ACS SYMPOSIUM SERIES

Immunoassays
for Trace Chemical Analysis
Monitoring Toxic Chemicals in Humans, Food, and the Environment

Martin Vanderlaan, EDITOR
Lawrence Livermore National Laboratory
Larry H. Stanker, EDITOR
Lawrence Livermore National Laboratory
Bruce E. Watkins, EDITOR
Lawrence Livermore National Laboratory
Dean W. Roberts, EDITOR
National Center for Toxilogical Research

Developed from a symposium sponsored
by the International Chemical Congress
of Pacific Basin Societies,
Honolulu, Hawaii,
December 17–22, 1989

American Chemical Society, Washington, DC 1990

Library of Congress Cataloging-in-Publication Data

Immunoassays for trace chemical analysis: monitoring toxic chemicals in humans, food, and the environment
    Martin Vanderlaan...[et al.] editors

    p.   cm.—(ACS Symposium Series; 451).

"Developed from a symposium sponsored by the International Chemical Congress of Pacific Basin Societies, Honolulu, Hawaii, December 17–22, 1989."

Includes bibliographical references and index.

ISBN 0–8412–1905–2

1. Immunoassay—Congresses.   2. Trace   analysis—Congresses.
3. Poisons—Analysis—Congresses.   I. Vanderlaan, Martin, 1948—.
II. International Chemical Congress of Pacific Basin Societies (1989: Honolulu, Hawaii).   III. Series   [DNLM: 1. Environmental Monitoring—congresses.   2. Environmental   Pollutants—analysis—congresses.   3. Food   Contamination—analysis—congresses.
4. Immunoassay—methods—congresses. QW 570 I333 1989]

RA1223.I45I47   1991
616.07'.56—dc20
DNLM/DLC
for Library of Congress   90–14504
                                                            CIP

The paper used in this publication meets the minimum requirements of American National Standard for Information Sciences—Permanence of Paper for Printed Library Materials, ANSI Z39.48–1984. ∞

PRINTED IN THE UNITED STATES OF AMERICA

ACS Symposium Series

M. Joan Comstock, *Series Editor*

1991 ACS Books Advisory Board

Foreword

THE ACS SYMPOSIUM SERIES was founded in 1974 to provide a medium for publishing symposia quickly in book form. The format of the Series parallels that of the continuing ADVANCES IN CHEMISTRY SERIES except that, in order to save time, the papers are not typeset, but are reproduced as they are submitted by the authors in camera-ready form. Papers are reviewed under the supervision of the editors with the assistance of the Advisory Board and are selected to maintain the integrity of the symposia. Both reviews and reports of research are acceptable, because symposia may embrace both types of presentation. However, verbatim reproductions of previously published papers are not accepted.

Contents

IMMUNOASSAYS FOR NATURAL TOXINS

Preface

THE TOXICOLOGICAL SIGNIFICANCE OF HUMAN EXPOSURE to low levels of trace toxic chemicals is unknown. Most human exposures occur as the result of incidental exposure to trace quantities of carcinogens, mutagens, and other toxicants in complex mixtures of chemicals present in such media as drinking water, food, and the occupational environment. Quantifying the levels of exposure is a challenge for toxicologists and epidemiologists. Yet, accurate, sensitive means of measuring chemicals are essential if we are to define current levels of human exposure, relate exposure to target tissue dose and ultimately to human disease, identify environmental sources, and limit future exposures. One long range goal might be to have each person act as his or her own dosimeter, with body fluid or tissue samples providing an index of exposure history. Before we see people checking their own exposures, we must improve analytical techniques for measuring the chemical traces left in the form of DNA and protein adducts and residual chemicals stored in adipose tissue.

Related to the above issues in toxicology are similar needs for improved analytical methods to monitor residues in foods and the environment. Public concern over chemical exposures and the quality of the food supply has increased dramatically in recent years. This concern translates into increasing pressure on regulatory agencies of all countries to do more sampling for more chemicals. This pressure occurs at a time when budgets are limited, so that the only alternative is increased cost-effectiveness of monitoring programs. In addition to reducing the cost per assay, there is a need to develop field portable assays that provide rapid, on-site data on chemical levels.

In the 1980s, we saw an explosion in the number of publications about using immunochemical methods for trace chemical analysis. Generally, these reports can be grouped into the broad categories of environmental monitoring for regulated synthetic compounds, food monitoring for pesticide and mycotoxin residues, and human monitoring for evidence of carcinogen exposure through the detection of adducts or metabolites. Representative work in all of these areas is included in this book. One goal of the book was to bring together a broad sampling of the applications of analytical immunochemistry to allow a cross-fertilization of techniques among those working in the field. The chapters show that immunochemical methods offer one approach to the complex problems presented by trace chemical analysis and should be part of the standard repertoire of techniques used in the analytical chemistry laboratory.

It is our hope that this volume will introduce many new scientists to the possibilities of immunoassays and provide a reference for this subject. With any meeting, support for travel for the attendees is essential to the success of the meeting. We express our appreciation for the generous support provided by the American Chemical Society, the U.S. Department of Agriculture, the U.S. Food and Drug Administration, and the following Corporations: Nestle, Inc., E. I. du Pont de Nemours and Company, and J & W Scientific. We would also like to thank the symposium participants for making the meeting lively and informative.

MARTIN VANDERLAAN
Lawrence Livermore National Laboratory
Livermore, CA 94550

LARRY H. STANKER
Lawrence Livermore National Laboratory
Livermore, CA 94550

BRUCE E. WATKINS
Lawrence Livermore National Laboratory
Livermore, CA 94550

DEAN W. ROBERTS
National Center for Toxilogical Research
Jefferson, AR 72079-9502

August 30, 1990

IMMUNOASSAYS FOR CHEMICAL RESIDUES IN FOOD AND THE ENVIRONMENT

Chapter 1

Immunochemical Techniques in Trace Residue Analysis

Martin Vanderlaan, Larry Stanker, and Bruce Watkins

Biomedical Sciences Division, Lawrence Livermore National Laboratory, University of California, P.O. Box 5507, L-452, Livermore, CA 94550

Immunoassays offer a sensitive, specific, cost-effective means of screening many samples for trace residues of toxic chemicals, their metabolites, and adducts. Antibodies can be used both as detectors to quantify the amount of a chemical present and in immunoaffinity chromatography to purify and concentrate material for subsequent analysis. Applications of these assays include detection of pesticide residues, mycotoxins, biomarkers of toxicity, and industrial chemicals.

The magnitude and frequency of adverse health effects resulting from exposure to trace residues in the environment, including known carcinogens, pesticide residues, food additives, and natural toxins, are not yet fully known. The detection and control of environmental pollution are important because human exposures to such substances may contribute to the incidence of neoplastic disease and other adverse health effects. These varied and frequently inadvertent exposures present special challenges to toxicologists, analytical chemists, and regulators. This volume presents the state-of-the-art of one approach, immunochemistry, to providing sensitive, specific, cost-effective dosimetry methods for monitoring human exposures to potentially toxic chemicals. The intent of this chapter is to provide an introduction to the concepts involved in trace-chemical analysis by immunoassay. To a large extent the recent advances and successes in this emerging field are the result of eclectic applications of prior developments in immunology, instrumentation, and analytical chemistry.
This book highlights recent developments in the novel application of immunoassay technology to food and environmental human exposure assessment (see *1-3* for reviews). The use of immunoassays for quantification of small organic compounds, is not itself a new technology, since it is widely used in clinical laboratories. Currently, more than half of the total clinical immunoassay market is aimed at low molecular-weight analytes, amounting to an estimated $580 million in sales in the United States in 1990 (4). Many of the immunoassay formats successfully applied in the clinic can be transferred directly to environmental analysis. Assays can be quantitative (e.g., serum dioxin levels) or yes-no screening assays (e.g., urinary illicit-drug tests). They can be part of a centralized pathology laboratory with the automated handling of hundreds of samples (e.g., AIDS testing), or they can be operated in the field on single samples by untrained personnel (e.g., home pregnancy tests).

What distinguishes clinical immunoassays from the assays presented in this book is that clinical assays most frequently quantify polar compounds in the matrices of blood and urine. In contrast, environmental and food applications of immunoassays often require development of antibodies to analytes that are lipophilic and that may occur in such diverse complex matrices as sediments, nuts, and animal fat.

"Analyte" is a term from analytical chemistry meaning the compound for which an assay is developed (e.g., a mycotoxin, its metabolites, and toxin-DNA adducts are all different analytes that might be detected by different immunoassays following exposure). "Hapten" is a term from immunology and refers to low molecular weight compounds (typically less than 2500 daltons) that do not illicit an immune response by themselves but that can be rendered immunogenic by conjugation to a higher molecular weight molecule, typically a protein. Antibodies raised to hapten-protein complexes can be used to detect a variety of analytes that are structurally related to the immunizing hapten, depending on the binding selectivity of the antibodies. A derivatized analog of the mycotoxin would be the hapten used to illicit antibodies to the analytes listed above. Classical analytes in trace residue analysis include the parent compound, its metabolites and degredation products. One can also develop immunoassays for genotoxic dose such as carcinogen-DNA adducts and carcinogen-protein adducts. These latter analytes are often termed "biomarkers" of exposure.

Why Consider Immunoassays for Analytical Chemistry?

There are three primary motives for considering immunochemical methods when surveying the range of analytical techniques available. First, for some chemicals, immunoassays allow measurements not possible by other means. For example, thermal lability and low volatility prevent gas chromatographic (GC) analysis of some compounds, while lack of a distinctive chromophore may hamper liquid chromatographic (HPLC) analysis. In contrast, immunochemical detection is based on the ability of an antibody to act as a receptor for the analyte of interest; binding occurs through ionic and van der Waals forces and is unrelated to properties of volatility, thermal stability, and chromogenicity. As examples, the thermal lability of pyrethroid insecticides can interfere with their analysis by GC, but they can be readily quantified by immunoassay (5). Plant, parasite, and fungus-derived toxins are usually too large for GC, but are well suited for immunoassay (see papers on mycotoxins in this volume). Many new, biotechnology-derived insecticides pose particularly difficult analytical problems for conventional approaches because of their large mass, but antibodies to avermectins and *Bacillus thuringiensis* toxins illustrate the applicability of immunoassays for insecticides in this upcoming class of compounds (6,7).

The high degree of selectivity in antibody binding allows quantification of trace chemicals in the presence of large excesses of similar chemicals. A prime example is the detection of modified DNA bases in occupationally exposed individuals, which typically represent only one base in 10^7 to 10^9 of the normal bases (see papers on human monitoring in this volume). The epidemiologist now may make quantitative biochemical measurements about the "dose" of a given carcinogen, such as aflatoxin or benzo[a]pyrene. In the area of pesticide detection, many recently developed pesticides derive their species selectivity by inhibiting specific enzymes in the target species. These substrates are often structural analogs of natural substrates, which means the chemist must determine trace levels of a pesticide in the presence of abundant amounts of the chemically-similar natural compound. With increasing public demands that pesticides not affect non-target species, the use of analogs of natural substrates will increase, and the selectivity of antibody binding offers an obvious solution to the problems of residue analysis.

The second motive for selecting immunoassays as an analytical method is their cost effectiveness. Immunoassays offer the promise of being inexpensive, rapid, and field-portable. They also allow for parallel sample processing, unlike the sequential sample processing required by chromatographic methods. In some cases immunoassays have been applied on site, resulting in additional savings in the costs and time involved in transporting samples back to a central laboratory (see papers by Rittenburg, Fleeker and Thurmand in this volume). As screening assays, immunoassays can eliminate the need for complete work-up of negative samples, freeing the analytical laboratory to focus on the more interesting, positive samples. Ultimately, these cost savings make studies economically feasible that otherwise would be prohibitively expensive; for example, the mass screening of individual wells for drinking-water contamination, or the quantitative analysis of many an individuals' exposures in epidemiologic studies. These properties of immunoassays commend them to regulatory agencies that must screen ever increasing numbers of samples on limited budgets. They also allow the introduction of monitoring programs in third world countries that lack the skilled personnel to staff sophisticated laboratories. It is important, however, not to overstate the cost savings. Adequate sample numbers and good sampling technique are still needed to ensure that the samples accurately reflect the area studied. For many applications, the analytes must be extracted from the matrix, and some level of sample preparation and cleanup is required. However, sample cleanup is usually significantly less than is required for GC/MS analysis.

The third motive for considering immunochemical methods derives from their use in conjunction with other systems to form hybrid analytical procedures. Hybrid systems, such as gas chromatography with electron capture (GC/EC) and liquid chromatography with ultraviolet absorption (HPLC/UV), are well accepted in analytical chemistry laboratories. Immunoassays can be combined with other techniques to exploit the advantages of both methods. As illustrated in Figure 1, antibody affinity columns can be used in sample preparation to selectively concentrate and purify chemicals. Affinity columns are ideally suited for concentrating dilute polar compounds that are often difficult to concentrate from environmental samples using conventional resins. Broad-selectivity antibodies can be used to concentrate both the parent compound and its metabolites from urine, for example. Individual compounds then can be quantified by separating them by HPLC following immunoconcentration or, as in the case of aflatoxin, immunopurification can then be followed by fluorescence detection, using the intrinsic fluorescence of the analyte.

The antibodies that function best in chromatography are not necessarily the same ones that are optimal in quantitative assays. Quantitative assays are competition assays, and for these assays, the highest affinity antibodies offer the greatest sensitivity, while for affinity chromatography the reversibility of the binding, allowing recovery of the analyte from the column, may be the most important feature. In addition, the literature for the immunoaffinity purification of proteins indicates that pH 3 or 2M KSCN buffers are often sufficient to recover bound material, but the recovery of organic compounds often requires more harsh conditions, such as 50% MeOH/H_2O or 50% DMSO/H_2O (8-9).

Alternatively, antibodies can be used as chemically-selective detectors following HPLC separation of material from a complex mixture of chemicals. HPLC/Immunoassay (HPLC/IA) in some cases offers a 1,000-fold increase in sensitivity over (HPLC/UV), and the selectivity to quantify compounds present as only trace fractions of the total. One current limitation of HPLC/IA is that there is no on-line immuno-detector allowing real time immunochemical sensing of the column effluent. Rather, fractions must be collected, the solvent exchanged to be compatible with the antibodies, and immunoassays run on the individual fractions from the HPLC. The automated, parallel process nature of immunoassays means that it is relatively simple to assay up to 100 fractions from a chromatogram. In spite of the

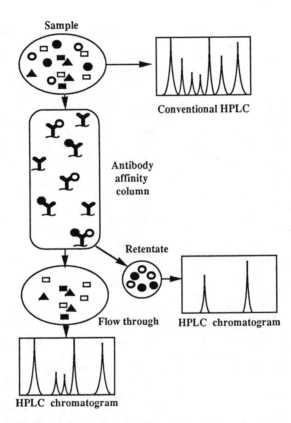

Figure 1. - Immunoaffinity chromatography.
A sample containing a complex mixture of chemicals, when analyzed by conventional HPLC chromatography, shows a number of peaks. Application of the sample to an affinity column containing immobilized antibody results in the pass through of material that is depleted of those compounds retained on the column. Subsequent elution of the column recovers the bound material (retentate) in a reduced volume (both concentration and purification have occurred). Two related chemicals, a parent compound and a metabolite, were retained on column and can be subsequently separated by HPLC.

somewhat cumbersome nature of this combination, the analytical power of retention time and immunochemical selectivity is impressive for the identification of unknowns in complex mixtures of chemicals (10, 11). While it has yet to be explored, supercritical fluid extraction (SFE) and chromatography (SFC) may be ideally suited for a hybrid analytical technique with immunoassay (e.g., SFC/IA) (12).

Antibody Development

The critical component in an immunoassay is the antibody. Antibodies are typically bivalent serum proteins that are part of the vertebrate immune defense system. Exposure of an animal to small organic molecules having molecular weights less than 2500 daltons has generally not lead to the production of antibodies. Rather the hapten must be presented to the animal in the context of a larger molecule to elicit antibodies (i.e., to be immunogenic). For example, covalently conjugating the hapten to a protein will, upon injection of the hapten-protein conjugate, elicit some antibodies directed against the hapten specifically. Other antibodies will be formed to the protein and to the protein-hapten complex, but these are irrelevant to the immunoassay. A small subpopulation of the total antibody response will bind with high affinity to the unconjugated analytes of interest, free in solution, and it is this subpopulation that forms the basis for specific chemical-selective assays.

The site of conjugation, and the chemical functionality at the site, influence the spectrum of antibodies produced (see Harrison, this volume). Ideally, the conjugation chemistry should be immunologically neutral. Usually, this is achieved by using four- to six-member carbon alkane linkers between the hapten and carrier protein. Antibodies will be most selective for that portion of the hapten farthest removed from the site of conjugation. Thus, if there is a series of structurally related pesticides, for example, the chlorinated cyclopentadienes, having a common ring structure, conjugation at a site distal from the common structure should elicit antibodies that cross react with all members of the chemical class (see papers by Harrison and Stanker, this volume).

Typically, conjugates are injected into mice or rabbits to generate an antibody response. Bleeding of the animals yields serum that contains a population of antibodies (polyclonal antisera). Cloning of the antibody-secreting cells from immunized animals yields monoclonal antibodies. In experienced hands, both polyclonal and monoclonal antibodies are readily prepared and suitable for immunoassays. In theory, there are some advantages to monoclonal antibodies, although these advantages are only occasionally of practical consequence, and it is often immaterial to the end user of the assay whether the antibodies are of monoclonal or polyclonal origin. Monoclonal antibodies can be more specific than polyclonal antibodies, and therefore may be more suited to detecting a single analyte rather than screening for a class of analytes. The amount of antibody becomes limiting only in immunoaffinity chromatography, where typically milligram quantities are used per sample. For these applications, monoclonal antibodies offer clear advantages because antibody production can be scaled up to produce gram quantities of antibody.

Enzyme Immunoassays

Immunoassays for small organic compounds usually are formatted as competition assays (13,14). There are many assay formats and choices of label for the eventual quantification of binding. An equilibrium is approached for formation of complexes between the antibody, labeled or immobilized hapten (hapten*), and free analyte in the sample, according to the following equation.

(analyte) + (antibody) + hapten* <-> [analyte-antibody] + [hapten*-antibody]

In the example illustrated in Figure 2A, a hapten-protein conjugate is immobilized in a plastic tube or in a well on a microtiter plate, and the sample or standard containing the analyte is added along with the antibody. The antibody partitions between the plastic-immobilized hapten-protein conjugate and the analyte in solution. Increasing the amount of competitor in the sample reduces the amount of antibody binding to the solid phase. Antigen-antibody binding reactions are reversible, and the competition approaches equilibrium in 15 to 30 min. Washing removes unbound analyte and antibody from the solid phase, effecting the separation of bound and free antibody. The amount of antibody binding to the plastic is measured by tagging the antibody with an enzyme, hence the name Enzyme Immunoassay. The amount of enzyme-antibody conjugate bound to the solid phase is measured by the generation of a chromogenic or fluorogenic reaction product.

An alternative format for enzyme immunoassays makes use of an enzyme-hapten conjugate, and is illustrated in Figure 2B. In this format, the antibody is immobilized onto the solid phase, and the antibody partitions between this conjugate and the analyte in the sample. The antibody may either have been immobilized on a solid phase to begin with, or may be trapped on a solid phase after the incubation period. The amount of enzyme-hapten conjugate retained by the antibody on the solid is then inversely proportional to the amount of free analyte in the sample, and may be quantified with a suitable substrate for the enzyme.

In both of the assay formats depicted in Figure 2, there is the possibility that the analyte, or co-extracted matrix material, may directly inactivate the enzyme or cause dissociation of the antigen-antibody complex, giving the false appearance that competition has occurred. This possibility can be removed by making the assay a two-step procedure, adding the enzyme as a second step after the analyte-antibody binding reaction has occurred. For example, in Figure 2A, the antibody is conjugated to biotin (a vitamin), and in Figure 2B biotinylated-hapten is used as the competitor instead of hapten-enzyme conjugate. In both cases, a conjugate of enzyme-avidin (a biotin-binding protein) is then added as the second step to reveal the presence of bound biotin. This two-step procedure involves extra pipetting and washing steps, but minimizes direct contact between sample and enzyme.

Using either competition format, the end result is the immobilization of enzymatic activity on a solid phase. The percentage inhibition of enzyme binding to the solid phase caused by the presence of the free analyte is then measured. By comparison to a standard curve, the amount of analyte in an unknown sample is determined. In the simplest assays, the amount of enzyme product is compared visually with color standards to give a qualitative estimate of the amount of analyte. Alternatively, the amount of enzyme product can be quantified spectro-photometrically. The measured competitive antibody binding curve is illustrated in Figure 3. The sigmoidal inhibition curve can be fit by computer using the four-parameter equation shown (15). Alternatively, the data can be "linearized" by plotting the optical density as a function of competitor on log-logit graph paper.

Data in the vicinity of 50% inhibition give the most precise estimate of the amount of competing material because the curve has the greatest slope in this region. The absolute amount of material required for 50% inhibition (I_{50}) is a function of the assay format, the size of mass action reaction, and the affinity of the antibody. The lower the amount of hapten-protein conjugate bound to the solid phase and lower the amount of antibody used, the less competitor needed for inhibition, and the more sensitive the assay. Two approaches can be used to improve the sensitivity, both of which operate by reducing the affinity of the antibody for the coating of the solid phase: (1) providing a disbersed hapten-protein coating by minimizing the absolute amount of hapten-protein used and diluting it in unlabeled protein, and (2) choosing an analog of the immunizing hapten for which the antibody has only weak binding, and using this analog to synthesize the protein-hapten conjugate.

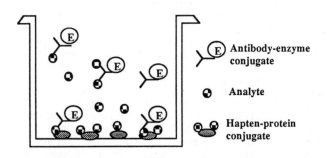

Figure 2A. - Enzyme-antibody conjugate immunoassay.
A hapten-protein conjugate is immobilized on the solid phase. Antibody
conjugated to enzyme is mixed with analyte. The antibody-enzyme combination
partitions between the bound hapten-protein and the analyte in solution.
Unbound material is then washed away, and the amount of enzyme-antibody
bound to the hapten-protein on the solid phase may be detected by providing a
substrate for the enzyme. The greater the amount of free analyte in the sample,
the more antibody-enzyme will bind to it, and the less will be available to bind to
the immobilized hapten-protein. Enzymatic activity retained on the solid phase is
therefore inversely related to the amount of analyte in the sample.

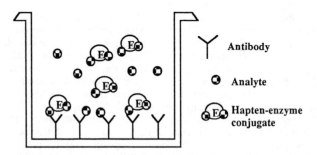

Figure 2B. - Enzyme-hapten conjugate immunoassay.
Antibody is immobilized on the solid phase, and then enzyme-hapten and sample
with analyte are added. The antibody will competitively bind either the enzyme-
hapten or analyte, and unbound material is washed away. Addition of substrate
reveals the presence of the enzyme-hapten that has been trapped on the surface by
the antibody. As in Figure 2A, the enzymatic activity retained on the solid phase
is inversely related to the amount of analyte in the sample.

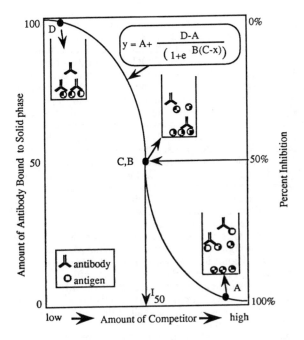

The equation shown in the figure:

$$y = A + \frac{D-A}{\left(1 + e^{B(C-x)}\right)}$$

Figure 3. - Sigmoidal competitive inhibition curve.
Because the mass action laws govern the partitioning of the antibody between the analyte and a hapten-protein conjugate, there is a sigmoidal shape to the dose-response curve. This shape may be fit with the four-parameter function shown. At high concentrations of analyte, there is little binding to the solid phase (100% inhibition, "A" in the function), while at low concentration of analyte there is maximal binding to the solid phase (0% inhibition, "D" in the function).

Because the assay has its maximum sensitivity at the limit of minimum antibody concentration and minimum labeled or immobilized hapten, the intrinsic affinity of the antibody for the analyte, and the capacity of the enzyme to produce detectable product at useful signal-to-noise ratio, govern the sensitivity of the assay. Typically, nanogram quantities of conjugates and antibodies are used, with I_{50} values in the range of 0.1 to 1.0ng (0.1 to 1 picomoles) per assay. Maximal sensitivity requires the highest possible affinity for the antibody, and highest activity reagents (enzymes, substrates, etc.) used for labeling. Ultimate sensitivity may not be required for regulated compounds, however, since allowable residue levels are often well in excess of the minimum immunochemically detectable levels. In such cases, it may be most appropriate to adjust the assay I_{50} value to the regulated residue level.

Example of an ELISA.   A representative enzyme linked immuosorbent assay (ELISA) is presented below. The assay illustrated in Figure 2A was conducted using monoclonal antibodies to assay meat for the aromatic amine carcinogen, 2-amino-3,8-dimethylimidazo[4,5-f]quinoxoline (MeIQx)(10). MeIQx-albumin was adsorbed (by drying) onto the surface of a 96-well microtiter plate, antibodies along with various levels of competitors were co-incubated in the coated wells, and the plate was washed to remove unbound antibody and competitors after 60 min. A peroxidase-conjugated second antibody (goat anti-mouse immunoglobulins) was then added to each well. Unbound second antibody was washed away and the presence of bound enzyme was detected using o-phenylenediamine (OPD) as the substrate for the peroxidase enzyme. After 30 min the reaction was stopped and the optical density in each of the 96-wells was quantified using a microtiter plate reader. The data were accumulated on a MacIntosh (Apple Computer, Cupertino, CA) using software we have written.

Figure 4 is the computer image of the 96-well microtiter plate. Each well is illustrated as a circle, and the thickness of the line used to draw the circle graphically displays the measured optical density in the wells. In the experiment illustrated, rows of wells "A" and "H" were unused and are blank. For each individual row, wells 3 through 10 illustrate a two-fold serial dilution of competitor. Wells in columns #1 and #2 for rows "B" through "G" are controls where the MeIQx-albumin was omitted. The absence of color in these wells demonstrates the lack of non-specific binding of the anti-MeIQx and peroxidase-anti-mouse antibodies, and defines experimentally "complete inhibition" for the competitive ELISA. Wells "B11" through "G12" define the other extreme experimental condition, that of no inhibition. In these wells, no competitor was added and all of the anti-MeIQx antibody was available to bind to the solid phase.

In the example shown, rows "B" through "G" are the analyses of six consecutive fractions from an HPLC chromatographic separation of well-done fried beef in the region of the chromatogram where MeIQx is expected. As illustrated, competitive inhibition of antibody binding is highest in well 3 for each lane, where the amount of competitor is greatest. Inhibition decreases (solid-phase binding increases) as one moves to the right on any given row and the competitor is diluted. As one moves from row B to D there is increasing inhibition in each row, and from D to G there is decreasing amount of inhibitor. This progression reflects the "peak" in the chromatogram is present in row D, with it having more inhibitor than the HPLC fractions on either side of it.

The results from row D are further illustrated in Figure 5. The fraction of inhibition in wells 3 through 10 was calculated by subtracting the average of the background in wells 1 and 2 for row D, and dividing by the average of the "no inhibition" values in wells 11 and 12 of row D. The fractional inhibition is plotted as a function of the dilution of the sample (well 3 corresponding to no dilution, i.e., 1.0). The four-parameter computer curve fit to the data is shown as open circles in Figure 5 and the figure shows that 50% inhibition occurred at a dilution of the sample of 0.02 (i.e., 1:50) and that the residual sum of squares (a measure of the quality of

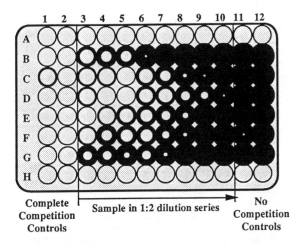

Figure 4. Computer image of a 96-well microtiter ELISA plate. Individual wells are shown as circles arranged in 12 columns and 8 rows. The thickness of the line for each circle is drawn in proportion to the measured optical density in the plate wells as determined by a plate-scanning spectrophotometer.

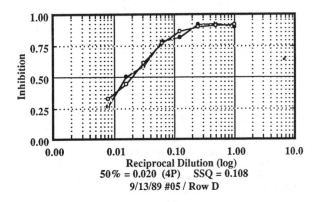

Figure 5. Computer analysis of the data from row D in the plate shown in Figure 4. Solid circles show the calculated fractional inhibition as a function of sample dilution for the wells D3 through D10. Open circles show the four-parameter curve fit to the data.

the fit of the computer equation to the data) is 0.108. The standard curve had a 50% inhibition value of 0.1 ng per assay well, so in this case the amount of material in the undiluted sample in row D was 50 x 0.1, or 5 ng. Similar fits to the data from other rows in the plate show that 50% inhibition occurs with less dilution, indicating less immunochemically positive material in the starting fractions.

<u>Assay Validation</u>

As with any analytical technique, it is important to have thoroughly validated assays and use confirmatory methods whenever possible to ensure accurate data. Many of the papers in this volume address issues of sample preparation and comparison of immunochemical assays with conventional analytical methods.

   With the exception of a limited number of assays that can be run directly in the sample matrix (e.g., water, juice, and urine), most other sample matrices must be extracted and treated prior to immunoassay. Typically, the amount of treatment required before immunoassays is considerably less, however, than is required for GC/MS analysis, thereby offering savings of time and money in sample preparation. The extraction and cleanup procedures for the immunoassay should be compatible with subsequent GC or HPLC analysis. This allows confirmation of the immunoassay results without the necessity of a separate sample extraction and processing. Our experience with immunoassays for screening of permethrin, heptachlor, and dioxin in foods and environmental samples is that many of the widely used extraction protocols are excellent places to start when developing extraction procedures for immunoassays. Often these can be applied with minor modifications, or only carried through the initial steps, to produce an extract suitable for immunoassay. Specifically, if the analyte is hydrophobic, the non-polar matrix material must be removed in order to achieve a single-phase system and allow the analyte to come in contact with the antibody. Because antibodies are water-soluble proteins, use of mixed solvent systems can simplify the analysis of hydrophobic analytes. Several of our antibodies work well in mixed solvent systems, such as 1:1 aqueous buffer:ethanol, or with the analyte dispersed in detergents (<u>16</u>). Others have reported functional immunoassays in organic solvents such as acetonitrile or DMSO, but the general utility of organic solvents has yet to be shown (<u>17</u>). One prospect for the future is that general methods will be developed to allow direct injection of organic extracts of samples into the immunoassay.

<u>Conclusion</u>

Immunoassays are finding increasing application in the chemical analysis of trace organic compounds, as illustrated by the publications in this volume. Applications include monitoring of residues in foods, in the environment, and in humans for residues of both synthetic and naturally occurring toxins. The primary motives for the development of these assays are their high sensitivity, high selectivity, portability, short analysis time, low cost, and potential for parallel processing samples. Parallel processing of samples means that immunoassays are highly applicable to mass screening studies either for monitoring regulatory compliance or for epidemiology studies. Particularly powerful analytical approaches use antibodies in conjunction with other methods, for example, the use of immunoaffinity columns to concentrate and purify the analyte before measurement by conventional means.

   In the past decade, environmental immunoassays have moved from novelties being developed in a few laboratories to widespread applications such as those described in this volume. Almost all Federal agencies involved with some aspect of public health (Food and Drug Administration, Department of Agriculture, National Institutes of Health, Environmental Protection Agency, and Department of Energy) are currently funding environmental immunoassay research and development. Several corporations also are marketing environmental immunoassays. These milestones illustrate the general maturation of this field that has ocurred in recent years. The next decade should see incorporation of some of these assays into the routines of environmental monitoring laboratories and the acceptance of immunochemical data by regulatory agencies. In addition, new application formats may be developed, such as the development of "immunosensors" to provide chemically selective detectors for passive monitoring.

Acknowledgments

Funding for this work was provided by the Environmental Protection Agency under interagency agreement IAG #DW89931433-01-3, the Department of Agriculture under interagency agreement 13-37-7-046, and the National Institutes of Health under grant CA48446-03. Work performed under the auspices of the U.S. Department of Energy by Lawrence Livermore National Laboratory under Contract W-7405-Eng-48.

Literature Cited

1. Hermann, B.W. In *Immunological Techniques in Insect Biology*; Gilbert, L.I.; T.A. Miller, Eds.; Springer-Verlag: New York, NY, 1988; pp 135-180.
2. Newsome, H.W. *J. Assoc. Official Anal. Chem.* 1986, **69**, 919-923.
3. Vanderlaan, M.; Watkins, B.; Stanker, L. *Environ. Sci. Tech.* 1988, **22**, 247-254.
4. Goodsaid, F., Syntex Corporation, 1989, Personal communication
5. Stanker, L.H.; Bigbee, C.; van Emon, J.; Watkins, B.E.; Jensen, R. H.; Morris, C.; Vanderlaan, M. *J. Agric. Food Chem.* 1989, **37**, 834-839.
6. Schmidt, D.J.; Clarkson, C.E.; Swanson, T.A.; Egger, M.L; Carlson, R.E.; Van Emon, J.M.; Karu, A.E. *J. Agric. Food Chem.* 1990, (in press).
7. Wie, S.I.; Hammock, B.D.; Gill, S.S.; Grate, E.; Andrews, R.E.; Faust, R.M.; Bulla, L.A.; Schaefer, C.H. *J. Applied Bact.* 1984, **57**, 447-454.
8. Groopman, J.D. ; Kensler, T.W. *Pharmac. Ther.* 1987, **34**, 321-334.
9. Turesky, R.J.; Forster, C.M.; Aeschbacher, H.U.; Wurzner, H.P.; Skipper, P.L.; Trudel, L.J.; Tannenbaum, S.R. *Carcinogenesis* 1989, **10**, 151-6.
10. Vanderlaan, M., Hwang, M., Knize, M.G.,Watkins, B.E., and Felton, J.S. In *Mutation and the Environment. Part E: Environmental Genotoxicity, Risk, and Modulation;* Mendelsohn, M; R. J. Albertini, Eds.; Wiley-Liss: New York, NY, 1990; pp 189-198.
11. Vanderlaan, M.; Watkins, B.E.; Hwang, M.; Knize, M.; Felton, J. S. *Carcinogenesis* 1989, **10**, 2215-2221.
12. McHugh, M.; Krukonis, V. *Supercritical Fluid Extraction;* Butterworths: Boston, MA, 1986.
13. Monroe, D. *Anal. Chem.* 1984, **56**, 920A.
14. Tijssen, P. *Practice and Theory of Enzyme Immunoassays;* Elsevier: New York, NY, 1985.
15. Rodbard, D. *Clin. Chem.* 1974, **20**, 1255.
16. Vanderlaan, M;Stanker, L.H.; Watkins, B.E.; Petrovic, P.; Gorbach, S . *Environ. Toxicol. Chem.* 1988, **7**, 859-870.
17. Russell, A.J.; Trudel, L.J.; Skipper, P.L.; Groopman, J.D.; Tannenbaum, S.R.; Klibanov, A.M. *Biochem. Biophys. Res. Comm.* 1989, **158**, 80-85.

RECEIVED August 30, 1990

Chapter 2

Hapten Synthesis for Pesticide Immunoassay Development

Robert O. Harrison[1], Marvin H. Goodrow, Shirley J. Gee, and Bruce D. Hammock

Departments of Entomology and Environmental Toxicology, University of California, Davis, CA 95616

The production of pesticide specific antibodies requires presentation of a conjugated form of the pesticide to an animal's immune system. This in turn demands the design of an appropriate chemical structure (a hapten) which can be covalently coupled to a carrier, but which will still elicit the production of antibodies recognizing the target analyte. The immunoassay literature does not adequately address the question of what chemical structure is required for the production of specific antibodies to low molecular weight compounds, especially in the pesticide area. Hapten design historically is considered an art, with minimal or no explanation for hapten structures which failed to produce the desired antibodies. Most failed attempts remain unpublished and thus have not shed light on the structural requirements for antibody production. We present here an examination of selected examples of successful and unsuccessful hapten designs from our laboratory and from the literature. These examples are used herein to illustrate several criteria deemed critical for successful hapten synthesis strategies.

The Development And Utilization Of Pesticide-Specific Antibodies in Immunoassays

The procedures for production of specific antibodies and their application in a competitive inhibition ELISA (Enzyme-Linked Immunosorbent Assay) are discussed in detail in the preceding chapter (Vanderlaan et al., this volume). In addition, other comprehensive overviews of the immunoassay development process in the pesticide field are available (5,7,8,19). In general, a

---

[1]Current address: ImmunoSystems, Inc., 4 Washington Avenue., Scarborough, ME 04074

synthetic antigen (immunogen or <u>immun</u>izing anti<u>gen</u>) is used to immunize an animal for antibody development. This immunogen is composed of a synthesized hapten (mimicking the structure of the target compound) which is covalently attached to a carrier protein. Antibodies are produced against many sites on the immunogen, including the conjugated hapten. Antibodies thus made may recognize either the carrier protein, the hapten, or a site combining parts of both. Because several quite different results are possible from an immunization, animals or cultured cells must be tested carefully to assess the usefulness of the antibodies produced. The more common undesired results include excessive recognition of the spacer arm, recognition of the immunizing hapten without adequate recognition of the target compound, and failure of the hapten to elicit a specific response despite a successful immunization procedure.

Antibodies thus produced must be evaluated for specificity to allow analysis of the success or failure of the hapten design. The problem of antibody screening requires special consideration due to the potential for error and misinterpretation of data. Careful design of the screening protocol is especially crucial for development of monoclonal or recombinant antibodies; critical decisions must be made rapidly to preserve growing cells, but one can potentially be overwhelmed by cells producing undesirable antibodies. Antibody screening must address the following questions for either polyclonal or monoclonal procedures.

a. Does the antibody bind to the immunizing antigen? (i.e. Did the animal respond specifically to the immunogen? This test is needed only in the rare retrospective analysis of a total failure. Our screening generally starts at b; if b is successful, a is moot.)

b. Does the antibody bind to the conjugated hapten on the ELISA antigen (plate coating antigen; see Vanderlaan et al., this volume, Figure 2a) , which has a different carrier protein than the immunogen, but is otherwise identical? (i.e. Are antibodies present which recognize the hapten portion of the immunogen?)

c. Does the free immunizing hapten inhibit the binding shown in b? (i.e. Are antibodies present which recognize the unconjugated hapten structure independent of attachment to a carrier protein?)

d. Does the target analyte inhibit the binding shown in b? (i.e. Are antibodies present which recognize the analyte portion of the hapten independent of the spacer arm?)

e. Does the antibody bind to a conjugated heterologous hapten (a plate coating antigen made with a hapten related to but not identical to the immunizing hapten), and can this binding be inhibited by the target analyte? (i.e. Do the antibodies against the immunizing hapten also bind other conjugated haptens having significant differences in structure from the immunizing hapten? If the antibodies give poor sensitivity in test d, can the

problem be overcome by the use of heterologous assay
systems?)

Some studies (10,21,22) have reported antibodies which meet
tests b and c above, but not test d. This result indicates much
stronger recognition of the conjugated or free immunizing hapten
than the target analyte. Such antibodies may still yield a useful
assay, as noted in test e above and the section below. Other
studies (6,17) have reported antibodies which meet tests b and c
above, but not tests d or e. This result indicates recognition of
the conjugated or free immunizing hapten without significant
recognition of the target analyte. In this case the antibodies are
not useful for analysis of the target compound. We suspect both of
these problems may be more frequent than is apparent from the
literature, especially for small target molecules. We also suspect
that more careful exploration of heterologous assay systems,
facilitated by more extensive hapten synthesis, would increase the
sensitivity of many existing assays. In some cases, as described
below, such an approach has salvaged a functional assay from what
otherwise would have been a total failure.

A common misconception is that monoclonal antibodies can be
used to circumvent problems of handle recognition. Although this is
conceptually possible, in practice most monoclonal antibodies to
small molecules demonstrate extensive handle recognition. Proper
and extensive screening of numerous cell lines, as outlined above,
can yield truly superior antibodies with excellent detection of the
analyte (23). However, it is unreasonable to consider such a major
expense unless there has been a parallel intellectual investment in
good hapten design. Excellent, rugged immunochemical assays result
from a combined investment in both chemistry and immunology.

Hapten Design And Utilization

The initial and critical step in the development of effective
immunoassays for pesticides and other low molecular weight
environmental chemicals lies in the selection of appropriate haptens
which will elicit the production of antibodies demonstrating maximum
specificity and sensitivity for the target molecule. It is
important to distinguish between the problem of antibody production
and the development of many hapten-protein conjugates for assay
optimization using existing antibodies. Assay specificity and
sensitivity are determined primarily by the antibody produced in
response to the immunogen; thus its design and preparation are
critical. However, for a given antibody, other (heterologous)
haptens used in the later stages of assay design may offer critical
improvements in sensitivity (10,20,22; below and e above). Recent
reviews on immunoassay techniques (5,7,8,19) emphasize the
importance of attention to hapten selection during assay
development.

The principles of good hapten design are simple and
straightforward, given a basic understanding of the process of
antibody production. However, this is not an exact science. For
instance, even hybridomas or inbred animals which yield antibodies
recognizing the same molecule may produce antibodies with very
different amino acids in their combining sites. In designing

haptens for antibody production, several general guidelines are
clear. It is preferable to avoid spacer attachment at or near
functional groups of the target molecule as this can reduce the
number of potential sites contributing to antibody binding (nitro,
amine, hydroxy, halide, etc.), either directly by chemical
modification or indirectly by sterically blocking access to these
groups. To enhance compound specificity, synthesis strategies
should maximize exposure of unique determinants by attachment of the
spacer at a position on the target molecule where members of a class
have identical (or very similar) structural features. Spacers
containing strong determinant groups, such as aromatic rings,
conjugated double bonds, or heteroatoms should be avoided if
possible, to minimize the production of spacer-specific antibodies.

The Importance Of Negative Results

It is difficult to completely understand the immune response to a
particular hapten. Development of strong data on the inability of a
hapten to produce specific antibodies requires considerable
resources, including many animals, time for immunization, and
extensive characterization to demonstrate lack of appropriate
antibodies. It is possible to reduce the difficulty of obtaining
negative data through the use of mice rather than rabbits (6), a
route not usually pursued except when planning monoclonal antibody
production. In any case, rigorous synthetic chemistry remains
critical to understanding of the final results, successful or not.
The few available studies which describe unsuccessful immunizations
provide valuable insight (6,17), as do those failures within our
group. Based on the literature and our own experience, we have
identified and examined several criteria for the successful
development of immunoassays for pesticides. We present here
examples illustrating and contradicting those criteria.

Criteria For Hapten Synthesis And Discussion Of Examples

The structures of the compounds discussed below are given in Figures
1 and 2, with references indicated under the compound names. Sites
of conjugation are indicated by arrows, while dotted polygons
enclose parent molecule structure which is not retained in the
immunizing hapten.

Spacer Arm Location: Exposure of Determinants. The location of the
spacer arm on the target molecule should be distal to important
haptenic determinants to maximize their exposure for antibody
binding. This point is supported by several examples. In the
development of molinate haptens (3), the unique hexamethyleneimine
ring was retained, leading to antibodies which were specific for
molinate and recognized EPTC poorly. This must be due to the
difference between the cyclized hexamethyleneimine ring of molinate
and the open *N*-dipropyl group of EPTC. Assays developed for the *s*-
triazine herbicides using a range of immmunizing haptens (4,
Harrison et al. J. Agric. Food Chem., in press) exhibited
specificity reflecting the structures of the immunizing haptens;
chloro and alkylamino groups were each important determinants of
specificity. In a series of papers detailing the synthesis of

Thiocarbamate Haptens (3)

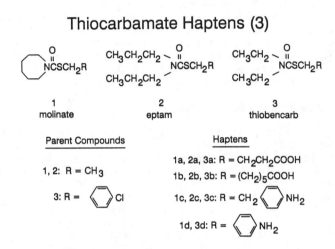

Parent Compounds   Haptens

1, 2: R = CH₃

1a, 2a, 3a: R = CH₂CH₂COOH
1b, 2b, 3b: R = (CH₂)₅COOH
1c, 2c, 3c: R = CH₂⟨⟩NH₂

1d, 3d: R = ⟨⟩NH₂

Triazine Haptens (4)

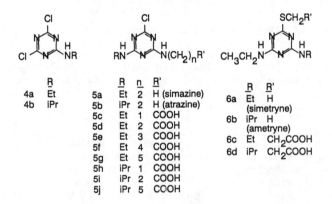

Figure 1. Structures of thiocarbamate and triazine haptens and parent compounds.

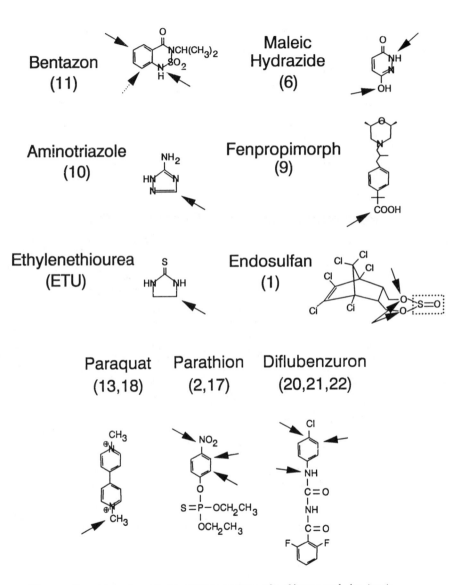

Figure 2. Structures of other compounds discussed in text. Arrows indicate sites of conjugation. Dotted polygons enclose parent molecule structure not retained in the immunizing hapten. The numbers under compound names refer to literature cited.

Continued on next page

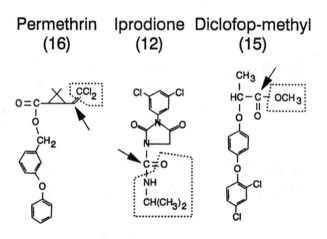

Figure 2. Continued

haptens and production of antibodies for diflubenzuron (20-22), several haptens which were derivatized distal to the 2,6-difluorobenzoylurea moiety yielded antibodies specific for this end of the molecule. An assay for fenpropimorph (9) produced highly specific antibodies because of preservation of the unique dimethylmorpholine ring. Conjugation of a permethrin hapten (16) distal to the 3-phenoxybenzyl group produced antibodies which crossreacted with several related pyrethroids because they share this portion of the permethrin structure. For iprodione (12), conjugation distal to the *m*-dichlorophenyl ring produced antibodies which were class specific, recognizing the parent, a metabolite, a hydrolysis product, and two related fungicides.

However, several exceptions to this generalization have been observed. The conjugation of an aminophenyl thiobencarb hapten by diazotization (Gee et al., this volume) produced antibodies which recognized the target compound. Conjugation of maleic hydrazide through the nitrogen of the distinctive hydrazide group (6) yielded antibodies recognizing the target compound. Preservation of the seemingly important nitro group on parathion's phenyl ring (by conjugation ortho or meta to the nitro group) led to antibodies which did not recognize parathion (17). However, useful antibodies were produced in response to diazotized aminoparathion (2; nitro group of parent compound was reduced to amino group for conjugation), but aminoparathion was also strongly recognized.

Spacer Arm Location: Preservation of Functional Groups.   The location of the spacer arm on the target molecule should avoid attachment to functional groups, including heteroatoms, which might lead to change in polarity of the group and/or a reduction in the number of potential sites contributing to antibody binding. Coupling through a carbon atom appears preferable. Several examples can be cited to illustrate this principle. In the development of assays for the triazines (4, Harrison et al. *J. Agric. Food Chem.*, in press), a sulfur atom was used in the spacer to mimic the size and other properties of the chloro group, allowing antibodies against thioether haptens to retain good recognition of chloro compounds while having excellent recognition of S-methyl triazines. In assays for diflubenzuron (20-22), attachment of the spacer through the ring NH produced specific antibodies, but the parent compound (target analyte) could not compete the antibody off the plate coating antigen. This problem was overcome by changing the hapten used in making the plate coating antigen. With this change, N-methylated diflubenzuron was still recognized more strongly than diflubenzuron, indicating the critical importance of the difference between secondary and tertiary nitrogens. The importance of this difference between secondary and tertiary nitrogens has also been observed for bentazon (11). Attachment of the spacer to the ring nitrogen produced specific antibodies, but these recognized N-methylated bentazon 100 to 1000 times better than bentazon. Similarly, in the production of antibodies to maleic hydrazide (6), conjugation through one oxygen prevented tautomerization of the remaining carbonyl to a phenol. This hapten produced antibodies which recognized the immunizing hapten, but not the phenolic parent compound. In two immunoassays for paraquat (13, 18), retention of

the positive charge on the nitrogen used for conjugation led to
antibodies which bound poorly to singly charged monoquat.

The following exceptions to the above principle must be noted.
In an assay for thiobencarb (Gee et al., this volume), removal of
the chloro group and conjugation through that position did not
prevent the production of specific antibodies. For maleic hydrazide
(6), conversion of the secondary nitrogen to tertiary by conjugation
through the nitrogen did not prevent the production of antibodies
recognizing the parent compound. In one parathion assay (17,
preservation of the nitro group on the phenyl ring (by conjugation
ortho or meta to the nitro group) led to antibodies which recognized
the immunizing haptens, but not parathion.

Selection of Spacer Arm. The spacer arm length and structure should
be chosen carefully to reduce spacer recognition, while retaining
specificity for the target molecule. Functional groups in the
spacer arm should be avoided, if possible, to minimize "spacer"
recognition. Alkyl spacers appear preferable; heteroatoms appear
undesirable. This is borne out by several examples. In the assays
for both molinate (3) and EPTC (Gee et al., this volume), antibodies
against aralkyl haptens produced high titer antibodies which could
not be used because spacer recognition was too strong. For
thiobencarb (Gee et al., this volume) and parathion (2), the
aromatic ring in the spacer was important to antibody binding, but
this was desirable because it constituted part of the target analyte
as well. Aminotriazole haptens (10) conjugated using the
heterobifunctional reagent MBS (maleimidobenzoic acid N-
hydroxysuccinimide) produced antibodies which bound the homologous
hapten-protein conjugate, but did not recognize the target analyte.
This result indicates recognition of the bulky and distinctive MBS
spacer group. The length of the spacer arm may also be an important
factor. Several examples are given below, primarily for the use of
alternative haptens in ELISA with antibodies against other haptens.
In one assay for the s-triazines (4, Harrison et al. J. Agric. Food
Chem., in press), the ELISA use of a hapten with reduced length
spacer, ($(CH_2)_5$ less than long alkyl spacer of immunizing hapten)
increased assay sensitivity 100 fold over the homologous system. In
our assays for molinate (3), thiobencarb (Gee et al., this volume),
and EPTC (Gee et al., this volume), antibodies against short chain
alkyl acid haptens produced no inhibition in homologous assay
systems, but gave acceptable sensitivities when the spacer length of
the ELISA antigen was increased by $(CH_2)_3$. Similar results were
obtained for diflubenzuron (20-22), where the use of site and spacer
heterologous haptens improved the assay sensitivity in most cases.

For many haptens, more complex spacers containing aromatic
rings or multiple heteroatoms do not prevent the production of the
desired antibodies. A bentazon hapten (11) containing a benzyl
group in the spacer produced antibodies recognizing the target
compound. Similarly, antibodies raised against diazotized
aminoparathion recognized parathion (2). This is not surprising,
since the phenyl ring of the diazotized hapten is shared by the
target parathion. We have also observed the usefulness of multiple
spacer types for the development of heterologous assays for
improving sensitivity. In the case of the thiocarbamates, this is
the only simple route for making heterologous assays. In the assays

for both molinate (<u>3</u>) and thiobencarb (Gee et al., this volume),
antibodies against alkyl acid haptens produced the best assays when
ELISA conjugates (plate coating antigens) used spacer heterologous
aminophenyl haptens.

<u>Selection of Coupling Chemistry.</u> In practice, the selection of
coupling chemistry is closely linked to the points discussed in the
previous section. The choice of chemical reaction for the
conjugation of hapten to protein must take into consideration the
reactivity of the hapten molecule's other functional groups. Most
of the compounds described in this review used alkyl COOH or aryl
NH_2 groups for conjugation.The differences in stability among azo,
amide, ester, and disulfide linkages illustrate the range of
possibilities available with different conjugation reactions. The
easiest (and most common) approach is to select the simplest and
most stable conjugation chemistry which is appropriate for the most
easily produced hapten(s). This is not always the best strategy.
The stability (chemical and biological) of the hapten molecule
during conjugation and subsequent use must allow for the *in vivo*
production of antibodies and *in vitro* production and use of hapten-
protein conjugates. Functional group protection strategies may be
required during conjugation. For example, in attempting to produce
stable aminotriazole conjugates (<u>10</u>), chromophore groups were used
to protect the reactive primary amine, providing colorimetric
monitoring of conjugation; these groups were then removed after
conjugation. In the preparation of bentazon conjugates (<u>11</u>),
dimerization of the activated hapten occurred during the conjugation
procedure, due to reactivity of the unprotected secondary nitrogen.
One possible solution would be to use a spacer which can cyclize
with the secondary nitrogen, temporarily protecting it while
providing another functional group for conjugation; after
conjugation, ring opening would restore the secondary nitrogen,
leaving an alkyl spacer. Our work on ethylenethiourea (ETU) is an
example where normal hapten conjugation protocols did not produce
verifiable conjugates. Because the unique portion of this molecule
is also highly reactive, special precautions were required for
verification of conjugation.

<u>Solubility of Hapten and Conjugate.</u> The solubility of hapten and
conjugate must also be considered, due to the nonpolarity of most
pesticides. Unusual measures may be required to overcome problems
of hydrophobicity. For endosulfan (<u>1</u>), periodate conjugation
(aqueous $NaIO_4$) of the diol was difficult due to poor water
solubility of the hapten. However, the N-hydroxysuccinimide active
ester method could be performed using over 50% dimethylformamide as
a cosolvent. In diflubenzuron antigen preparation (<u>20-22</u>), the use
of diazotization resulted in a reactive intermediate which is
charged and thus very soluble in aqueous systems. In addition,
differences exist among the possible carrier proteins. In assays
for the s-triazines (<u>4</u>, Harrison et al. <u>J. Agric. Food Chem.</u>, in
press), the solubility of all haptens was improved by using 20%
dimethylformamide as a cosolvent in aqueous conjugation reactions.
The solubility of all bovine serum albumin (BSA) conjugates was
better than keyhole limpet hemocyanin (KLH) conjugates, due to
solubility differences intrinsic to the two proteins. Thus KLH

conjugates were used for immunization (emulsification for injection mitigates solubility problems) and BSA conjugates were used for ELISA.

Ease of Synthesis. The ease of hapten synthesis is also important, especially in minimizing the number of synthetic steps required. For most of the assays covered here, generally two or three steps were required from commercially available starting materials. It is also convenient to use readily available starting materials if possible. For examples, the direct conjugation of reactive dichlorotriazines yielded ELISA antigens useful for analysis of the s-triazines (4, Harrison et al. J. Agric. Food Chem., in press). Direct conjugation of acid metabolites produced specific antibodies against both fenpropimorph (9) and diclofop-methyl (15). We must emphasize however that synthetic expediency at the expense of the other considerations we have discussed above is likely to create more problems than it solves.

Further Discussion of Thiocarbamates and Triazines

The haptens shown in Figure 1 were synthesized for the thiocarbamates molinate, EPTC, and thiobencarb. Two haptens for thiobencarb (3a-3b) contained S-alkyl spacers (in place of the p-chlorobenzyl group) with a terminal carboxylic acid group for conjugation, while two haptens contained aralkyl spacers (3c-3d) and were conjugated via a diazonium salt. The thiobencarb haptens with alkyl spacers elicited adequate antibodies, but the aralkyl haptens produced antibodies with superior specificity for thiobencarb, since hapten 3d is nearly identical to the target compound. This antibody could be used in both heterologous and homologous assays for the parent thiobencarb with acceptable sensitivity.

Sometimes options for spacer arm location are limited, as is the case for EPTC (Figure 1). Many thiocarbamates contain the S-ethyl moiety; thus only the dipropylamino group is unique to EPTC. Hence the spacer attachment for all our EPTC haptens (2a-2c) was at the sulfur, distal to the dipropylamino group. These haptens elicited good titer antibodies, but the best EPTC assay was 100 fold less sensitive than the best assays for molinate or thiobencarb. It appears that this deficiency is due to the EPTC structure itself. In light of this result, it is interesting to note the structural similarity of molinate and EPTC; symmetric opening of the molinate hexamethyleneimine ring yields EPTC. We suspect that the critical difference between these two compounds is the added rotational freedom of the two n-propyl groups of EPTC. While the relative assay sensitivities suggest that the alicyclic hexamethyleneimine group is superior to the aliphatic dipropylamino group for the production of antibodies, to our knowledge no systematic study of the advantages of ring systems exists.

In the triazine series (Figure 1), the best immunizing hapten for making triazine-class specific antibodies was compound 6d. This hapten has the spacer attached at the 2-position and contains 4-ethylamino and 6-isopropylamino groups. Hapten 6c, the 6-ethylamino analog, was slightly less effective, but this may be related to the decreased solubility and accompanying differences in conjugation and in vivo antigen presentation. It is not clear at present which is

the best immunizing hapten for producing simazine specific
antibodies. The most sensitive monoclonal antibody (1-2 ppb for
atrazine) was made in response to compound 6d. The immunizing
hapten producing the best specificity for atrazine was 5j
(containing 2-chloro and 4-isopropylamino groups, with the spacer
arm at the 6-position). Only 8% cross reactivity with simazine was
noted for the best rabbit antibody made against this hapten, a
surprising result considering that the structures differ only by one
methylene group. Antibodies against compound 5g (containing 2-
chloro and 4-ethylamino groups, with the spacer arm at the 6-
position) were also very specific for atrazine, as expected.
However, their cross-reactivity for simazine was approximately 100%,
but with reduced sensitivity compared to the antibodies against
compound 5j.

The recognition of variable spacer arm length triazine haptens
was evaluated using competitive ELISA. Using antibodies made
against haptens 5j and 6d, a clear decrease in recognition was
observed as the spacer arm was shortened. Based on these results,
the n = 1, 2, and 3 derivatives of simazine (5c-5e) and the n = 1
and 2 derivatives of atrazine (5h-5i) were chosen for conjugation to
alkaline phosphatase and BSA. The ultimate goal of this approach is
to produce more sensitive heterologous assays by exploiting the
reduced affinity of the antibodies for the conjugated haptens, while
retaining specificity for the triazine class (4, Harrison et al. J.
Agric. Food Chem., in press).

Conclusions

1. The importance of the above criteria, especially spacer
    recognition and preservation of parent molecule functional groups,
    appears to increase as the size of the target molecule decreases
    and as the number of clearly recognizable functionalities
    decreases (6,10).
2. Some molecules may be inadequate for the production of specific
    antibodies due to size, reactivity, or structure. Limitations due
    to size may be manifested only as decreasing assay sensitivity
    with decreasing molecular size.
3. Sometimes it is impossible or very difficult to avoid
    sacrificing a useful determinant group in hapten synthesis;
    exploration of multiple haptens is crucial in such cases (6).
4. Multiple haptens should be prepared for each target compound,
    including different conjugation positions and spacer lengths and
    structures, if possible. While it may be unnecessary to use all
    of the synthesized haptens for antibody production, heterologous
    systems should be carefully explored to optimize assay sensitivity
    while retaining specificity and ruggedness.
5. The strategy of synthesizing a library of potential haptens
    during the early phase of agricultural chemical product
    development would be a valuable corporate policy because it would
    facilitate the later development of immunoassays. An added
    benefit is that these derivatives would offer new compounds for
    screening as potential pesticides or serve as metabolite
    standards.
6. The ability to generate class or compound specific antibodies
    depends greatly on the class/compound structure and the number of

closely related compounds which might be encountered in routine analysis. The use of multiple haptens employing different conjugation positions and substitution patterns allows exploration of class/compound specificity.

7. When designing assays for larger molecules, immunizing haptens containing less than the complete parent structure may yield antibodies which adequately recognize the target molecule (12, 16). Such haptens are more likely to produce class specific antibodies due to non-recognition of the omitted structure.

8. Study of the literature on low molecular weight drugs (14 for review) may offer further insight (valproic acid, phenobarbital, caffeine, nicotine, etc.).

9. Understanding of hapten-antibody interaction would benefit from physical-chemical study of binding (x-ray, NMR, etc.), with the realization that for most molecules there will be many possible antibody combining site structures.

Acknowledgments

This work was supported in part by NIEHS Superfund grant PHS ES04699, EPA grant CR-814709-02-0, and a grant from the California Department of Food and Agriculture. BDH is a Burroughs-Wellcome Scholar in Toxicology.

Literature Cited

1. Dreher, R.M.; Podratzki, B.; J. Agric. Food Chem. 1988, 36, 1072-1075.
2. Ercegovich, C.D.; Vallejo, R.P.; Gettig, R.R.; Woods, L.; Bogus, E.R.; Mumma, R.O. J. Agric. Food Chem. 1981, 29, 559-563.
3. Gee, S.J.; Miyamoto, T.; Goodrow, M.H.; Buster, D.; Hammock, B.D. J. Agric. Food Chem. 1988, 36, 863-870.
4. Goodrow, M.H.; Harrison, R.O.; Hammock, B.D. J. Agric. Food Chem. 1990, 38, 990-996.
5. Hammock, B.D.; Mumma, R.O. In Pesticide Analytical Methodology; ACS Symposium Series No. 136; Zweig, G., Ed.; American Chemical Society: Washington, DC, 1980; pp 321-352.
6. Harrison, R.O.; Brimfield, A.A.; Nelson, J.O. J. Agric. Food Chem. 1989, 37, 958-964.
7. Harrison, R.O.; Gee, S.J.; Hammock, B.D. In Biotechnology in Crop Protection: ACS Symposium Series No. 379; Hedin, P.A., Menn, J.J., Hollingworth, R.M., Eds.; American Chemical Society: Washington, DC, 1988; pp 316-330.
8. Jung, F.; Gee, S.J.; Harrison, R.O.; Goodrow, M.H.; Karu, A.E.; Braun, A.L.; Li, Q.X.; Hammock, B.D. Pest. Sci. 1989, 26, 303-317.
9. Jung, F.; Meyer, H. H. D.; Hamm, R. T. J. Agric. Food Chem. 1989, 37, 1183-1187
10. Jung, F.; Szekacs, A.; Hammock, B.D. International Chemical Congress of Pacific Basin Societies, Honolulu, 1989, Abstract 01-210.
11. Li, Q.X.; Hammock, B.D.; Seiber, J.N. International Chemical Congress of Pacific Basin Societies, Honolulu, 1989, abstract 01-217.

12. Newsome, W.H. In <u>Pesticide Science and Biotechnology,</u> <u>Proceedings of the Sixth International Congress of Pesticide</u> <u>Chemistry</u>; Greenhalgh, R., Roberts, T.R., Eds.; Blackwell: Oxford, 1987; pp 349-352.
13. Niewola, Z.; Hayward, C.; Symington, B.A.; Robson, R.T. <u>Clin.</u> <u>Chim. Acta</u> **1985**, <u>148</u>, 149-156.
14. Oellerich, M. <u>J. Clin. Chem. Clin. Biochem.</u> **1980**, <u>18</u>, 197-208.
15. Schwalbe, M.; Dorn, E.; Beyermann, K. <u>J. Agric. Food Chem.</u> **1984**, <u>32</u>, 734-741.
16. Stanker, L.H.; Bigbee, C.; Van Emon, J.; Watkins, B.; Jensen, R. H.; Morris, C.; Vanderlaan, M. <u>J. Agric. Food Chem.</u> **1989**, <u>37</u>, 834-839.
17. Vallejo, R.P; Bogus, E.R.; Mumma, R.O. <u>J. Agric. Food Chem.</u> **1982**, <u>30</u>, 572-580.
18. Van Emon, J.; Hammock, B.D.; Seiber, J.N. <u>Anal. Chem.</u> **1986**, <u>58</u>, 1866-1873.
19. Van Emon, J.; Seiber, J.N.; Hammock, B.D. In <u>Analytical Methods</u> <u>for Pesticides and Plant Growth Regulators: Advanced</u> <u>Analytical Techniques, Vol. XVII</u>; Sherma, J., Ed.; Academic Press: New York, **1989**, pp 217-263.
20. Wie, S.I.; Sylwester, A.P.; Wing, K.D.; Hammock, B.D. <u>J. Agric.</u> <u>Food Chem.</u> **1982**, <u>30</u>, 943-948.
21. Wie, S.I.; Hammock, B.D. <u>J. Agric. Food Chem.</u> **1982**, <u>30</u>, 949-957.
22. Wie, S.I.; Hammock, B.D. <u>J. Agric. Food Chem.</u> **1984**, <u>32</u>, 1294-1301.
23. Weiler, E. W. In <u>Chemistry of Plant Protection</u>; Bowers, W. S., Ebing, W., Martin, D., Wegler, R., and Yamamoto, I., Eds.; Springer-Verlag: New York, **1990**, pp. 145-220.

RECEIVED August 30, 1990

Chapter 3

Rapid On-Site Immunoassay Systems

Agricultural and Environmental Applications

J. H. Rittenburg, G. D. Grothaus, D. A. Fitzpatrick, and R. K. Lankow

Agri-Diagnostics Associates, 2611 Branch Pike, Cinnaminson, NJ 08077

A simple immunoassay system has been developed for rapid on-site analysis and quantitation of chemical residues. The system consists of an assay device with an absorbant core to which antibody can be immobilized, and a small handheld reflectometer for quantitation of the assay results. The competitive immunoassay is performed by adding reagents to the surface of the device from dropper bottles. As each solution is absorbed into the device, it passes through the surface zone of immobilized antibody allowing the antibody-antigen reactions to proceed. The immunoassay can be completed within 10 minutes with a visually observable color endpoint. Each assay device contains a negative control reference zone that is used for comparison to the sample zone. Results are quantitated using a handheld, dual beam, reflectometer that compares the color intensity of the sample zone to that of the reference zone. The development of both a multiwell and field usable immunoassay for quantitation of alachlor in the low to sub ppb range is described.

The accurate and precise analysis of pesticides is a critical requirement for the registration and use of pesticides throughout the world. Parent molecules, key metabolites and chemical breakdown products must be identified and studied in well designed laboratory and field research trials. Environmentally sound management practices rely on significant amounts of information about the levels and movements of pests, pathogens and specific chemical treatments within the environment. The methods available for such analysis have become extremely sophisticated and sensitive in response to the need to detect lower and lower levels of contaminants in crops, water, soil, and farm animals.

Despite the tremendous sophistication of pesticide residue and environmental chemical analysis, there remain a number of serious limitations to certain aspects of classical analysis. A number of those limitations can be addressed through the application of immunoassay technology to residue analysis (1-4). Immunoassays rely on highly specific antibody proteins and relatively simple analytical apparatus to detect and quantify a wide variety of target materials in a broad range of analytical matrices. Since the reagents are specific, immunoassays can generally be performed with relatively crude sample preparations. Reduced sample preparation, simple

0097–6156/91/0451–0028$06.00/0

assay procedures, high throughput capabilities, and relatively inexpensive automation make immunoassay procedures much less expensive on a per sample basis than conventional methods. Additionally, immunoassays can be readily adapted to simple and rapid on-site testing methods that generate timely information enabling better informed decisions to be made.

Immunoassay Technology

Immunoassays are analytical techniques based on the specific, high affinity binding of inducible animal-derived proteins called antibodies with particular target molecules called antigens. Antibody formation by higher vertebrates is remarkable both in the specificity of the induced antibody to its target and in the variety of organic molecules and macromolecules that are able to induce a specific antibody response. The primary binding between the antibody and the target antigen forms the basis of the immunoassay, and a wide variety of immunoassay "formats" have been developed to allow either visual or instrumental measurement of this primary binding reaction. The tremendous variety of immunoassay formats and reagent configurations presently being used in medical, veterinary, food and agricultural immunodiagnostics all represent different ways of visualizing the primary antibody-antigen reaction.

Over the past 25 years this methodology has been successfully applied to many of the analytical challenges of the medical health care industry for rapid and accurate measurement of analytes such as hormones, microorganisms, therapeutic drugs, drugs of abuse, and tumor markers . The past 10 years has seen a rapid expansion of immunoassay techniques into forensic, veterinary, food and agricultural analyses. The potential of rapidly measuring a very minute quantity of a specific analyte from within a complex sample matrix, often with little or no sample clean-up, is one of the attractive features that has led to the widespread application of immunoassay techniques.

Antibody-producing cells generally respond to macromolecules and compounds with a molecular weight greater than 10,000 Daltons when those materials are recognized as foreign by the immune system. In general, compounds of the molecular weight of most pesticides will not independently elicit an immune response. If those compounds are covalently attached to a carrier macromolecule, however, the immune system will respond and produce antibodies to the small molecule portion of the complex (hapten) as well as to other regions of the carrier.

The specificity of an antibody to a small molecule such as a pesticide can be influenced to a large degree by the design of the immunogen used to induce antibody formation. The immunogen is constructed by covalently coupling the small molecule or a related analogue to a carrier protein. In general, antibody specificity is highest for the part of the molecule furthest from the carrier protein. Thus in synthesizing the immunogen it is possible to orient the small molecule in ways that will favor antibody specificity to particular portions of the molecule. Through selective construction of the immunogen, it is possible to induce antibodies that may or may not, for example, differentiate a parent pesticide from its major metabolite, or a specific pesticide from a related family of pesticides. Additional levels of specificity may also be achieved through selection of appropriate monoclonal antibodies.

Most small molecules, including many of the pesticides, in the less than 1000 Dalton molecular weight range only have one antigenic determinant and thus must be analyzed using a competitive immunoassay format. A wide variety of competitive immunoassays have been developed for analysis of small molecules such as antibiotics (5), pesticides (6), toxins (7), and hormones (8). Further information concerning basic immunoassay technology is covered in detail in Chapter 1.

Alachlor Immunoassay Development

The herbicide alachlor is one of the most widely used pesticides in North America. It is used primarily to control grassy weeds in corn and soybeans and can be found as a groundwater contaminant. Some laboratory testing services routinely analyze water samples for alachlor residues using chromatographic methods. The availablility of an appropriate immunoassay would reduce the turnaround time and costs incurred in current residue analyses for alachlor and enable larger numbers of samples to be analyzed. Development of a laboratory immunoassay for alachlor analysis was first described by Wratten and Feng (9).

This paper describes work at Agri-Diagnostics toward the development of a standardized and stabilized multiwell immunoassay kit and development of a simple, rapid, and quantitative on-site immunoassay kit for analysis of alachlor in groundwater.

Experimental Methods

Protein conjugates of alachlor were synthesized by two different methods for use as immunogens and antibody screening conjugates. Conjugates were prepared using a carbodiimide driven condensation reaction of the 2-(4-aminophenylthio)-2',6'-diethyl-N-methoxymethylacetanilide and bovine serum albumin (BSA) as shown in Figure 1. Alachlor was also coupled directly to thiolated chicken albumin and BSA through an alkylation reaction. Acetyl homocysteine thiolactone was used to first thiolate the protein which was subsequently mixed with alachlor under alkaline conditons resulting in alkylation of the thiol by the herbicide to form the conjugate (Figure 2). These conjugates were used to immunize sheep and mice and also to coat 96 well polystyrene plates for use in antibody screening. Alachlor was also covalently coupled to horseradish peroxidase for use as an enzyme-hapten tracer in a direct competitive assay format.

Alachlor, metolachlor, and butachlor were obtained from Chem Service (West Chester, Pa). Acetanalide anlogues used in the cross-reativity studies were provided by Ricerca, Inc. (Painesville, Ohio).

Indirect and direct competitive multiwell assay formats, and a rapid field usable format were developed and performed as described below:

Indirect competitive multiwell assay:
1. Add 50μl of negative control, standards, and test samples to wells of multiwell plates coated with the chicken albumin-alachlor conjugate.
2. Add 50μl of alachlor antibody to each well and mix for 10 minutes.
3. Rinse out wells five times and add 100μl of anti-sheep globulin peroxidase to each well.
4. Rinse out wells five times and add 100μl of enzyme substrate (ABTS) to each well. Mix for 10 minutes.
5. Add 50μl stop solution to each well and mix for 10 seconds.
6. Read absorbance at 405nm.

Direct competitive multiwell assay:
1. Make a 1:1 mixture of the water sample or standard or negative control with the peroxidase-alachlor tracer.
2. Add 100μl of each mixture from step 1 into respective wells of multiwell plates coated with purified sheep anti-alachlor antibody and mix for 10 minutes.

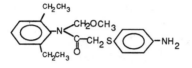

ALACHLOR AMINOBENZENE

+

BSA-COOH

+

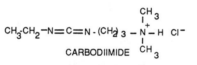

CARBODIIMIDE

$-H_2O$

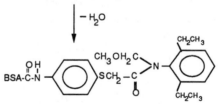

BSA - ALACHLOR CONJUGATE

+

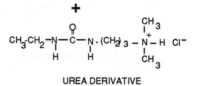

UREA DERIVATIVE

Figure 1.   Conjugation of alachlor aminobenzene to bovine serum albumin.

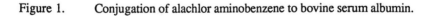

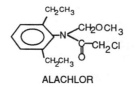

ALACHLOR

+

OA-NH$_2$

+

ACETYL HOMOCYSTEINE THIOLACTONE

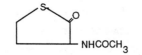

− HCL

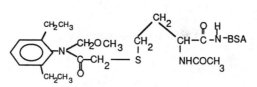

**CHICKEN ALBUMIN - ALACHLOR
CONJUGATE**

Figure 2.   Conjugation of alachlor to chicken albumin using acetyl homocysteine thiolactone.

3. Rinse out wells five times and add 100μl of enzyme substrate (ABTS) to each well. Mix for 10 minutes.
4. Add 50μl stop solution (1.5% NaF) to each well and mix for 10 seconds.
5. Read absorbance at 405nm.

Rapid field usable assay:
1. Make a 1:1 mixture of the water sample or standard or negative control with the peroxidase-alachlor tracer.
2. Add 1 drop of the negative control peroxidase-alachlor tracer mixture and one drop of the sample peroxidase-alachlor tracer mixture to respective antibody coated zones on the surface of the porous plastic device.
3. Allow all liquid from each drop to completely drain into the porous device.
4. Add 1 drop of rinse solution to each zone and allow to drain.
5. Add 1 drop of enzyme substrate solution to each zone and allow to drain.
6. Add 1 drop of stop solution to each zone and allow to drain into device
7. Quantitate results using hand-held reflectometer.
    Total test time approximately 5-10 minutes

Cross-reactivity is expressed as the ratio of the concentration of alachlor to each test compound at level that gives 50% inhibition of the immunoassay maximal binding level.

Results and Discussion

An indirect competitive assay was developed using the chicken albumin-alachlor conjugate as the solid phase antigen and antisera obtained from a sheep immunized with the BSA-alachlor aminobenzene immunogen. A sensitivity limit of approximately 1 ppb was observed. Cross-reactivity with two other acetanalide herbicides, metolachlor and butachlor, was 2.3% and 6.4% respectively, indicating that the methoxymethyl region of the alachlor molecule plays a significant role in antibody specificity (Figure 3).
    Additional sensitivity was obtained by formatting the assay in a direct competitive configuration where the antibody was immobilized to the solid phase and a peroxidase-alachlor conjugate was used in a simultaneous competition reaction with the sample. A sensitivity limit of approximately 0.1ppb was observed with an IC50 of about 2ppb. The dynamic range of the assay allows for quantitation as high as 50ppb (Figure 4). Cross-reactivity of the direct competitive assay with metolachlor and butachlor was 0.5% and 1.0% respectively. Cross-reactivity analysis of a variety of other chloroacetanalides indicated that changes to alachlor at either the methoxy methyl side chain or to the ethyl groups on the ring result in a major loss of antibody recognition (Table 1).

Field Usable Format

On-site immunoassay testing formats are regularly used in the medical area for doctors office and home testing applications. One type of on-site format that has become very popular for home pregnancy testing (10) and one which Agri-Diagnostics has successfully used for plant disease diagnostics (11) and is now applying to chemical analysis, is the flow through assay device. In this format the antibody or antigen is immobilized within a microporous surface that is in contact with an absorbant reservoir. The test sample is added to the surface of the device and flows through the activated area into the absorbant reservoir. The movement of the sample and the subsequent reagents across the immobilized antibody or antigen

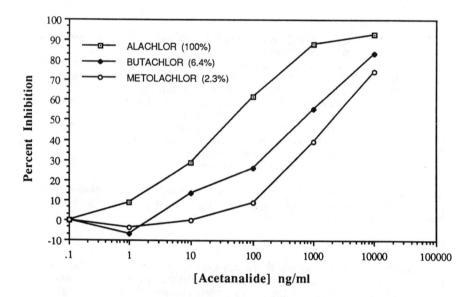

Figure 3.   Comparative dose response curves of alachlor, butachlor, and metolachlor using an indirect competitive ELISA.

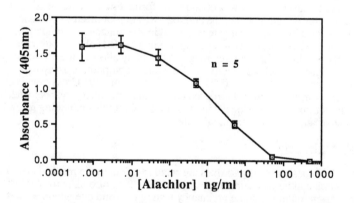

Figure 4.   Alachlor dose response curve using a peroxidase-alachlor conjugate in a direct competitive assay configuration.

Table 1

CROSS-REACTIVITY OF ACETANALIDE ANALOGUES
IN THE DIRECT COMPETITIVE ALACHLOR IMMUNOASSAY

Compound	I 50 Value (ug/ml)	% Cross-Reactivity	Structure
Alachlor	0.002	100.0%	
Butachlor	0.20	1.0%	
Metholachlor	0.40	0.5%	
SDS-023018	0.05	4.0%	
SDS-023946	5.50	< 0.1%	
SDS-024048	60.00	< 0.1%	
SDS-024742	20.00	< 0.1%	
SDS-025664	0.03	6.7%	
SDS-025665	7.00	< 0.1%	

zone enhances the reaction rates and improves the efficiency of rinsing away the unbound reagents. The reagents can be easily added to the device from dropper bottles and all the waste liquid from the assay accumulates within the device reservoir providing safe containment for toxic analytes.

The rapid, field usable assay format developed detects alachlor at concentrations between 1 and 500 ppb. The device and meter system is illustrated in Figure 5. The competitive immunoassay is performed by sequentially adding sample, enzyme-labeled reagent, and enzyme substrate to the surface of the device from dropper bottles. As each solution is absorbed into the device, it passes through the surface zone of immobilized antibody or antigen. The immunoassay can be completed within 10 minutes resulting in a visually observable color endpoint. Each assay device contains a negative control reference zone that is used for comparison to the sample zone. The assay device couples to a handheld dual beam reflectometer which compares the color intensity of the sample zone to that of the reference zone. The results can be displayed as percent inhibition or the analyte concentration can be extrapolated from a pre-programmed curve.

The immunoassay device is comprised of an absorbant, porous plastic reservoir having a microporous surface with a controlled pore size (Figure 6). The microporous surface is treated with a hydrophobic mask to leave two circular zones exposed to which antibody or antigen bound latex can be entrapped. To prepare the device for use in the direct competitive assay, antibody sensitized latex particles are immobilized to the two zones on the device surface. After drying, the sensitized devices can be stored for up to 12 months prior to use. To perform the test a mixture of the sample and enzyme-hapten conjugate is applied to one zone and a mixture of a negative control or reference and enzyme-hapten conjugate is applied to the other zone. As these solutions are absorbed into the device they move through the antibody coated latex layer enabling the competition of any free analyte with the enzyme-hapten tracer for binding to the solid phase antibody. Enzyme-hapten conjugate that has not been specifically bound by the antibody is rinsed through the surface of the device using drops of a rinse solution. A chromogenic peroxidase substrate, 4 chloronaphthol, is then applied to the surface of the device. In the presence of peroxidase a visible blue precipitate will develop at the surface of the device as the substrate passes through the antibody zones. The amount of color that develops in each zone is inversely proportional to the amount of analyte in the sample. A battery operated handheld dual beam reflectometer was developed to quantitate the color endpoint reactions on the immunoassay device. The meter can be programmed to take the ratio between the reflectance of the sample zone and the negative control zone and display the percent inhibition which can then be extrapolated to concentration values from a dose response curve. Alternatively analyte concentration can be extrapolated from either a pre-programmed dose response curve or from a single point standard curve (in this case a specific level of analyte would be applied to the control zone) and displayed as either absolute concentration (eg. ppb) or as numbers or ranges (eg. LO, MED, HI). The meter memory can store up to 84 readings with an optional 4 digit label to identify each reading. An Rs 232 interface enables contents of the memory to be copied directly to a printer or computer. The use of the reflectometer removes the subjectivity from operator interpretation and provides data for documentation.

The development of a rapid on-site immunoassay system with a versatile and easy to use handheld meter provides an objective means of screening for levels of agricultural and environmental chemicals in either a remote site or laboratory setting and will enable better monitoring of the levels and movement of chemicals through the environment.

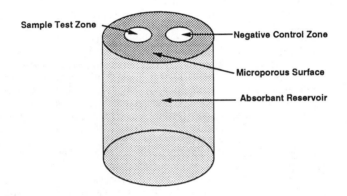

Sample Test Zone ● ── Negative Control Zone
── Microporous Surface
── Absorbant Reservoir

Figure 5.   Agri-Diagnostics on-site assay device and meter system.

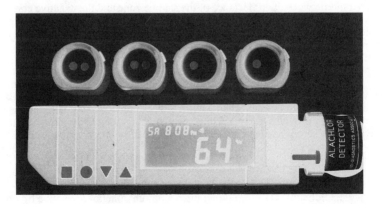

Figure 6.   Schematic diagram of flow through immunoassay device.

Summary and Conclusions

Agri-Diagnostics has developed and successfully marketed rapid on-site immunoassay systems for monitoring the levels of a variety of plant pathogens (12,13). The information provided by these tests aids in the selection and timing of the appropriate chemical treatments to be used in an agricultural management program. These same types of immunodiagnostic systems are now being developed for the rapid on-site analysis and quantitation of agricultural and environmental chemicals.

   Development of a simple, rapid, and quantitative on-site immunoassay system was undertaken to enable better monitoring of the levels and movement of the chemicals in the environment. Alachlor was selected as an initial target compound for immunoassay development. Immunogenic conjugates of alachlor with bovine serum albumin(BSA) were synthesized through a carbodiimide driven condensation reaction of the 2-(4-aminophenylthio)-2',6'-diethyl-N-methoxymethylacetanilide and BSA and also through an alkylation reaction between alachlor and thiolated BSA. A 40 minute indirect competitive multiwell immunoassay was developed using an chicken albumin-alachlor conjugate as the solid phase antigen. A sensitivity limit of approximately 1 ppb was observed with 2.3 % metolachlor and 6.4% butachlor cross-reactivity. A 25 minute direct competitive multiwell format was also developed which demonstrated sensitivity to alachlor at the 0.1ppb level and cross-reactivities of 0.5% and 1.0% with metolachlor and butachlor repectively.

   A quantitative, field usable system was developed for detection of alachlor at levels down to 1ppb in less than 10 minutes. The immunoassay device is comprised of a porous plastic reservoir having a microporous surface with a controlled pore size. The microporous surface is treated with a hydrophobic mask to leave two circular zones exposed to which antibody bound latex can be entrapped. A battery operated handheld dual beam reflectometer was developed to quantitate the color endpoint reactions on the immunoassay device. The meter can be programmed to take the ratio between the reflectance of the sample zone and the negative control zone and display the percent inhibition. Work is currently in progress to increase the on-site assay sensitivity to 0.5ppb through additional optimization of the assay parameters and possibly incorporation of a simple solid phase extraction and concentration step. Further antibody development is also in progress for use in a group specific acetanalide immunoassay.

   The development of a rapid on-site immunoassay system with a versatile and easy to use handheld meter provides an objective means of screening for levels of agricultural and environmental chemicals in either a remote site or laboratory setting. This technology provides a cost effective way to obtain the timely information needed in many crop management and environmental monitoring programs.

Acknowledgments

We thank Tom Magee and Ricerca, Inc., 7538 Auburn Road, Box 1000, Painesville, Ohio 44077, for providing the 4 aminothiophenol derivative of alachlor and the acetanaline analogues used in this work.

Literature Cited

1.   Hammock, B.D.; Mumma, R.O. In Pesticide Analytical Methodology; Zweig, G., Ed.; ACS Symposium Series No. 136; American Chemical Society: Washington, D.C., 1980; pp 321-352.
2.   Mumma, R.O.; Brady, J.F. In Pesticide Science and Biotechnology; Greenhalgh, R. and Roberts, T.R., Eds.; Blackwell: Oxford, 1986; p 341.

3.   Harrison, R.O.; Gee, S.J.; Hammock, B.D. In <u>Biotechnology for Crop Protection</u>; Hedin, P. A.; Menn, J.J.; Hollingworth, R.M., Eds; ACS Symposium Series No. 379; American Chemical Society: Washington, D.C., 1988; pp 316-330.

4.   Vanderlaan, M.; Watkins, B.; Stanker. (1988). Environ. Sci. Technol., **22**, (3), pp 247-254.

5.   Yao, R.C.; Mahoney, D.F. (1984). J. Antibiotics, **37**, (11), pp 1462-1468.

6.   Van Emon, J.M.; Hammock, B.D.; Seiber, J.N. (1985). Anal. Chem. **58**, pp 1866-1873.

7.   Dixon-Holland, D. E., Pestka, J. J., Bidigare, B. A., Casale, W. L., Warner, R. L., Ram, B. P., and Hart, L. P. (1988). J. Food Protection, **51**, (3), 201-204.

8.   Meyer, H. H. D. and Hoffmann, S. (1987). Food Additives and Contaminants, **4**, (2), 149-160.

9.   Wratten, S. J.; Feng, P. C. (1989). In <u>Development and Application of Immunoassay for Food Analysis.</u>, Rittenburg, J. H., Ed.; Elsevier Applied Science Publishers, London., 1990.

10.   Valkirs, E. G., Owen, N. C., and Levinson, P. A. (1986). U.S. Patent No. 4,632,901, Dec. 30

11.   Miller, S.A., Rittenburg, J. H., Petersen, F. P., and Grothaus, G. D. (1988). Brighton Crop Protection Conf. Proc. 795-803; Lavenham Press Limited, Lavenham U.K. Publ.

12.   Miller, S.A.; Rittenburg, J.H.; Petersen, F.P., Grothaus, G.D. (1990). In <u>Techniques for Detection and Diagnosis in Plant Pathology.</u>, Torrance, L.; Duncan, J.R., Eds.; Blackwell Scientific Publications, in press.

13.   Steele, W., (1989). Turf, July.

RECEIVED August 30, 1990

Chapter 4

Testing of Food and Agricultural Products by Immunoassay

Recent Advances

William P. Cochrane

Laboratory Services Division, Agriculture Canada, Ottawa, Ontario K1A 0C6, Canada

The evaluation of a number of immunoassay diagnostic
kits was undertaken to determine their usefulness in
a regulatory analytical laboratory environment in the
food, feed and pesticide areas. Four rapid enzyme
immunoassay tests for the detection of aflatoxin
residues at the 20 ppb level in animal feeds were
compared to the official HPLC procedure. In the
pesticide area, a commercial pentachlorophenol
competitive inhibition assay for residues in water was
investigated as to its applicability to poultry and
pork liver matrices. In addition, an ELISA screening
procedure for the herbicide fusilade was developed.
Modifications were incorporated into the rapid
immunoband 1-2 Test procedure for the detection of
motile Salmonella in various food and animal feed
products resulting in quicker analysis than the
standard culture method. Also, a comparative
evaluation of a Quik-Card Test for sulphamethazine
drug residues in pork urine, liver and muscle tissue
is described.

New diagnostic tests have been developed based on monoclonal and
polyclonal antibody technology and have become commercially
available in the last two to three years for a wide range of
applications in agriculture. Economical, rapid, sensitive and
easy to use, these diagnostic tests are making an impact on three
major analytical areas, namely:
- Testing for pathogenic organisms.
- Detection and identification of toxin and drug contamination
  in animal feeds and foods.
- Monitoring for pesticide residues in water, soil, foods and
  agricultural crops.

These diagnostic products have appeared in the form of
hand-held cards, visual tests, affinity columns that have the
potential of replacing the standard clean-up columns in residue
analysis, and new ELISA tests.

It is, therefore, essential to evaluate these new products

0097–6156/91/0451–0040$06.00/0
© 1991 American Chemical Society

and to determine their strengths and limitations. To this end a number of rapid test kits in the animal feed and food areas have been assessed, where possible, in tandem with the normal conventional methods currently in use.

Animal Feeds

The presence of fungal toxins or mycotoxins in foods and animal feeding stuffs has become one of the more important concerns in human and animal health (1-2). Mycotoxin contamination of animal feeds occurs as a result of invasion of crop by field fungi or the growth of fungi in crops stored under less than ideal conditions. The toxicity of mycotoxins in general and the carcinogenic potential of aflatoxins in particular are well documented (3). Significant aflatoxin-contamination can occur in crops such as corn and cereals, which are major constituents of animal feeds. The potential, therefore, exists that aflatoxin in feed can be left as residue in meat, milk or eggs (4). Agriculture Canada has an ongoing program of testing feeds from the domestic market and export certification for aflatoxin (B1, B2, G1, G2) residues using an HPLC technique sensitive at the 2.5 ppb level for each aflatoxin with results being reported as "total" aflatoxin. To help increase sample throughput and reduce analysis time, four commercially available enzyme immunoassay tests for aflatoxin residues were evaluated.

The test kits studied were: Aflatest, Agri-Screen, E-Z Screen, and Oxoid.

E-Z Screen. Manufactured with a proposed use for aflatoxins B1, B2, G1, G2 at 10 ppb, was found to be effective for B1 and G1 in spiked corn at the 10 ppb level. Sample preparation was as prescribed by the kit manufacturer except the sample was shaken for one hour, instead of blending for one minute. A 50 g sample was shaken for one hour with 100 mL 80% methanol and allowed to settle. Then a 100 uL of extract was diluted with 200 uL buffer and 50 uL of this mixture was applied to the sample port of the EZ card. Aflatoxin B2 was detected at the 15 ppb level and G2 at the 20 ppb level. This card test was less sensitive for mixed feeds which contain more pigments. Due to the lack of demand, the manufacturer has stopped producing cards sentitive at the 10 ppb level, but control cards at the 20 ppb are available. In one study 22 samples were tested along with standards and formulated product. No false negatives were encountered and only one false positive. No false positives were encountered on 40 samples tested in another laboratory. Recently cards sensitive to 5 ppb were introduced. Preliminary research indicates the cards will perform to specifications. As the presently used HPLC procedure processes 6 samples in a 2-day period and 99% of the samples were negative, the aflatoxin E-Z Screen could provide the necessary screening capability.

Agri-Screen. Here an antibody fixed in wells can either be used for fast visual screening or used in conjunction with instrumentation for quantitation of aflatoxin B1. Normally, standards or controls are used, one of which must be at the zero ppb level and

one at the level of detection - these being prepared by adding
the appropriate aflatoxin B1 amount to 55% MeOH/H_2O extracts of
aflatoxin-free samples of the feed being tested. The proposed
use is for aflatoxin B1 at the 2 ppb level but the antibody had
reduced cross reactivity levels for B2, G1 and G2 at 31%, 25% and
12% respectively. The test was found to detect aflatoxin
residues in spiked corn and mixed feed samples at the following
levels: B1 - 5 ppb, G1 - 10 ppb, B2 and G2 - 20 ppb.
Approximately 400 samples (feeds, feed ingredients and cereal
grains) were qualitatively analyzed. About 1% of the samples
gave a positive result; HPLC analysis showed these results to be
false positives. In a collaborative study with the manufacturer,
results for B_1 correlated well to expected values; results for
B_2, G_1 and G_2 correlated to the relative cross reactivities.

Oxoid. This affinity column detection procedure proved to be the
most sensitive of all the kits, 0.25 - 0.5 ppb for each aflatoxin
with the limit of detection for total aflatoxin being 1 ppb. The
test involved extraction of 10 g sample with 20 mL MeOH/water
(60/40) using a blender for two minutes or a one hour shake. A
portion of the centrifuged extract is diluted with phosphate
buffered saline (PBS), this is added to an affinity column, which
is washed with water and the aflatoxins then diluted with 2 mL
methanol. For qualitative analysis sensitive to 1 ppb total
aflatoxin, 6 mL water is added to the methanol eluate then the
aflatoxins are extracted into chloroform. The chloroform extract
is passed through a mini column which is then viewed under a UV
light. The aflatoxin is evident as a fluorescent band at the
florisil-silica gel interface. Quantitation can be done by HPLC,
but not to the same sensitivity. The kit was evaluated on 20
spiked feed and corn samples. Although the manufacturer
recommends a 2:1 extractant/sample weight ratio, this was not
adequate for some matrices and increased volume of extractant was
required. There was no sample extraction, dilution, or filter
kit available, therefore, one would have to be designed for use
at field level.

Aflatest. This immunoaffinity column also met the lower
detection requirement for the USFDA at a permissible limit of 20
ppb for aflatoxin. Qualitative results can be obtained using a
florisil tip, however, the florisil tip was found difficult to
interpret. Evaluation showed the quantitative analysis to be as
accurate as other HPLC procedures, but much more rapid. Using
the Aflatest for sample extract purification, and HPLC with post
column derivatization using iodine, a limit of quantitation of
2.5 ppb for B_1 and G_1 is easily achieved. Sensitivity to B_2 and
G_2 is lower due to less specificity of the Aflatest antibodies to
these toxins.
    The time required for complete analysis using any of the
above kits was about 30 minutes depending on the amount of time
required for the filtration of the sample. In summary, any of
the above immunoassay tests could be use for screening of animal
feeds down to a level of 20 ppb. Since the above study, the USDA
recently completed a comparative study on six commercially
available aflatoxin test kits including the above four with

essentially the same findings (5) except that the Agri-Screen kit
was not originally recommended. The Agri-Screen Test was
subsequently retested and found to be satisfactory.

Foods

Salmonella. The health and safety of the nation's food supply is
a prime concern to regulatory agencies. One of the key analyses
performed in microbiological testing laboratories is the
detection of Salmonella. All known Salmonella spp are pathogenic
to man and/or animals and cause the well-known Salmonellosis
infection. Even though federal regulations prohibiting the
presence of Salmonella in foods are in place, the recovery of
this organism from a particular food commodity may be quite
difficult.
    Surveillance of salmonellae by the 6-step conventional
culture method (6) is both costly and time-consuming - a negative
test requiring a minimum of 96 hours and a positive test at least
an additional 24 hours. Although, a number of prospective rapid
methods have been developed, none however, have equalled the
accuracy and reliability of the culture method.
    One of these rapid methods, namely the Salmonella 1-2 TEST
was compared to the conventional procedure. The 1-2 TEST is a
small plastic disposable device with an inoculation and a
motility chamber. The test utilizes the motility of the
Salmonella organism and a reaction with flagellar antibodies to
form a distinguishable ImmunoB and seen 8-16 hours after
inoculation. The Salmonella 1-2 TEST received official first
action status at the 1988 meeting of the Association of the
Official Analytical Chemists (AOAC) (7) but subsequently
unfavourable evaluations of the 1-2 TEST were published (8-9).
In our initial evaluation (Table I, Phase 1), 196 food and feed
samples were analysed by both methods. Thirty-four samples were
found positive by the wet culture method but only 26 were found
positive by the 1-2 TEST (10). The procedure was modified to
include an enrichment stage using tetrathionate brilliant green
broth. Of the 314 subsequent samples analysed (Table I, Phase
2), 84 samples were found to be positive. There were 79 samples
found positive by both methods while of the remaining 5, 3 tested

Table I: Salmonella 1-2 TEST Results

Product	Phase 1			Phase 2			Phase 3		
	# Sam	Cult Meth	1-2 TEST	# Sam	Cult Meth	1-2 TEST	# Sam	Cult Meth	1-2 TEST
Animal Feeds	96	29	23	208	80	81	225	64	62
Cheese Environ.	20	0	0	-	-	-	-	-	-
Egg Environ.	38	5	3	59	0	0	40	2	2
Cheese & Gums	32	0	0	-	-	-	-	-	-
Milk (dried pdr)	10	0	0	45	0	0	-	-	-
Proc. Egg Prod.	-	-	-	-	-	-	18	7	6
Fertilizer Sludge	-	-	-	2	1	1	-	-	-

positive only with the 1-2 TEST and 2 positive by the culture method. The selective enrichment modification enhanced its reliability for Salmonella detection in foods and feeds. The manufacturer's current procedure specifies a selective enrichment step.

Although the reliability of the 1-2 TEST had been improved greatly, when inoculated from the enrichment broth, it unfortunately prolonged the time for analysis by 24 hours. A further protocol modification examined the effect of reducing the incubation period from 24 hours to 7 hours and increasing the amount of inoculum introduced into the 1-2 TEST inoculation chamber to 1.5 mL, made possible by previously emptying the chamber of its contents (11). In this study, 73 of the 283 samples (Table I. Phase 3) were found to be contaminated with Salmonella by the wet culture method while 70 were detected as being positive by the 1-2 TEST. Therefore the rate of false-negative reactions with the 1-2 TEST was approximately 1%. There were no false-positive reactions observed in this study.

The three isolates that were detected with the culture method but not with the 1-2 TEST System were S. worthington, S. senftenberg and S. heidelberg. These three serotypes were isolated more than once in these studies, and were detected with the modified 1-2 TEST in at least one other sample (Table II). Therefore failure of the 1-2 TEST to detect Salmonella in these samples was not due to its inability to react with these specific serovars.

It is concluded that the modified 1-2 TEST system can be viewed as a reliable rapid screening method for the detection of Salmonella in a variety of naturally contaminated animal feeds, feed ingredients, frozen egg products and environmental samples.

Table II: Listing of Salmonella Serovars
            Detected by the 1-2 TEST

S. senftenberg (13)	S. infantis (3)	S. lingstone (2)
S. mbandaka (11)	S. agona (3)	S. kentucky (2)
S. worthington (5)	S. newington (3)	S. bareilly (2)
S. schwarzengrund (4)	S. heidelberg (2)	S. new brunswick (2)
S. johannesurg (4)	S. anatum (2)	S. saint paul (2)
S. typhimurium (3)	S. havana (2)	S. tennessee (1)
S. ohio (1)	S. hadar (1)	S. bredeny (1)
S. cerro (1)	S. drypool (1)	S. poona (1)
S. westhampton (1)		

Note - Figures in brackets are the number of samples

SULPHAMETHAZINE RESIDUES. Sulphonamides are used in veterinary practice as antibacterial agents. The use of sulphamethazine in swine has made it the single most prevalent residue of all drugs used (12). Sulphamethazine (SMZ), at levels of 0.011% may be added to swine starter and pre-starter feeds to be used in the control of atrophic rhinitis, maintenance of growth rate, and feed efficiency. Regardless of use of SMZ, a strict withdrawal of at least 10 days must be adhered to by the farmer. Pork liver

and muscle samples must not exceed a violation residue level of
0.10 ppm sulphamethazine. The current method of analysis of
sulpha residues in pork utilizes thin-layer chromatography and
fluorometric scanning (13) with a limit of quantitation of 0.02
ppm. The procedure, however, is time-consuming, labour intensive
and requires extensive sample clean-up, which has led to a keen
interest in rapid immunological tests.

A study was undertaken to evaluate the performance of the
EZ-Screen Quik-Card Test for sulfamethazine in pork urine, liver
and muscle tissue (Table III). Initially, the EZ-Screen Quik
Card was compared with the current USDA Sulfa-on-Site TLC method
(14) for the screening of 569 pork urine samples at the plant
level. Eighty-one urine samples were found positive by the
EZ-Screen and only 36 by SOS. The corresponding 81 pork liver
samples were then analyzed by both EZ-Screen and TLC.
Subsequent analysis of corresponding pork muscle samples (only
78) were analyzed by EZ-Screen and positive results above 0.1
ppm were reanalyzed by TLC.

Table III   Analysis of Pork Urine, Liver and Muscle Tissues
for Sulfamethazine Residues by the EZ-Screen Test,
Thin-Layer Chromotography (TLC) or the Sulpha-on-
Site (SOS) Procedures

Sample Type	Samples No.	EZ-Screen ppb		TLC ppb		SOS ppm[d]	
		Neg. 0.1	Pos. 0.1	Neg. 0.1	Pos. 0.1	Neg. 0.4	Pos. 0.4
Initial Survey							

Pork Urine	569	488	81	–	–	533	36
Pork Liver	81	53[a]	28	64	17	–	–
Pork Muscle	78	49[a]	29[b]	(22)	(7)	–	–
Follow-up Survey							

Pork Muscle	160	95[a]	65	124	36	–	–
Pork Liver	63	46	17	39	24[c]	–	–

a   no false negatives
b   reanalyzed using TLC, results shown in brackets
c   50% false negatives when re-analyzed by E-Z Screen
d   For urine samples only: corresponds to 0.1 ppb in tissue
samples

In a follow-up survey, SOS screening was conducted at the
plant level on pork urine and all muscle and liver samples
corresponding to any positive urine results were reanalyzed by
EZ-Screen and TLC.

To obtain clear liver and muscle extracts, a centrifugation
step using a Centriflow Membrane Cone was used to obtain a clear

colourless liquid suitable for application directly onto the card.

The results for liver tissues showed that the EZ-Screen using the centrifugation sample preparation is not reliable for screening purposes. It is possible that the liver tissue contains interfering enzymes which compete for the active sites on the card resulting in negative results. However, there is excellent correlation between results obtained by the official method and the EZ-Screen Quik-C ard test for muscle tissue containing sulfamethazine residues over the regulatory tolerance level of 0.1 ppm. Statistical analysis of the data verified that the two methods are equivalent with the confidence limit of 99%. In addition, the results for urine show that the EZ-Screen Quik-C ard test is too sensitive for use as a screening tool for regulatory use, as too many positives are generated. It should be noted that these were not false positives as sulfamethazine was found in the liver by the official method.

It was concluded that the use of the EZ-Screen Quik-C ard on muscle tissue, as a screening test, would result in a reduction of the number of samples analysed by the official method by 60%. Twelve samples can be analysed per day by the official TLC method compared to one hour by using the EZ-Screen Quik-C ard in tandem with the centrifugation step.

Pesticides

Pentachlorophenol. The evaluation of an immunoassay kit for the determination of pentachlorophenol (PCP) residues in water (15) was carried out in conjunction with the current gas chromatography/electron capture detection procedure. The kit is based on a monoclonal antibody-based competitive inhibition immunoassay and has a shelf-life of about one year. PCP is the second most heavily used pesticide in the USA. In C anada, it is registered as a wood preservative, insecticide and herbicide with the sodium salt of PCP often being used as a general disinfectant for trays in mushroom houses and wood preservative in crates. Since toxic pesticides of this type can pollute streams and ground water, it is necessary to monitor its presence.

A standard 96-well microtiter plate is used with the developed colour being measures by optical density. A standard curve was generated over the range 0.3 ppb - 1.2 ppm together with spiked samples. In water, recoveries of 97 - 112% were found at the 100 ppb level. Since the method is based on disappearance of colour, with 100 ppb corresponding to full-scale absorbance, the limit of detection is about 10 ppb without pre-concentration. The method permitted rapid analysis of large number of samples.

Since PCP is frequently found in pork livers, the applicability of the kit to these and poultry liver matrices was investigated. As the kit tolerates significant levels of organic solvent, standard clean-up procedures could be used. Two main approaches were tried - partition with concentrated sulphuric acid and automated gel permeation. Sulphuric acid proved the best procedure (linear over about 2 orders of magnitude). The interferences that remained following clean-up were not sufficient to destroy the ability of the assay to detect added quantities of PCP in a predictable manner. The assay had a high cross reactivity with tri- and tetra- chlorophenols and there was a significant substrate dependence. Violative levels of PCP (100 ppb) were readily detected for systems in which PCP was added (10 to 100 ppb range) to reference blank livers. However, the test could not be applied to real liver samples as a screening technique at the current level of development of the methodology as the kit was only accurate within a factor 2 or 3.

Fusilade. Fusilade is the trade name for the active ingredient fluazifop-butyl which is a potent selective herbicide used to control grass weeds in many crops such as sugarbeets, soybeans, rapeseed etc. Residues in most crops are less than 0.1 ppm, with the exception of strawberries and soybeans, where a Maximum Residue Limit (MRL) of 1.0 ppm has been established in Canada.

Since the current methodologies are long and time consuming (16-18) and concern as to teratology studies is still an important ongoing regulatory issue an ELISA screening procedure was developed. The preparation of the serum involved reaction with N-Hydroxysuccinimide and conjugation with bovine serum albumen (BSA) prior to injection into rabbits. The rabbit serum was purified by using a BSA-Affigel filtration technique to separate out the antibodies specific to BSA from those specific to fusilade. Initial sensitivity of the competitive assay was 1.2 ppm which can be lowered to 100 ppb by optimisation of the detection system. Studies are continuing to further lower the limit of detection by using different labelling enzymes for application to agricultural (rapeseed, potatoes, etc), food and animal feed samples.

Conclusion

Immunoassay procedures can readily complement existing analytical procedures. It has been estimated that immunoassay kits for pesticides and fungal toxins could obtain a 41% and 75% share respectively of the toxin monitoring market by the year 2000. In addition, antibodies bound to a solid support to form "off-the-shelf" affinity columns will be used increasingly to concentrate drugs, pesticides and other toxins from various food, plant and environmental substrates. These techniques are being readily accepted due to their low cost and ease of use.

Suppliers of Immunodiagnostic Kits

1. E-Z Screen Aflatoxin and Sulfamethazine Quick Card Tests: Environmental Diagnostic Inc., PO Box 908, 2990 Anthony Road, Burlington, North Carolina, USA 27215

2. Agri-Screen Aflatoxin Kit: Neogen Corp.; 620 Lesher Place, Lansing, Michigan, USA 48912

3. Oxoid Toxin Detection Kits:

   - Unipath Inc., 217 Colonade Road, Nepean, Ontario, Canada, K2E 7K3

   - Vicam, 29 Mystic Avenue, Sommerville, Massachusetts, USA 02145

4. Aflatest 10 Aflatoxin Kit: 29 Mystic Avenue, Sommerville, Massachusetts, USA 02145

5. Pentachlorophenol Field Analysis Kit: Westinghouse, Bio-Analytic Systems, Inc., 15225 Shady Grove Road, Suite 306, Rockville, Maryland, USA 20850

6. 1-2 Test for Salmonella: Biocontrol Systems Inc., 19805 North Creek Parkway, Bothell, Washington, USA 98011

7. Sulfa-on-Site Test Kit: Environmental Diagnostic Inc., PO Box 908, 2990 Anthony Road, Burlington, North Carolina, USA 27215

Acknowledgments

The cooperation of H. Cohen, T. Kuevers, H. Campbell, J. Oggel, C. Barry and G. Labelle is greatly appreciated.

The use of trade names does not imply endorsement by Agriculture Canada.

Literature Cited

1. Lynch, G.P., J. Dairy Sci., 1971 55, 1244-1255.
2. Stoloff, L., JAOAC, 1980, 63, 1067-1073.
3. Campbell, T.L. and Stoloff, L., J. Agri. Food Chem., 1974, 22, 1006.
4. Rodericks, J.V. and Stoloff, L., in "Mycotoxins in Human and Animal Health", Rodericks, J.V., Hesselton, C.W. and Mehlman, M.A. (Eds), Pathotox Publishers Inc., Park Forest South, IL, 1977, pp. 67-69.
5. Keoltzow, D.E. and Tanner, S.N. "Comparative Evaluation of Commercially Available Aflatoxin Test Methods", USDA, Federal Grain Inspection Service, June 1989.
6. Methods for the Isolation and Identification of Salmonella from Foods. Method MFHPB-20, Health Protection Branch, Health and Welfare Canada, Ottawa, Ontario, 1978.
7. Flowers, R.S. and Klatt, M.J., JAOAC, 1989, 72, 303-311.
8. D'Aoust, J.Y. and Sewell, A.M., J. Food Prot., 1988, 51, 853-856.
9. Allen, G., Satchell, F.B., Andrews, W.H. and Bruce, V.R., J. Food Prot., 1989, 52, 350-355.
10. Nath, E.J., Neidert, E. and Randall, C.J., J. Food Prot., 1989, 52, 498-499.
11. Oggel, J.J., Nundy, D.C. and Randall, C.J., in press, 1990.
12. Dixon, D.E., Dessertation Abstracts International, December 1986, 47(6), 2301-B.
13. Thomas, M., Soreka, K.E. and Thomas, S.H., JAOAC, 1983, 66, 881-883.
14. USDA-FSIS Sulfa-On-Site, TLC Method 5.018, March 1986.
15. "WBAS Pentachlorophenol Immunoassay Kit" Westinghouse Bio-Analytic Systems Co., Rockland, Maryland, U.S.A., 1988.
16. Imperial Chemical Industries Ltd., Plant Protection Division Residue Analytical Method No. 62, August 1981.
17. Clegg, B.S., J. Agric. Food Chem., 1987, 35, 269-273.
18. Worobey, B.L. and Shields, J.B. JAOAC, 1989, 72, 368-371.

RECEIVED August 30, 1990

Chapter 5

Pesticide Residues in Food

U.S. Food and Drug Administration's Program for Immunoassay

Marion Clower, Jr.

Pesticide and Industrial Chemicals Branch, Division of Contaminants
Chemistry, Center for Food Safety and Applied Nutrition, U.S. Food and
Drug Administration, 200 C Street S.W., Washington, DC 20204

The U. S. Food and Drug Administration's (FDA) enforcement and
monitoring programs for pesticide residues in foods are based primarily
on five multiresidue methods which use traditional quantitative techniques, such as gas or liquid chromatography, to provide coverage for a
large number of residues in a single analysis. A two–pronged assessment of the applicability of immunoassays as a supplemental technique
has been initiated: (1) FDA is examining the potential utility of selected
commercial kits by evaluating the kits as rapid detection systems when
combined with appropriate extraction and cleanup steps to form complete methods. (2) Six methods based on monoclonal antibody immunoassay technology are being developed under contract. Each method will
provide information necessary to evaluate the applicability of immunoassays to the Pesticide Program. Methods are being developed for
paraquat in potatoes; fenamiphos and its sulfoxide and sulfone metabolites in oranges; carbendazim in apples; benomyl in apples; thiophanate-methyl in apples; and glyphosate in soybeans. The methods will
be evaluated by the same criteria applied to traditional residue methods.
Results of studies by FDA and others will allow the agency to determine
the role of immunoassay technology in its Pesticide Programs.

Federal regulation of pesticides in the United States is accomplished by the joint
effort of three agencies. The Environmental Protection Agency (EPA) registers
(approves) the use of pesticides and establishes tolerances (maximum allowable residue concentrations) for potential residues that may occur in or on foods. The Food
and Drug Administration (FDA) enforces these tolerances for residues in foods,
except for meat and poultry, which are the responsibility of the U. S. Department of
Agriculture (USDA).

This paper briefly describes FDA's pesticide monitoring activities including the
goals of the program and the nature of the analytical methodology used to generate
the residue data on which the program is founded. The nature of enzyme immunoassay technology and the potential benefits offered by incorporation of this analytical
technique into monitoring activities are discussed. FDA is committed to use of
immunoassay wherever monitoring can be enhanced through use of properly validated methods. Potential applications of such methods and FDA's research in this
area are presented.

FDA Monitoring Program

FDA monitors the food supply using two different but complementary approaches: regulatory monitoring and the Total Diet Study analysis. *Regulatory monitoring* is specifically designed to enforce tolerances and other regulatory limits for residues in domestic foods in interstate commerce and in foods offered for import into the United States. Regulatory monitoring is focused on analysis of a wide variety of raw agricultural commodities for large numbers of pesticide residues. Information on the incidence and level of pesticide residues in the general food supply is also developed as part of this activity. Sampling and analysis are conducted by FDA offices and laboratories across the country.

FDA conducts the *Total Diet Study* to monitor dietary intake of pesticide residues and to identify trends in residue occurrence and levels. Foods collected from retail markets in representative areas of the country are analyzed in a single laboratory. In contrast to regulatory monitoring, foods in the Total Diet Study are analyzed after being prepared ready to eat.

Accurate determination of residue levels is clearly necessary for tolerance enforcement and regulatory monitoring, but it is also essential for calculation of dietary intakes even when residue levels are well below applicable tolerances. Therefore, both monitoring approaches require quantitative analytical methods. These methods must be applied uniformly in all laboratories.

Types of Analytical Methods

The analytical methods used in both types of monitoring fall into three categories. General purpose *multiresidue methods*, or MRMs, provide coverage for a large number of residues of different chemicals and are usually applicable to many different commodities. The large number of pesticides available for use on foods and the unknown treatment history of many samples collected in FDA's monitoring program have caused FDA to rely primarily on MRMs because of their ability to identify and measure more than one residue in a single analysis. In this manner, FDA makes the most efficient use of resources while providing the quantitative results required from its monitoring activities.

Selective multiresidue methods, can identify and measure a small number of structurally similar pesticides and are used when MRMs cannot determine potential pesticide residues of interest. Recently registered pesticides, as well as many pesticides that have been registered for some years, frequently have chemical and physical properties that are radically different from those for which the MRMs were developed. A selective MRM usually can determine fewer than a dozen residues. These methods are also frequently more complicated than MRMs and often require more expertise, or they may require recently developed, sophisticated instrumentation.

Single residue methods or SRMs, are used by FDA when no general or selective MRM is available. These methods are usually capable of determining the level and identity of the residue of only one pesticide from a limited number of commodities. SRMs are usually submitted during the registration process to meet EPA's requirement for an analytical method capable of determining compliance with the requested tolerance.

Steps in Residue Analytical Methods

Traditional methods (1-3) of analysis for pesticide residues in foods use a four-step approach:

- compositing of the laboratory sample—A homogeneous analytical sample must be obtained if results of the analysis are to be representative. Twenty pounds or more of food is ground or chopped, and then thoroughly mixed before analytical portions are taken.
- extraction of residues—Organic solvents of moderate polarity are typically used to extract pesticide residues from foods. The least polar solvent that satisfactorily extracts the analytes is used to avoid extraction of unwanted food materials that would require additional cleanup procedures.
- cleanup of the extract—Test sample extracts usually contain undesired food materials that have been co–extracted with the analyte(s). Cleanup procedures, often consisting of both a liquid-liquid partitioning step and a column chromatographic step, are used to preclude interference with detection and measurement of the residue.
- determination of residues—Residues are identified and amounts quantitated by gas or liquid chromatography, which provides additional selectivity by separation of multiple components.

Most pesticide residue methods are limited in one or more of the following areas: coverage (residues and foods), speed, complexity, and expense. SRMs are the most limited since they determine only one residue, usually in only a few commodities. SRMs often take as long or longer than an MRM, require specialized equipment, and in many cases are difficult to perform. This is not unexpected since SRMs are usually developed for a specific purpose.

Selective and general purpose MRMs have broader coverage, compared to SRMs, but also suffer to varying extents from the same limitations. The cleanup step generally requires a large portion of the total analysis time because these methods use traditional detection techniques, such as gas chromatography (GC) and high pressure liquid chromatography (HPLC). Although both GC and HPLC provide the advantage of additional separation after cleanup, the detectors employed are often not highly specific. Many co-extractives, if not removed by extensive cleanup, have properties that will produce a detector response and thus may interfere with detection or quantitation of the analyte of interest.

Immunoassay Considerations in Regulatory Analysis

Application of enzyme immunoassay (EIA) methods to regulatory analysis of foods in a pesticide residue monitoring program is a new area with little definitive documentation. In addition to FDA's activities described here, USDA's Food Safety and Inspection Service has issued proposed regulations describing the procedure for approval of immunoassays for use in its program for monitoring residues in meat and poultry (4). EPA is investigating immunoassays for environmental monitoring (5). Canada's Health Protection Branch has published several EIA methods for analysis of foods for pesticide residues (6, 7) and a review of immunoassay application to food analysis (8).

As an analytical approach to residue analysis, immunoassay methods are not well characterized, and no validation protocols have been established. The Association of Official Analytical Chemists, whose primary purpose is validation of analytical methods, established a "Task Force on Test Kits and Proprietary Methods" (9), which has addressed some of the issues relating to immunoassay methods. The International Union of Pure and Applied Chemistry's Commission on Food Chemistry has established a "Working Group on Immunochemical Methods," whose first project is to develop draft guidelines on criteria for evaluation, validation, and quality control for radio-immunoassay methods (10). Similar guidelines for EIAs will also be developed. These documents will assist in development and standardization of requirements for precision for both between-laboratories and within-laboratory analyses, accuracy, and ruggedness, and—for qualitative methods—false positive and false negative rates.

Since the absolute specificity of EIAs is unknown, investigation and documentation of the degree of crossreactivity of an EIA will have a great effect on acceptance and regulatory use of EIA methods. After appropriate study, a given EIA method may be classified as suitable for analysis of a certain food for a specific residue, but crossreactivity with matrix components may make it unsuitable for use with a similar food and thus significantly restrict the utility of the method.

The antibody used in an EIA method is analogous to the standard reference material of traditional residue methods; analytical results are severely compromised by impure standards. The greater purity and specificity of monoclonal antibodies relative to polyclonal antibodies suggest that the resulting method is less susceptible to false positive and negative findings. The capability of raising more monoclonal antibodies identical to those used in previous assays satisfies the requirement for a continuing supply of pure, characterized standard reference material. Without such a standard (i.e., with polyclonal antibodies), sensitivity and selectivity might vary from assay to assay depending on the lot of antibody used. The resulting variability in method performance would be unacceptable in regulatory residue analysis unless validated procedures to assure analytical reproducibility were followed.

Requirements for refrigerated storage may limit use of EIA methods. Although laboratories are typically capable of low temperature storage, field personnel often have limited facilities in this area. Shelf life may require careful planning of EIA method usage in either locale.

Perhaps the most often overlooked aspect of the application of EIA methods to pesticide analysis of foods is the size of the analytical portion taken for analysis. As mentioned above, about 20 pounds of food is usually collected. After a compositing step to assure homogeneity, traditional methods typically require analysis of 100 g of this mixture to obtain a representative portion and to provide the quantitation limits needed for regulatory monitoring. Analytical portion size is being studied to determine the minimum quantity that is statistically representative of the composite (11). Preliminary indications are that the analytical portion must be at least 10 g.

EIA techniques are often sensitive enough that analytical portions weighing only several grams can be analyzed. Although a single apple contains several 10 g analytical portions, it is unrealistic to assume that one apple is representative of an entire lot. The typically small analytical portion of many EIA methods, therefore, may lead to inaccurate analytical results. Moreover, collection, shipment, and storage of the large numbers of laboratory samples that can be conveniently analyzed by an EIA method would require significant additional staff and would greatly increase shipment and storage costs.

Potential Benefits of Immunoassays in Regulatory Analysis

Enzyme immunoassay methods offer a biological-based detection technique which is complementary to that of traditional residue analysis. Potential advantages must be evaluated in light of the intended application—use in a nationwide program for quantitatively measuring pesticide residues in foods. Ideally, each EIA method would contain simple, rapid extraction and cleanup (if required) steps, require minimal or portable equipment, be usable in the field by individuals with little scientific training, and produce quantitative results.

Most attractive of the potential advantages of EIA methods is the small amount of time, relative to traditional methods, required for analysis. EIA detection procedures usually require several minutes to an hour to complete. Although traditional chromatographic detection steps require a similar amount of time, extraction and cleanup steps, which must accompany such detection, are usually quite lengthy. The most widely used detectors in chromatographic systems respond to chemical properties of the analyte that are common to numerous chemicals which may be present in the extract examined. For most foods, EIA methods must contain extraction and cleanup steps, but these procedures can often be simple and require less time than those of traditional methods because of the nature of the immunoassay itself.

The specificity of each EIA derives from the antibody upon which the assay is based. Ideally, this antibody reacts with only the desired target chemical. However, in practice this is not always the case. Therefore, studies to determine the extent of crossreactivity must be conducted to assure suitability of the assay for its intended purpose. EIAs typically useful for pesticide residue analysis employ antibodies having a sufficient degree of specificity so that extensive cleanup is unnecessary. Overall, residue detection through antibody reaction with a specific portion of the analyte molecule, rather than detection of a chemical property common to many molecules, results in the need for less cleanup of the extract and a shorter, less complicated analytical method.

In addition to laboratory glassware and equipment necessary for cleanup of the extract, traditional pesticide residue methods require expensive chromatographic instrumentation for identification and quantitation of residues. EIA methods require minimal amounts of glassware, disposable plasticware, or other supplies. Quantitative EIAs often make use of 96-well microtiter plates for multiple simultaneous assays of about a dozen extracts and associated reference standards. Major equipment consists of a plate reader, which automatically measures the absorbance of each well. Plate readers can be used alone or in conjunction with a personal computer, which can correlate replicate measurements, construct the calibration curve, calculate results, and provide a complete statistical analysis. Such an EIA workstation can be obtained for roughly half the cost of the GC or HPLC system typically used for pesticide residue analysis.

Extremely polar pesticides, pesticide metabolites, and other chemicals not readily soluble in organic solvents currently provide the greatest challenge to traditional pesticide residue methodology because of difficulties in residue extraction and cleanup, chromatography, and detection. EIAs work best in aqueous media and may greatly simplify methodology for these chemicals since chromatography is not used and cleanup is usually minimal. Use of EIA methods in the monitoring program could significantly expand coverage of pesticide chemicals since most multiresidue chromatographic methods focus on chemicals of low or intermediate polarity and recover polar chemicals only with great difficulty.

The foregoing discussion assumes use of quantitative EIA methods by a trained chemist in a laboratory facility. Simple EIA methods requiring minimal equipment offer the potential for use in the field by nonlaboratory personnel. Portable, battery-operated microtiter plate readers are commercially available and could be used when electricity is not available. Development of qualitative versions of quantitative methods eliminates the need for a plate reader or complicated electronic hardware, if a suitable visual detection step is included in the method. Potential application of such EIA methods to residue analysis in nonlaboratory settings offers the prospect of analyzing increased numbers of test samples and eliminates the cost and time required for shipment of foods to a laboratory for analysis. It must be recognized that some means of using qualitative results in a program based on quantitated residue levels would need to be developed for qualitative methods to be broadly applicable. At a minimum, follow-up quantitative analyses would be required for foods with residues above applicable tolerances.

EIA methods also offer the potential for determination of multiple residues through the use of mixed antibody cocktails. Extensive crossreactivity studies would have to be conducted to assure independent reaction of each antibody-analyte pair, and a subsequent separate analysis would be required to determine the presence of each particular analyte.

Potential Applications of EIA Methods in FDA Monitoring

Qualitative EIA methods may play an important role in screening of food shipments for residues. In this context, screening methods are defined as those used to ascertain the presence (detectability) or absence (nondetectability) of a particular analyte (or group of analytes) at a predetermined concentration in the food. Screening methods may also provide a semiquantitative estimate of the analyte level. In all cases, qualitative EIA methods need to be well characterized and to have documented detection limits. The frequency of false positive and false negative findings must also be documented so that test results reasonably reflect actual residue occurrence. Ideally, false negative results would never be produced so that all foods containing residues above the method's detection limit would be identified. A small percentage of false positive results may be acceptable; however, each false positive result requires shipment of the food to a laboratory for quantitative analysis and increases analytical costs. Validation procedures to document these qualitative methods and to assure the reliability of the results must be defined before actual use in any monitoring program.

Screening methods could significantly increase the number of analytical samples examined in "contamination incidents" when a known pesticide is accidentally, or through misuse, applied to a crop. Availability of a suitable EIA method for aldicarb would have greatly facilitated analysis during the contamination of watermelons several years ago. The unpredictability of these incidents implies EIA methods for all pesticides would be required for adequate preparation for the next incident; however, the cost and resources needed for such an FDA method development effort would be prohibitive.

General screening of foods for residues may also be a suitable application of qualitative EIAs. Such screening would be especially advantageous in situations when data are needed for a pesticide for which traditional analysis is difficult. Since qualitative EIA methods are usually simpler than quantitative methods, they may be suitable for use in the field by investigators or inspectors.

In both situations, the incidence of residue contamination could be estimated. Laboratory analysis by a quantitative method would be required for foods giving a positive response to the EIA so that residue concentration could be determined. The occurrence of too many false positive results, however, could potentially overburden existing laboratory capability and escalate shipping and storage costs.

Quantitative EIA methods, in contrast to qualitative methods, could potentially be inserted directly into the monitoring program. Even though most EIA methods recover only a single residue, this disadvantage is offset by the relatively rapid, simple nature of the EIA determination itself. Moreover, method development costs may be recovered if EIA methods are developed for residues that are difficult to determine by traditional methodology. For unrestricted use, quantitative EIA methods must meet the validation requirements of traditional methods for acceptable precision, accuracy, reproducibility, repeatability, and ruggedness. Additional factors such as reagent stability and storage conditions must also be well documented. The most productive application of such methods appears to be for analysis of samples already in the laboratory. Increased sample handling costs are thereby avoided while residue coverage is expanded.

The principal current barrier to incorporation of quantitative EIA methods into FDA's monitoring program is the lack of validated quantitative methods. Furthermore, in spite of their simplicity, quantitative methods will require new equipment, training of analysts, and experience in their application before they can be widely used in the monitoring program.

Immunoassay Research at FDA

FDA's research and evaluation of immunoassay techniques in its monitoring program involve both commercial immunoassay detection kits and new EIA methods developed for selected pesticides. Although no complete immunoassay methods are commercially available, several detection kits have been marketed. Development of new methods designed to meet FDA monitoring program requirements will provide valuable information under conditions of actual use.

Commercial Immunoassay Kits. Most commercially available immunoassay kits have been developed for determination of pesticides in water. At a minimum, appropriate residue extraction procedures must be developed before these kits can be applied to pesticide residue analysis of foods. Encouraging results have been obtained in preliminary FDA evaluations of several kits (12). For example, in studies of a kit for detection of triazine herbicides, a typical residue extraction solvent (acetonitrile) was used and then diluted with water to levels tolerated by the immunoassay. Visual comparison of color developed for extract, standard, and reagent blank was made for qualitative analysis. Spectrophotometric readings of the color were made for quantitative analysis.

Several limitations of this kit were evident from these simple experiments. Visual comparison of color differences in extract and standard was difficult. For quantitative analysis, a calibration curve must be prepared from assays of standard solutions, each containing different amounts of analyte. Physical manipulation and timing involved were so cumbersome with more than 4 or 5 tubes that simultaneous analysis of several extracts and sufficient standards to prepare a calibration curve was not feasible.

EIA Method Development. A contract with Research Triangle Institute, Research Triangle Park, NC, was initiated to develop six EIA methods based on monoclonal antibody technology. Target pesticides were selected to provide EIA methods that would allow evaluation of their potential application to FDA's monitoring program as well as to produce methods of maximum utility. Crops were chosen after examination of the major uses of each pesticide. The desired method limit of quantitation was fixed at roughly half of the applicable tolerance for each pesticide-crop combination. Documentation of method behavior (*e.g.,* recovery, limit of quantitation, and crossreactivity) was required for each method developed.

The first phase of the contract focuses on paraquat, a small, ionic, aromatic pesticide, in potatoes. Several immunoassay methods for paraquat have been reported in the literature (13-16). Successful development of an EIA method for paraquat has obvious advantages for use in the monitoring program, since traditional chromatographic methods are very difficult, time-consuming

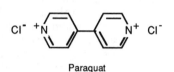

Paraquat

SRMs. Although FDA is just beginning its investigation of this method, the importance of cleanup of the extract in EIA methods for foods was demonstrated during method development—a potato coextractive interfered in the immunoassay determination unless it was removed by a short cleanup column.

In the second phase of this contract, a single method is being developed for the nematocide fenamiphos and its most important metabolites, fenamiphos sulfoxide and fenamiphos sulfone, in oranges. No immunoassay methods have been developed for any of these chemicals. As with the other pesticides in this contract, analysis for residues of these chemicals is difficult. Moreover, this method was designed to provide some evaluation of the selectivity of monoclonal antibodies since a single antibody will be used to detect all three chemicals. A number of antibodies with high, but not equal, affinity to all three chemicals were readily raised. The final method will rely on a simple oxidation to convert all three chemicals to a single moiety before residue measurement.

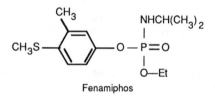

Fenamiphos

Phase three of the contract will produce a separate EIA method for each of three structurally related pesticides in apples: benomyl, thiophanate-methyl, and carbendazim. Both benomyl and thiophanate-methyl are fungicides registered for use in the United States and both degrade to carbendazim. Although carbendazim is not registered in the United States, it is registered in foreign countries for use on crops imported into the United States. Beyond the great degree of difficulty of traditional methods of analysis for these chemicals, it is often impossible, because of their degradation in the solvents used, to determine which pesticide was applied and therefore the applicable tolerance. Successful development of EIA methods for these chemicals would not only greatly simplify residue analysis but also permit identification of the pesticide used. An EIA method which detects benomyl as carbendazim has been reported (17), but it cannot distinguish between the two chemicals.

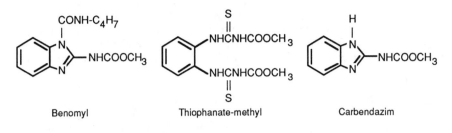

Benomyl   Thiophanate-methyl   Carbendazim

In phase four, a method will be developed for determination of glyphosate, a herbicide, in soybeans. Glyphosate is an extremely small, very polar molecule containing an amino acid moiety, which may hamper EIA detection. Development of a method for soybeans will provide an indication of the degree of difficulty in EIA determination of residues in fatty foods, which require significantly more cumbersome and time-consuming extraction and cleanup steps for determination by chromatographic approaches.

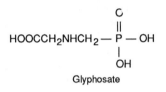

Glyphosate

Summary

FDA's program for monitoring the nation's food supply for pesticide residues is based on quantitative multiresidue analytical methods using traditional determinative techniques such as GC and HPLC. FDA will continue to evaluate available commercial immunoassay kits, develop complete immunoassay methods, and incorporate them into the pesticide residue monitoring program. Immunoassay methods have the advantages of high speed, low cost, and analytical simplicity. The cost of developing these methods and their ability to analyze for only one residue at a time will restrict their use to specific analytical situations. Multiresidue methods will remain the principal analytical approach used in monitoring the food supply for pesticide residues.

Studies of commercial kits will concentrate on expanding monitoring coverage. Kits may be used as screening techniques when applied to extracts from existing methods or when combined with rapid extraction procedures developed especially for the kits. Potential applications include screening of foods in cases of pesticide misuse or contamination, and in other situations when residue data are needed and residue identity is known. Such screening would provide an evaluation of residue level relative to a documented analytical concentration based on method capability. This "go-no go" analysis would increase monitoring coverage by requiring traditional quantitative analysis for only those foods found positive. Validation, documentation of method performance, and assurance of efficient, cost-effective application will be completed before a method is used in monitoring.

In a similar manner, development of complete quantitative EIA methods will continue. Quantitative methods could be incorporated into the monitoring program as soon as validated methods are available. The greatest benefit is to be derived from methods that recover residue chemicals that are difficult to determine by traditional methods, thereby exploiting the inherent simplicity of the immunoassay. Six such enzyme immunoassay methods are now under development. Investigation of the method for paraquat is under way. Initial research will verify method documentation and performance. Extension of this and subsequent EIA methods for other residues to crops beyond those for which they were developed will have high priority since this extension, if feasible, will significantly increase the number of foods covered at low cost.

Literature Cited

1. Pesticide Analytical Manual, Vol. I, Food and Drug Administration, Washington, D. C., 1986.
2. Pesticide Analytical Manual, Vol. II, Food and Drug Administration, Washington, D. C., 1986.
3. Official Methods of Analysis, Williams, S., Ed., Association of Official Analytical Chemists: Arlington, VA, 1984.
4. Federal Register 54, #158 (8/17/89) 33920-33923.
5. Von Emon, J. Paper AGRO 46, 198th ACS National Meeting, September 10-15, 1989, Miami Beach, FL, USA.
6. Newsome, W. Bull. Environ. Contam. Toxicol., 1986, 36, 9-14.
7. Newsome, W. J. Agric. Food Chem., 1985, 33, 528-530.
8. Newsome, W. J. Assoc. Off. Anal. Chem., 1986, 69, 919-923.
9. Anon in The Referee, Cassidy, C., Ed.; 12, #11; Association of Official Analytical Chemists: Arlington, VA, 1989; p 1.
10. Pohland, A., Chairman, Commission on Food Chemistry, IUPAC, Food and Drug Administration, Personal communication, 1989.
11. Trotter, W.; Young, S. Food and Drug Administration, Personal communication, 1989.
12. Pohland, A.; Trucksess, M.; Thorpe, C.; Clower, M. Paper AGRO 45, 198th ACS National Meeting, September 10-15, 1989, Miami Beach, FL, USA.
13. Van Emon, J.; Hammock, R.; Seiber, J. Anal. Chem., 1986, 58, 1866-1873.
14. Niewola, Z.; Walsh, S.; Davies, G. Int. J. Immunopharmacol., 1983, 5, 211-218.
15. Niewola, Z.; Hayward, C.; Symington, B.; Robson, R. Clin. Chim. Acta, 1985, 148, 149-156.
16. Niewola, Z.; Benner, J.; Swaine, H. Analyst, 1986, 111, 399-403.
17. Newsome, W.; Collins, P. J. Assoc. Off. Anal. Chem., 1987, 70, 1025-1027.

RECEIVED August 30, 1990

Chapter 6

Monoclonal Immunoassay of Triazine Herbicides

Development and Implementation

A. E. Karu[1], Robert O. Harrison[2,4], D. J. Schmidt[1], C. E. Clarkson[1], J. Grassman[1], M. H. Goodrow[2], A. Lucas[2], B. D. Hammock[2], J. M. Van Emon[3], and R. J. White[3]

[1]Department of Plant Pathology, University of California, Berkeley, CA 94720
[2]Department of Entomology and Environmental Toxicology, University of California, Davis, CA 95616
[3]Environmental Monitoring Systems Laboratory, U.S. Environmental Protection Agency, Las Vegas, NV 89193

This paper summarizes a three-laboratory effort to develop a sensitive, reliable enzyme immunoassay (EIA) for triazine herbicides using monoclonal antibodies (MAbs). Simazine and atrazine haptens with mercaptopropionic acid and aminohexanoic acid spacers were synthesized and conjugated to proteins via N-hydroxysuccinimide active esters. MAbs derived from mice immunized with these conjugates had I_{50} values of 3 ppb to 4 ppm for various triazines in standard and simazine-enzyme conjugate competition EIAs. The EIAs are compatible with simplified methods for triazine extraction and concentration from soil and water. The limit of detection for atrazine was approximately 0.05 to 0.1 ppb, similar to that obtained with gas chromatography. EIA and GC results agreed closely for 75 groundwater samples, with no "false negatives." Gas-liquid chromatography and EIA data for simazine in 48 soil extracts had a correlation of 0.97. The EIA has also been used to monitor groundwater from beneath a toxic waste pit and water from agricultural evaporation ponds.

The s-triazines, first developed in the early 1950s (1), are among the most effective and widely used herbicides known. They are of 3 major types, based on the substituent at R1 (Figure 1): the chlorotriazines, of which simazine and atrazine are the most-used, methoxytriazines, such as prometon, and the methylthio triazines, of which ametryne and prometryne are representative. Atrazine has been cited as the second most-used pesticide in the United States, with an estimated annual usage on the order of 79 million lbs (2). Roughly 3 million lbs. of triazines — mostly atrazine, simazine, and prometon — were applied in California from 1983 through 1987, with the largest percentages used in non-agricultural applications, such as industrial soil sterilization, landscape maintenance, and clearing of rights-of-way (3).

[4]Current address: ImmunoSystems, Inc., 4 Washington Ave., Scarborough, ME 04074

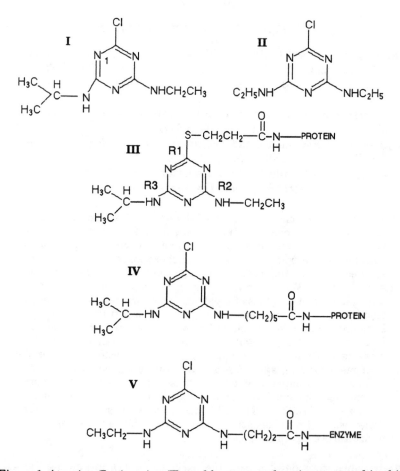

Figure 1. Atrazine (**I**), simazine (**II**), and haptens and conjugates used in this work. **III** — atrazine-mercaptopropionic acid hapten, and **IV** — atrazine aminohexanoic acid hapten, which were conjugated to BSA, CON, or KLH, and used as immunizing and EIA coating antigens. **V** — simazine-alkaline phosphatase "haptenated enzyme," used as the detector in competition EIAs where the monoclonal antibody was immobilized on the solid phase. Atoms on the triazine ring are numbered clockwise from N_1 shown in structure **I**.

Because of their different solubilities and modes of action, the triazines are selective to varying degrees in their effects on weeds and agriculturally important crops (4). Resistant plants dealkylate these compounds. Corn, sugarcane, and many other crops are naturally resistant, making triazines ideal for weed control on these crops. Various triazines can be used for pre- or post-emergence weed control, alone or in combination with other pesticides. Persistence of these compounds varies, and is a function of the soil properties and microbial ecology, and the climate. The triazines vary widely in their retention in various soils and their potential for leaching, and their mobility in groundwater is a good index of movement of other pesticides.

We undertook development of monoclonal antibodies and an immunoassay for triazines with sponsorship from the Environmental Monitoring and Pest Management Branch of the California Department of Food and Agriculture (CDFA), as part of a long-term plan to augment or replace more costly analytical methods with immunoassays, for regulatory purposes. The primary concern of the Environmental Monitoring and Pest Management Branch is groundwater. There are on the order of 40,000 domestic and municipal wells in California, and the state regulatory agencies analyze about 2,000 groundwater samples annually — primarily by gas chromatography (GC, 5). The number of wells CDFA must monitor will continue to increase, due to to recent legislation and increased public interest in water quality.

This report describes the initial results of a cooperative effort, in which haptens and conjugates were synthesized at UCD, monoclonal antibodies were derived and characterized at UCB, sample recovery methods were developed and initial feasibility tests with various types of field samples were conducted at UCD, UCB, and EMSL. The antibodies and assay methods were provided to the CDFA Analytical Laboratory in Sacramento, CA, in August 1989. Staff of that laboratory are in the process of validating the assay and acquiring data and experience that will be used to integrate the triazine EIA into their repertoire of tests for regulatory monitoring.

Methods

Details of the synthesis of haptens and conjugates, the production and characteristics of the MAbs, and optimization of the immunoassays, will be published separately (6; Schmidt et al., in preparation; Jung, et al., in preparation).

Synthesis of triazine haptens and hapten-protein conjugates. Simazine and atrazine were derivatized with mercaptopropionic acid (mpa) at R1, or aminohexanoic acid (aha) at R2, and these haptens were covalently linked to keyhole limpet hemocyanin (KLH), conalbumin (CON), or bovine serum albumin (BSA), by forming active esters with N-hydroxysuccinimide (6) (Figure 1, structures III and IV). This technique was also used to couple simazine-aminohexanoic acid to calf intestine alkaline phosphatase (Figure 1, structure V), for use as the "haptenated enzyme" in the EIA format described below.

Preparation of triazine-specific MAbs. Pairs of Swiss Webster, Biozzi, and
B10.Q mice were immunized with 4 doses of one of the triazine-protein
conjugates in Ribi adjuvant (MPL + TDM Emulsion, Ribi Immunochem Research,
Hamilton, Montana) over 3 months. The sera showed wide variations in
triazine-specific serum titers, limiting detectable dose and I_{50} (the dose giving
half-maximal inhibition) in a competition EIA, using conjugates with a carrier
and linker different from those of the immunizing antigen. Splenocytes from the
four best-responding mice (two Swiss Webster and two B10.Q) were fused with
P3X63AG8.653 myelomas, essentially as described by Fazekas de St. Groth
and Scheidegger (7). 15,936 cultures were seeded (166 96-well culture plates),
from which 3,156 colonies developed, and were screened for triazine-directed
antibodies, again using conjugates with a carrier and linker different from those
of the immunizing antigen. Of 232   triazine-specific antibodies, 74 were
inhibited by free atrazine or simazine, and 36 of these proved to be genetically
stable after several passages in culture. The 15 most sensitive MAbs had I_{50}
values of 3 to 15 ppb for atrazine and 35 to 60 ppb for simazine, and all were of
the IgGκ subclass. By contrast, the sera from the mice used to derive the
hybridomas had I_{50} values of 100 to 200 ppb for atrazine and simazine.

The 5 most sensitive MAbs were subcloned by limiting dilution, and at
least 12 clones of each cell line were frozen. Cultures were expanded to produce
pools of 500 to 750 ml of antibody-containing culture fluid, which were used
without purification in the assays. (Figure 2)

Enzyme Immunoassays. We carried out these studies with 3 variations of the
competition EIA. Initial surveys of the responses in mice, screening and initial
characterization of the hybridomas, and some of the method development studies
were performed using a "classical" competition EIA, in which triazine in
solution competed with atrazine-protein conjugate immobilized on the EIA
plates, for binding a limiting amount of antibody, which was in solution. Most of
the studies to optimize the quantitative EIA with soil and water extracts, and
many of the specificity studies were carried out using a "haptenated enzyme"
format, in which the MAb was immobilized on the EIA plate by trapping it with
a goat anti-mouse antibody, and triazine in solution competed with a simazine-
alkaline phosphatase conjugate for binding to the MAb. We recently perfected a
more rapid and convenient version of this format, which was done as follows:

EIA wells (Immulon 2, Dynatech) were coated overnight at 4°C with 0.1 ml
(approx. 200 ng) of affinity-purified goat anti-mouse IgG+IgM (Boehringer-
Mannheim no. 605 24) 1:1,000 in "coating buffer" (0.015 M Na_2CO_3 — 0.035 M
$NaHCO_3$ — 0.003 M NaN_3 pH 9.6). The wells were washed 3 times with "PBS-
Tween" (0.01 M KH_2PO_4-K_2HPO_4, pH 7.4 — 0.15 M NaCl — 0.02% NaN_3 —
0.05% Tween 20), 0.1 ml of triazine MAb AM7B2 (hybridoma culture fluid,
diluted 1:400 with PBS-Tween containing 0.5% bovine serum albumin) was then
added to each well, the plates were incubated for 1 hr at room temperature, and
then stored (with the fluid left in the wells) at -20°C in a sealed container to
prevent evaporation until they were needed. At the time of assay, the EIA plates
were thawed and washed 3 times with PBS-Tween. Standards and unknowns
were diluted in PBS-Tween in microplates or polypropylene tubes, and mixed

with a limiting amount of simazine-N(C2)- alkaline phosphatase in PBS-Tween to give a final volume of 0.24 ml per well. Aliquots of 0.05 ml of these mixtures were then transferred to the EIA plates. The competition reaction was complete after 30 min at room temperature (Figure 3), at which time the plates were washed 3 times with PBS-Tween, and dried by rapping on lint-free paper towels. Substrate solution (1 mg/ml p-nitrophenyl phosphate in 10% (w/v) diethanolamine-HCl, pH 9.8 — 0.4 mM $MgCl_2$ — 3 mM NaN_3) was then added, and color development at 405 nm was monitored on an EIA reader.

Data Analysis. Standard curves (generally 11 dilutions in triplicate from a spectrophotometrically standardized stock solution) were fitted by iterative regression to the 4-parameter logistic equation (10) using Passage II™ (Passage Software, Inc., Fort Collins, CO) on a Macintosh computer, or Softmax™ software (Molecular Devices, Menlo Park, CA) on an IBM PC. Sample concentrations were determined by interpolation from the best-fit curves. Values that fell outside of the "working range," defined as 20% to 70% of the maximum normalized response, were not used.

Solid-phase Extraction of Atrazine from Water. Water samples of 100 to 220 ml were divided in two aliquots, one of which was spiked with atrazine standard to 0.2 ppb. C_{18} solid-phase extraction (SPE) columns (Analytichem "Bond-Elut") containing 100 mg or 300 mg resin were conditioned successively with 2 column volumes of pesticide analysis grade hexane, ethyl acetate, methanol, and glass-distilled water. The water samples were filtered through two layers of Whatman No. 4 paper to remove solids, and the filtrates were applied to the columns at 8 to 15 ml/min, followed by a wash with 2 column volumes of glass-distilled water. Triazines were eluted into glass tubes with a total of 2 ml of ethyl acetate. The eluates were evaporated to near-dryness under nitrogen, and dissolved in 1 ml of PBS-Tween. Generally, 5 dilutions of each sample were assayed in triplicate on each of two EIA plates, which also included atrazine standards in triplicate. This procedure shown schematically in Figure 4.

Solvent Extraction of Simazine from Soil. Soil samples of 10 grams (sandy loam with low organic carbon content) were dried at 80 °C, suspended in 10 ml of ethyl acetate, and shaken or sonicated at low power for 30 min. Solids were allowed to settle, and the extract was decanted. The soil was resuspended in 10 ml of ethyl acetate, and this second extract was added to the first one, and filtered through Na_2SO_4. These extracts were used directly for gas chromatography. For EIA, the ethyl acetate was evaporated to dryness, the eluate was reconstituted in 1 ml PBS-Tween, and aliquots were taken directly into the EIA.

Solid-phase Extraction of Atrazine from Soil. This procedure was modified from the method described by Hill and Stobbe (8). For studies involving spiked samples, atrazine standards in methanol were added to give the desired ng of atrazine per gram of dry soil, and the samples were dried again before extraction. Samples of 5 grams of "U.S. Army Standard Soil" were suspended in 10 ml of acetonitrile: water :: 9:1, and the slurry was sonicated (30 min,

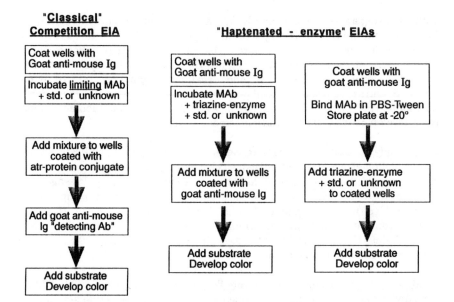

"Classical" Competition EIA

Coat wells with Goat anti-mouse Ig

Incubate <u>limiting</u> MAb + std. or unknown

Add mixture to wells coated with atr-protein conjugate

Add goat anti-mouse Ig "detecting Ab"

Add substrate Develop color

"Haptenated - enzyme" EIAs

Coat wells with Goat anti-mouse Ig

Incubate MAb + triazine-enzyme + std. or unknown

Add mixture to wells coated with goat anti-mouse Ig

Add substrate Develop color

Coat wells with goat anti-mouse Ig

Bind MAb in PBS-Tween Store plate at -20°

Add triazine-enzyme + std. or unknown to coated wells

Add substrate Develop color

Figure 2. Flow diagrams of 3 competition EIA procedures. The "classical" competition EIA (left panel) was used to monitor the immunizations, select the hybridomas, and for several of the demonstration projects described in this paper. The haptenated-enzyme EIA in the center panel was used for most of the method development. This is the assay that is presently being evaluated by the CDFA analytical laboratory. The simplified haptenated-enzyme EIA in the right panel is described in Methods.

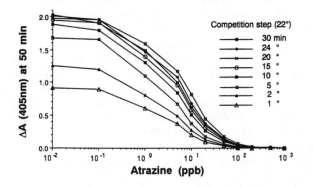

Figure 3. Kinetics of the competition step in the simplified "haptenated-enzyme" EIA. The assay diagrammed in the rightmost panel of Fig. 2 was conducted at room temperature as described in Methods. Mixtures of atrazine standards and simazine-alkaline phosphatase conjugate in PBS-Tween were added to rows of EIA wells coated with MAb AM7B2, which was "trapped" on the wells by affinity-purified goat anti-mouse IgG. At the times indicated, the wells were rinsed, substrate solution was added, and the absorbance was read 50 min later.

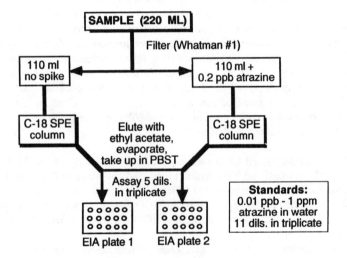

Figure 4. Flow chart for recovery and EIA measurement of atrazine in water.

Branson B12 sonic bath). The extracts were decanted, centrifuged (10 min, 10,000 x g) to remove particulate material, and 0.01 volume of glacial acetic acid was added. These solutions were applied to SCX aromatic sulfonic acid SPE columns (Analytichem) containing 300 mg of resin. The columns were washed with 5 ml of 1 M K_2HPO_4, and atrazine was then eluted with 2 ml of acetonitrile: 0.1 M K_2HPO_4 :: 1:1. The eluates were diluted to 5 ml with PBS-Tween, and dilutions were analyzed by EIA.

Results

Derivation of MAbs   We used 3 strategies to obtain MAbs with the greatest sensitivity and specificity: First, to maximize the chances of evoking different repertoires of antibodies, we tested simazine and atrazine haptens with two different linker groups (aha or mpa) on each of 3 different carriers (BSA, CON, and KLH) as immunizing antigens in pairs of 3 strains of mice (Swiss Webster, Biozzi, and B10.Q). The responses to the triazine were quantified by EIA on wells coated with a conjugate that had a linker and carrier different from the immunizing antigen. Second, for hybridoma production we selected only the best responding mice, with respect to serum titer, lowest detectable dose, and I_{50} for atrazine and simazine. Third, we prepared and screened a large number of hybridomas. Although all of the immunizing antigens evoked good triazine-directed responses in most of the mice, the statistics cited in the Methods section demonstrate that the most sensitive MAbs were only a small percentage of all of the triazine-directed MAbs.

Specificity of the MAbs. At UCB we compared the specificity of the MAbs using the "classical" competition EIA, which measured the ability of various triazines to compete with atrazine conjugates (immobilized on the EIA wells) for binding the MAbs. A similar set of experiments at UCD was done using the "haptenated enzyme" EIA format, with simazine-N(C2)-alkaline phosphatase as the competitor.

Table I summarizes the relative recognition of 37 triazine analogs and haptens, by MAbs AM7B2 and AM5D1. The results using the two different EIA formats and simazine-N(C2)-alkaline phosphatase were essentially the same for 7 of the most-used triazines. These results can be summarized as follows: (a) Propazine, procyazine, and cyanazine were recognized better than atrazine. Atrazine-mercaptopropionic acid, which was the hapten used to elicit the antibodies, was also recognized better than atrazine by both MAbs. This indicated that the MAbs bound better to analogs with isopropyl, cyclopropyl, or cyanoisopropyl groups at R2 or R3. (b) Both MAbs were much less reactive with prometon, which is used in substantial amounts in California and elsewhere, than they were for atrazine and simazine. (c) Hydroxyatrazine and hydroxysimazine reacted only 1% to 5% as well as atrazine.   (d) The mono-dealkylated triazines reacted 0.1% to 0.2% as well as atrazine, and (e) these MAbs did not measurably (< 0.2%) recognize di-dealkylated triazines. Thus, the MAbs are not effective probes for these triazine metabolites.

Table I. Relative reactivity of triazine MAbs AM7B2 and AM5D1 with various triazines and triazine haptens.

	Compound	R1	R2	R3	% cross-reactivity AM7B2	AM5D1
1	procyazine	Cl	$NHCH(CH_2)_2$[a]	$NHCCN(CH_3)_2$	526	583
2	atrazine-mpa	$S(CH_2)_2COOH$	$NHCH_2CH_3$	$NHCH(CH_3)_2$	261	181
3	propazine	Cl	$NHCH(CH_3)_2$	$NHCH(CH_3)_2$	196	161
4	cyanazine	Cl	$NHCH_2CH_3$	$NHCCN(CH_3)_2$	106	116
5	atrazine	Cl	$NHCH_2CH_3$	$NHCH(CH_3)_2$	100	100
6	dipropetryne	SCH_2CH_3	$NHCH(CH_3)_2$	$NHCH(CH_3)_2$	95	68
7	simazine-mpa	$S(CH_2)_2COOH$	$NHCH_2CH_3$	$NHCH_2CH_3$	66	76
8	simazine	Cl	$NHCH_2CH_3$	$NHCH_2CH_3$	31	31
9	prometryne	SCH_3	$NHCH(CH_3)_2$	$NHCH(CH_3)_2$	30	16
10	tertbutylazine	Cl	$NHCH_2CH_3$	$NHC(CH_3)_3$	23	22
11	terbutryne	SCH_3	$NHCH_2CH_3$	$NHC(CH_3)_3$	21	17
12	atr-N(C5)-COOH	Cl	$NH(CH_2)_5COOH$	$NHCH(CH_3)_2$	21	24
13	sim-N(C5)-COOH	Cl	$NH(CH_2)_5COOH$	$NHCH_2CH_3$	16	19
14	ametryne	SCH_3	$NHCH_2CH_3$	$NHCH(CH_3)_2$	14	14
15	sim-N(C4)-COOH	Cl	$NH(CH_2)_4COOH$	$NHCH_2CH_3$	8.2	12
16	cyanazine amide	Cl	$NHCH_2CH_3$	$NHCCONH_2(CH_3)_2$	6.5	6.2
17	hydroxyatrazine	OH	$NHCH_2CH_3$	$NHCH(CH3)_2$	5.7	4.1
18	prometon	OCH_3	$NHCH(CH_3)_2$	$NHCH(CH3)_2$	5.1	3.3
19	terbumeton	OCH_3	$NHCH_2CH_3$	$NHC(CH_3)_3$	5	4
20	simetryne	SCH_3	$NHCH_2CH_3$	$NHCH_2CH_3$	4.4	4.7
21	sim-N(C3)-COOH	Cl	$NH(CH_2)_3COOH$	$NHCH_2CH_3$	3.8	4
22	atratone	OCH_3	$NHCH_2CH_3$	$NHCH(CH_3)_2$	2.3	2.3
23	trietazine	Cl	$NHCH_2CH_3$	$N(CH_2CH_3)_2$	1.8	1.7
24	atr-N(C2)-COOH	Cl	$NH(CH_2)_2COOH$	$NHCH(CH_3)_2$	1.5	1.1
25	hydroxysimazine	OH	$NHCH_2CH_3$	$NHCH_2CH_3$	1.3	1.1
26	desmetryne	SCH_3	$NHCH_3$	$NHCH(CH_3)_2$	1.2	1.1
27	sim-N(C2)-COOH	Cl	$NH(CH_2)_2COOH$	$NHCH_2CH_3$	1.2	1.5
28	desethyl simazine	Cl	NH_2	$NHCH_2CH_3$	0.9	1
29	desethyl atrazine	Cl	NH_2	$NHCH(CH_3)_2$	0.7	0.8
30	desethyl simetryne	SCH_3	NH_2	$NHCH_2CH_3$	0.2	0.3
31	atr-N(C1)-COOH	Cl	$NHCH_2COOH$	$NHCH(CH_3)_2$	<0.2	<0.2
32	sim-N(C1)-COOH	Cl	$NHCH_2COOH$	$NHCH_2CH_3$	<0.2	<0.2
33	didesethyl simazine	Cl	NH_2	NH_2	<0.2	<0.2
34	ammelide	NH_2	OH	OH	<0.2	<0.2
35	ammeline	NH_2	NH_2	OH	<0.2	<0.2
36	melamine	NH_2	NH_2	NH_2	<0.2	<0.2
37	cyanuric acid	OH	OH	OH	<0.2	<0.2

NOTE: The assays were conducted using the haptenated-enzyme competition EIA shown in the middle panel of Figure 2, with simazine N(C2)-alkaline phosphatase as the competitor. Stocks of each analog were prepared by weight, and their molar concentrations were calculated from the molecular weight. The concentrations of each analog giving half-maximal inhibition (I_{50}) were calculated from multi-point dose-response curves, and the "percent cross-reactivity" is the ratio of the I_{50} for the analyte to the I_{50} of atrazine.

[a] (cyclopropyl)

Quantitative EIA for triazines. The MAbs were compatible with the 3 variations of the competition EIA shown in Figure 2. For the "haptenated-enzyme" format a simazine-enzyme conjugate proved to be better than an atrazine-enzyme conjugate for detection of atrazine. Simazine is recognized only about 30% to 40% as well as atrazine. This enables free atrazine to compete better than the simazine-enzyme detecting conjugate, making the assay more sensitive. The optimized "haptenated-enzyme" EIA done with simazine-alkaline phosphatase conjugate proved to be about 5-fold more sensitive than the conventional competition EIA using atrazine-protein conjugates as competitor.

The working range of the EIA standard curves was generally from 0.7 to 70 ppb, with an I_{50} of about 13 ppb. Thus, using a sample concentration step of about 100-fold for the EIA brought the limit of detection (an inhibition of 2 standard deviations from the signal with zero analyte) down to, or below that of gas chromatography, i.e., 0.01 to 0.02 ppb. Regardless of format, the EIA is very economical; it requires less than 50 ng of triazine conjugate and less than 1 μl of MAb culture fluid (which could be used as filtered hybridoma culture fluid without additional processing) per well. The culture fluid could be freeze-dried and reconstituted with no significant loss of triazine binding capacity, and EIA plates coated with captured MAb could be stored frozen until they were needed. The maximum response of the EIA in these was lower, but the I_{50} and slope values were the same as with plates prepared the night before use. These properties lengthen shelf-life and improve quality control. Furthermore, the MAbs developed for this study tolerated at least 20% (v/v) methanol in the PBS-Tween buffer used for the competition step. This makes it easier to use solvent- and solid-phase extracts in the assay, and may reduce sequestration of analytes in lipid micelles from various sample matrices.

Variability between assays is an important consideration for regulatory applications of EIA. Figure 5 is a plot of the I_{50} values and slopes of the standard curves for 17 consecutive "classical" competition EIAs performed at UCB during June and July 1989, using plates coated with atrazine-aha-BSA. The I_{50} values remained in the same range through December 1989, indicating that there was no apparent deterioration of this conjugate.

Variation when the same assay is done by different analysts is also a concern. At UCD, a study was conducted in which an immunochemist with several years' experience and a graduate student newly trained in EIA each analyzed 56 well water samples containing 0 to 0.25 ppb, using the haptenated enzyme EIA shown in Figure 2 (center panel). The results these persons obtained correlated with a slope of 1.08 and r = 0.98 (data not shown).

Sample Extraction Methods. Solid-phase recovery of atrazine and simazine on C_{18} columns proved to be convenient and efficient, using the method described above. Recovery of atrazine as a function of sample volume was assessed by gas chromatography. As Table II indicates, quantitative recovery of a 200 ng spike was achieved for samples of up to 1 liter. Efficiency was measured by recovery of [^{14}C]atrazine, as well as by gas chromatography (using a nitrogen-phosphorus detector). The results, summarized in Table III, indicate that recovery is nearly 100% from 0.1 to 100 ppb.

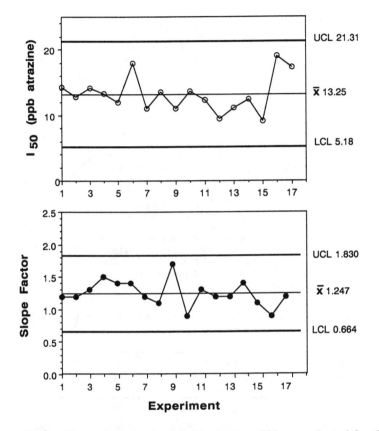

Figure 5. Quality control charts for 17 atrazine EIAs conducted by the "classical" method (Fig. 2, left panel) at U.C. Berkeley. Wells were coated overnight at 4° with 50 ng of atrazine-aha-BSA. Mixtures of standards and MAb AM7B2 in PBS-Tween were incubated overnight at room temperature, and then applied in 0.1 ml to the coated wells after they were rinsed 3 times with PBS-Tween. After 2 hr at room temperature, the wells were rinsed 3 times, and 0.1 ml of a 1:1,000 dilution of alkaline phosphatase-conjugated goat anti-mouse immunoglobulin (Sigma) was added. The plates were incubated for 2 hr at room temperature, rinsed again, substrate solution was added, and the color development was read on an EIA reader. The standard curves were fitted as described in Methods, to derive the I_{50} and slope values. The upper and lower confidence limits are one standard deviation from the mean.

Table II. Recovery of atrazine by C_{18} solid-phase extraction, as a function of sample size.

Sample vol. (ml)	atrazine recovered (ng)	%
100	220	110
250	210	105
500	200	100
1,000	200	100

NOTE: Duplicate samples of distilled water as indicated were spiked with 200 ng of atrazine from a reference standard in methanol, and applied to C_{18} columns (Analytichem Bond-Elut, 100 mg resin) at 8 to 15 ml/min. Columns were washed and eluted as described in Methods, and the eluates in ethyl acetate were analyzed by gas chromatography, using a nitrogen-phosphorus detector. Data were quantified as peak areas, relative to reference standards.

Table III. Recovery of atrazine from water using Analytichem C_{18} Bond-Elut™ columns.

Atrazine spike (ppb)	[14C]atrazine recovered (cpm ± s.d.)	%	Recovered (GC analysis) (ng)	%
0	25 ± 2	—	nd	nd
0.01	2,172 ± 67	96.6	nd	nd
0.1	2,321 ± 60	103.3	nd	nd
1	2,307 ± 43	102.7	45	60.2
10	2,265 ± 55	100.8	746	99.5
100			7,350	98.0
1,000			74,100	98.6

NOTE: Samples of 75 ml of distilled water were spiked with the indicated amounts of atrazine, from reference standards in methanol. Where indicated, 2,266 ± 67 cpm of ring-labeled [14C]atrazine was added to each spike. Samples were applied and recovered from 3 separate C_{18} columns as described in Methods, and the eluates were counted by liquid scintillation, or analyzed by gas chromatography (GC) using a nitrogen-phosphorus detector. Because of co-eluting contaminants, the limits of detection and quantification for GC were 2 ppb (3 std. dev.) and 7 ppb (10 std. dev.) respectively. The low (60.2%) recovery determined by GC for a 1 ppb spike appeared to be due to detector suppression by co-extracted material.

(nd = not determined)

EIA and GC results were compared for the analysis of simazine in soil, using ethyl acetate for extraction of dried samples of sandy loam from a site contaminated by an experimental simazine spill. For 24 samples that had simazine content between 0 and 350 ppb by GC analysis, the results by EIA correlated with r = 0.93 and a slope of 1.26 (Figure 6). For an extended data set of 48 samples containing simazine from 0 to 3 ppm, the correlation between EIA and GC determinations was 0.97, with a slope of 0.81 (A. Lucas, unpublished data). To ensure solubilization of the simazine recovered from the most heavily contaminated samples, methanol was added to the PBS-Tween to 5% (v/v) as cosolvent. Atrazine residues were recovered by extraction with 90% acetonitrile and concentration on SCX solid phase columns. For samples spiked with 10, 25, 50, and 100 ppb (ng atrazine per gram of soil) recoveries of 80%, 82%, 79%, and 93%, respectively, were obtained.

Demonstration Projects. During 1989 we conducted several studies to determine the accuracy, precision, and robustness of the EIA for quantifying triazines in various sample matrices. The U.C. Davis and U.C. Berkeley laboratories collaborated in EIA tests of well water samples that had been analyzed for triazine by gas chromatography at CDFA. Figure 7 is a bar chart of the triazine content of 75 of these samples, determined by the haptenated enzyme EIA. Three major results were evident from this study. First, the limit of detection of the EIA (the SPE blank in Figure 7) was below the limit of approx. 0.05 ppb for GC. Second, all of the samples that showed detectable amounts of triazine by GC also registered positive by EIA; in other words, there were no "false negatives" in the survey by EIA. Thirty-six of these samples were also analyzed at UCB, using the conventional EIA. Again, there were no "false negatives," and the results obtained by the two laboratories using different EIA methods correlated with a coefficient of 0.87. Third, the precision of the EIA was slightly better than that of the GC method, as shown by the coefficients of variance for the paired samples in Figure 7.

The UCB group also conducted a survey of groundwater from test wells sunk to different levels in and around a toxic waste site contaminated with atrazine. The data from this study are summarized in Table IV. Two groundwater samples from test wells contained high concentrations of triazines. Confirmatory values for these samples were obtained by the remediation site contractor using EPA Method 619. Two points can be made from these data. First, measurements obtained by recovering the triazines on C_{18} SPE columns were higher that those obtained when the samples were assayed directly after filtration through Whatman #1 paper. Second, the remediation contractor's gas chromatographic analysis of the samples from wells 4 and 5a revealed ppb to ppm levels of other contaminants, including 2,4,-D, 2,4,5,-T, MCPA, and xylene. These did not interfere appreciably with the triazine immunoassay.

The UCB group also performed triazine EIAs on samples from agricultural evaporation ponds in the San Joaquin Valley. These ponds collect drainage from fields where triazines may be used, and thus have the potential for accumulating these, and other pesticides. Metal ions, salts, and suspended solids accumulate in amounts up to 20 times those found in sea water, and various species of bacteria, algae and even brine shrimp may propagate in water from these

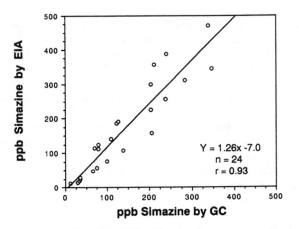

Figure 6. Comparison of soil analyses by gas chromatography and monoclonal EIA. Simazine-contaminated soil samples were extracted with methanol as described in Methods. Portions of the extract were analyzed by gas-liquid chromatography (GLC) using a nitrogen-phosphorus detector. The remainder of each extract was analyzed using the haptenated-enzyme EIA diagrammed in the center panel of Figure 2. The solid line was obtained by linear regression.

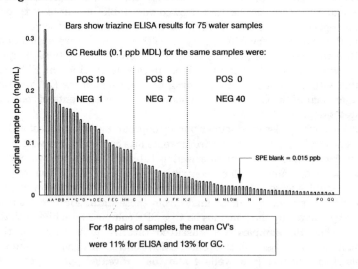

Figure 7. Summary of EIA results for 75 samples of well water. These samples were analyzed by gas chromatography (GC) at the CDFA analytical laboratory, and by the haptenated enzyme EIA (Figure 2, center panel) at U.C. Davis. For the EIA, triazine was recovered from the water samples essentially as diagrammed in Figure 4. Samples indicated by the same letter were quadruplicates taken at the wellhead, and analyzed in duplicate by EIA and GC. These samples are the basis for the precision comparison noted. Samples marked (*) are replicate determinations of one sample (mean ± SD, 0.16 ± 0.01 ppb; CV = 8.8%). Its duplicate sample is marked (+).

Table IV. Application of the triazine EIA to analysis of contaminated groundwater at a toxic waste pit.

Description	Appearance of C18 extract	Atrazine Spike (0.2 ppb)	Assay A	C18 Column Extracts				EPA Method 619†
				Assay B		Assay C		
				Plate 1	Plate 2	Plate 1	Plate 2	
			ppb triazine (atrazine equivalent, mean ± std. error) *					
Groundwater, test well approx. 100 ft east of toxic site	clear	-		≤ 0.08		< 0.1		
		+		0.18±0.03	0.29±0.05			
Groundwater, well MA-4	turbid	-	18.3	> 7.1		< 45	> 7, < 34	29
		+				< 45	> 7, < 34	
Groundwater, well MA-5a	yellow, with odor	-	465±36	> 7.1		633±74	541±58	510
		+				536±13	513±28	
Groundwater, domestic well about 150 ft west of toxic site	clear	-	< 0.1	< 0.08		< 0.1		
		+		0.19±0.03	0.26±0.01			
glass-distilled H2O		-	< 0.1	< 0.08		< 0.1		
"		+ 0.2 ppba		0.23±0.02	0.28±0.01			
"		+ 0.2 ppbb		0.19±0.01	0.23±0.03			
"		+ 100 ppb	104±18					

NOTE: Samples were collected at a remediation site in Northern California, concurrent with a scheduled quarterly sampling by the EPA-certified remediation contractor. Groundwater samples were obtained from two 3 inch test wells, MA-4 (150 feet deep) and MA-5a, (49 feet deep) at the edge of the toxic waste pit. To ensure that groundwater seepage was being measured accurately, ten well volumes were pumped from these pipes into a waste tank before the test samples were taken. The samples were transported to UCB in foil-covered glass bottles with foil-lined caps, on ice. Three assays were performed. In each case, half of the sample volume was unspiked, and the other half was spiked to a concentration of 0.2 ppb with atrazine before concentration or cleanup. In Assay A, the samples were filtered through 2 layers of Whatman #1 paper and analyzed directly. For Assays B and C, the filtrates were applied to C18 solid-phase columns (Fisher Prep-Sep) and the triazine was recovered and analyzed as described in Methods. After assay B was run, it was apparent that some samples exceeded the anticipated range. Assay C was then run using greater dilutions of the samples recovered from the C18 columns. Controls consisted of 0.2 or 100 ppb of atrazine in glass-distilled water. Four or 5 dilutions of each sample were measured by EIA with MAb AM7B2, and only the values falling between 0.2 and 0.8 of the full range of the standard curve were used. The data are the mean ± standard error for triplicate samples — i.e., 3 separate C18 column eluates, in Assays B and C.

* Values preceded by < are below the indicated minimum detection limit.
Values preceded by > are above the indicated maximum amount measurable in this assay.
Data without standard errors represent only one value in the working range.
† Data reported by EPA contractor to Calif. Regional Water Quality Control Board, July 1989
a Data from Assay B, plates processed 6/30/89   b Data from Assay C, plates processed 7/5/89

ponds. Table V shows results of EIAs on water from 3 evaporation ponds that were potential accumulators of triazines, and samples of sump and canal water from an area not subject to triazine application. Unspiked and spiked samples of the evaporation pond water showed an extreme matrix effect when they were added directly to the EIA. This effect was greatly reduced when the samples were subjected to the C_{18} solid-phase extraction protocol described in Methods. However, a significant matrix effect remained, as was evident from the values obtained for spikes of 0.5 ppb recovered from these samples. We speculate that the high metal and salt content in these samples may create an inhibitory "matrix effect" that could account for all of the "triazine" estimated in the unspiked samples. Additional studies are under way to determine and eliminate the cause of this bias in samples of this type.

Summary and Conclusions

The thiocarbamate herbicides described by Gee, et al., (this volume) and the triazines discussed in this paper are the first of several herbicides for which we plan to develop monoclonal antibodies and sensitive immunoassays for CDFA. Monoclonal antibodies offer the advantages of defined affinity and specificity, adaptibility to virtually any immunoassay format, and potentially unlimited supply. These advantages are of particular importance to agencies such as CDFA, that intend to configure and validate the immunoassays for regulatory purposes.

Although we can not draw many conclusions about structure-activity relationships from these data, the antibodies we generated showed a preference for binding to triazines that have isopropyl groups at R2 and R3. The substituents are clearly the major determinants of specificity, as shown by the very poor recognition of the mono- and di-dealkylated triazines and the hydroxytriazine metabolites. Detection of hydroxytriazines may be important for some environmental monitoring applications, because they are indicators of exposure of plants and soil microorganisms to the parent compounds. However, the hydroxytriazine metabolites are not herbicides, and they are not defined as hazardous pollutants. A recent paper by Schlaeppi, et al. (9) described the production of MAbs that were specific for hydroxyatrazine, using hydroxyatrazine conjugates as immunizing antigens.

Results with the three EIA formats used in this study were very similar. However, use of the simazine N(C2)-enzyme conjugate as the competitor in the haptenated enzyme format gave this method a more sensitive limit of detection than the classical competition EIA. The specificity of the EIA is primarily characteristic of the MAb that is used, although it may vary slightly with different EIA formats. The major advantage of the haptenated enzyme format was that its lower detection limit enabled one to work with smaller amounts of environmental samples. However, this format was also more sensitive than the classical competition EIA to inhibition by organic solvent in the incubation solution.

Tables II and III demonstrate that recovery of atrazine was quantitative from the small C_{18} solid-phase columns used for the analysis of groundwater and soil samples. Quantitative recovery was obtained from up to 1 liter of water,

Table V. Analysis of triazines in water from agricultural evaporation ponds.

	Atrazine spike (ppb) *		
Sample	0	1	10
Evap. pond A	8.7	10.8	NT
Evap. pond B	6.2	9.8	19
Sump T4	<0.2	NT	9.6
San Luis Canal	<0.2	NT	7.4

NOTE: Samples of water from 3 agricultural evaporation ponds in the San Joaquin Valley, and from a drain sump (T4) and the San Luis water delivery canal, were analyzed without concentration or cleanup. The samples and analyses of their ionic content were provided to us by the California Central Valley Regional Water Quality Control Board.

*NT = not tested

Sample	Matrix	Atrazine spike (0.5ppb)	ppb triazine (atrazine equiv., *mean ± std. error*) *	
			plate 1	plate 2
Evap. pond A	Mo > 6 mg/l	-	0.20 ± 0.03	0.13
	As > 1 mg/l	+	0.92 ± 0.06	0.82 ± 0.02
Evap. pond B	Se > 1.5 mg/l	-	0.13 ± 0.03	≤ 0.09
	SO$_4$ ~ 18 g/l	+	0.83 ± 0.08	0.74 ± 0.01
Evap. pond C	Cl$^-$ = 16 g/l	-	0.09	≤ 0.09
	SO$_4$ ~ 21 g/l	+	0.87 ± 0.08	0.82 ± 0.02
glass-distilled water		-	≤ 0.09	≤ 0.09
		+	0.66 ± 0.07	0.64 ± 0.03

NOTE: Residues recovered from the samples described above, and one other evaporation pond, using the SPE procedure for groundwater described in Methods.

* Values preceded by ≤ are below the indicated minimum detection limit.

Data without standard errors represent only one value in the working range.

and the detection limit of the assay was less than 0.1 ppb. To date, we have not examined the efficiency with which other triazines can be recovered, primarily because of the preference of the MAbs for atrazine. We speculate that if a method could be identified for selectively recovering triazines other than atrazine, it could be interfaced with the EIA using our MAbs, for single-analyte analysis.

The methods we adapted for recovery of atrazine and simazine from water and soil are faster and less involved than the recovery and cleanup procedures used for GC analysis. The recovery study in Table III demonstrated that our protocol for C_{18} solid-phase extraction recovered 100% of the atrazine from ordinary groundwater samples. In this experiment, data for spikes less than 1 ppb could not be obtained by GC, due to limitations of the detector and interference from co-extracted material. However, in other experiments, such as our studies on the 75 well water samples, the EIA was able to precisely quantify levels over 0.1 ppb. For example, 6 replicates of one sample in Figure 7 gave 0.16 ppb with a coefficient of variation of 8.8% (data not shown). Thus, the minimum detection limit of the EIA for atrazine appeared to be lower than that of GC. The recovery of atrazine from methanol extracts of soil was similarly efficient for the experiment of Figure 6. These results and the results of the toxic site groundwater study (Table IV) demonstrate that the monoclonal EIA is useful for surveys of highly contaminated soil and water, as well as for surveys of groundwater containing atrazine or simazine at the limit of detectability.

Our studies with solid-phase extraction also revealed differences in the types of errors it can introduce to GC or EIA analysis. The extreme metal and salt content of agricultural evaporation water is one example of a matrix that may interfere with the triazine EIA, and necessitate additional sample preparation steps. EIA and GC are likely to be sensitive to different sets of interfering factors, so EIA may not be as sensitive as GC to differences between manufacturers and different lots of SPE columns. However, material from improperly conditioned columns can interfere with EIA, and we found that this inhibitory effect was manifested as a bias toward higher estimates of the analyte. We found that to avoid this effect, the C_{18} columns must be scrupulously washed with hexane, ethyl acetate, methanol, and water before the sample is applied.

To facilitate the development of methods and test the immunoassay on the widest variety of samples, it has been our policy to distribute our antibodies and conjugates to all investigators who request them. We believe that this will help to more quickly reveal any shortcomings of the assay. It is enabling some investigators to conduct projects for which the cost and time for instrumental analysis would be prohibitive, and it is encouraging evaluation of the MAbs in new formats, such as sensors and field-portable kits. The availability of these MAbs should give more environmental chemists experience with immunoassay, help to establish the usefulness of the EIA as a screening method, and its validity as a quantitative research tool.

In summary, a coordinated effort between our three laboratories and the CDFA has resulted in development of MAbs , a sensitive, economical EIA, and simple, efficient residue recovery methods that CDFA analytical chemists are now validating, for integration into their repertoire of tests for regulatory monitoring of groundwater on a large scale in California. The assay and sample recovery protocols are sensitive, reproducible, fast, inexpensive, and amenable to automation. The limit of detection is comparable to that of gas chromatography. The monoclonal antibodies will provide a continuing source of the critical immunoprobe, with assured quality. These methods will initially be used as screening tools to reduce the number of samples submitted for instrumental analysis. However, future work at CDFA and in our laboratories will focus on identifying and eliminating sample matrix effects, and rigorously validating the entire procedure, so that it will be highly reliable, fully quantitative, and certifiable for regulatory purposes.

Acknowledgments

This research was supported by contracts from the CDFA to the co-authors at UCB, UCD, and EMSL. We are especially indebted to Dr. P. Stoddard of CDFA for his encouragement, assistance, and valuable guidance, and to S. Dumford of Ciba-Geigy Corp. for providing triazine reference standards. We also acknowledge the following CDFA scientists who are validating and implementing our triazine EIA for regulatory purposes: C. Cooper, H. Beermann, A. Braun, S. Powell, S. Richman, J. Melvin, and V. Quan. Numerous persons contributed to this work by offering help and expertise for the demonstration projects. These included J. Biggar (UCD), and J. Menke, J. Chillcott, and B. Niblack (Calif. Central Valley Regional Water Quality Board).

Literature Cited

1. Knüsli, E. In Residue Reviews; Gunther, F.A. and Gunther, J.D., Eds.; Springer-Verlag: New York 1970; vol 32, pp. 1-9.
2. Bushway, R.J.; Perkins, B.; Savage, S.; Lekousi, S.J.; Ferguson, B.S. Bull. Env. Contam. Toxicol. 1988, 40, 647-654.
3. California Dept. of Food and Agriculture; California Pesticide Use Reports, 1983 through 1987
4. Gast, A. In Residue Reviews; Gunther, F.A. and Gunther, J.D., Eds.; Springer-Verlag: New York 1970; vol 32, pp. 11-18.
5. Cardozo, C. et al.; 1988 Update, Well Inventory Database; California Dept. of Food & Agriculture: Sacramento, 1988; Document EH88-10.
6. Goodrow, M.H.; Harrison, R.O.; and Hammock, B.D. J. Agr. Food Chem. 1990, (in press).
7. Fazekas de St. Groth, S.; Scheidegger, D. J. Immunol. Meth. 1980, 35, 1-21.
8. Hill, B.D.; Stobbe, E.H. J. Agr. Food Chem. 1974, 22, 1143-1144
9. Schlaeppi, J.-M.; Föry, W.; and Ramsteiner, K. J. Agr. Food Chem. 1989, 37, 1532-1538.
10. Canellas, P.; Karu, A.E. J. Immunol. Methods. 1981, 47, 375-385.

RECEIVED August 30, 1990

Chapter 7

Reliability of Commercial Enzyme Immunoassay in Detection of Atrazine in Water

James R. Fleeker and Leonard W. Cook

Department of Biochemistry, North Dakota Agricultural Experiment Station, North Dakota State University, Fargo, ND 58105

Surface, municipal, and ground water (117 samples) from six midwestern states were tested for levels of eight triazine herbicides by a commercial enzyme immunoassay (EIA) and by gas chromatography (GC). The herbicides measured by GC were ametryn, atrazine, prometryn, propazine, simazine, terbuthylazine, terbutryn, and trietazine. Chromatographic analysis detected triazine levels of ≥0.1 µg/L. A limit of detection of 0.4 µg/L was found for the EIA. Atrazine was present in 34 of 36 samples testing positive for triazines by GC. Eight samples had triazine levels ≥0.4 µg/L by GC analysis and all tested positive by the EIA. Twenty-eight samples had triazine levels of 0.1–0.3 µg/L by GC and twelve of these gave EIA responses ≥0.4 µg/L. All samples having <0.1 µg/L of triazines by GC gave negative EIA responses.

Enzyme immunoassays (EIA) have been developed which offer a quick and inexpensive method to detect in water low levels of several environmental pollutants. The simplicity of some EIA protocols allow personnel with minimal training to screen samples. The use of an EIA as a screening method and chromatography for confirmation appears to be a cost effective approach to monitor large numbers of samples.

This paper reports on the reliability of a commercial EIA which detects atrazine and certain other triazine herbicides. The tests were designed to determine the reliability of the assay in detecting low levels of atrazine in surface, municipal and well water. The ability of the EIA to quantitate atrazine levels was not evaluated. The s-triazine herbicides are among the most widely used herbicides in the United States (1) and atrazine has been found to be a common ground water contaminant in the midwestern United States (2).

0097–6156/91/0451–0078$06.00/0

Materials and Methods

Reagents and Materials. Deionized water was obtained from a
Millipore Milli-Q water purification system. Triazine standards
were received from the U.S. EPA Pesticides and Industrial Chemicals
Repository, Research Triangle Park, NC 27711. Immunoassay kits were
supplied by ImmunoSystems, Inc., 4 Washington St., Scarborough, ME
04074.

Sample Collection. Samples from Minnesota, Iowa, and Nebraska were
supplied by the Minnesota Department of Agriculture, the U.S.
Geological Survey, and the Nebraska Department of Health, respectively. All samples were stored at 4°C in glass jars sealed with
lids lined with aluminum foil until analysis. Particulate material
was allowed to settle out before analysis. Just before extraction,
the pH and conductivity were determined (3).

Extraction for GC Analysis. The method was adapted from Muir and
Baker (4). Sodium chloride (5 g) was dissolved in 500 mL of the
water sample. The water was extracted with one 100-mL portion and
two 50-mL portions of ethyl acetate. After each extraction, the
organic layer was transferred to a 10 x 300 mm chromatography column
containing 10 g of anhydrous sodium sulfate. The column effluent
was collected in a 250-mL round-bottom flask and was concentrated to
1-2 mL on a rotary evaporator with the bath at 35°C. The concentrate was transferred to a small culture tube with three 1-mL rinses
of acetone. The solvent was removed with a stream of N_2 gas while
warming at 30°C. The residue was dissolved in 1.0 mL of toluene and
stored at 4°C until analysis by GC.

Gas Chromatographic Analysis. The quantitation method was adapted
from Roseboom and Herbold (5). Analysis was done with a Tracor 550
GC and a Tracor 702 thermionic detector. Inlet, outlet, and
detector temperatures were 200, 210, and 230°C, respectively.
Detector gas flows were air at 120 mL/min, hydrogen at 3.0 mL/min,
and helium makeup at 20 mL/min. Separation was done at 220°C on a
20 m x 0.53 mm Stabilwax (1.0 µm) column (Restek Corp.) with a
helium flow of 10 mL/min. Injection volume was 3.0 µL. Samples
giving a response >0.1 µg/L were confirmed with a 15 m x 0.53 mm
DB-17 (1.0 µm) column (J & W Scientific) programmed at $140-190^{\circ}$C at
5°C/min.

Immunoassay Protocol. The manufacturers recommendations were
followed for use of the EIA kits. EIA measurements were done just
prior to GC analysis. Each water sample was also spiked with
1.0 µg/L of atrazine and tested by EIA. Each sample set consisted
of a deionized water blank, and up to five samples.
     The water sample (160 µL) was incubated 5 min with 160 µL of
peroxidase-triazine conjugate in a plastic tube coated with antibody. This was followed by a washing step and then a 2-min incubation with hydrogen peroxide and tetramethyl-benzidine. The reaction
was stopped by the addition of dilute sulfuric acid. Deionized
water was added to the solution to bring the color intensity at 450
nm to a level appropriate for measurement in a spectrophotometer.

Usually 2.0 mL of water was added, but as the color development was done at ambient temperature, which varied day-to-day, the amount of water added was adjusted accordingly (1.5-2.5 mL).

Absorbance Measurement. A Beckman model 24 spectrophotometer was used unless otherwise specified. Solutions were read at 450 nm against a deionized water blank. A Hach model DR100 portable spectrophotometer was also evaluated with some of the samples.

Results and Discussion

The EIA system examined here is available on microtiter plates or small polystyrene tubes. The latter was used here. The tube assay has short incubation periods and is suitable for field work and a small number of samples.

Water Sources. Surface water was taken from lakes, ponds, rivers, and streams in the upper midwestern United States (Figure 1). These samples ranged from very clear to very cloudy and colored and had very little to large amounts of sediment. Municipal water was collected from towns and cities where the water was treated for human consumption. Ground water was gathered from rural areas and included private wells and test wells used by various government agencies. These samples ranged from very clear to colored and with very little to large amounts of sediment. The pH of the water ranged from 5.3 to 9.2 and there was no apparent effect on the EIA in this pH range. Conductivity varied from 0.1 to 180 mMHO's, with no apparent effect on the immunoassay in this range.

Several s-triazine herbicides besides atrazine are detected by this assay (Table I). While atrazine is the s-triazine usually found in ground and surface water in the midwestern United States, it was necessary to check by GC for the presence of other common triazine herbicides in order to establish false responses by the EIA. Prometron, simetryn, and cyanazine were unavailable as GC

Table I. Cross Reactivity of Triazine Herbicide With the
EIA and Recovery of the Compounds From Water
Prior to GC Analysis

Substance	50% Inhibition of EIA[a] (μg/L)	Recovery From Water (%)	
		0.1 μg/L	1.0 μg/L
Atrazine	0.04	104 ± 4	92 ± 11
Prometryn	0.4	125 ± 16	97 ± 9
Propazine	0.5	93 ± 16	90 ± 8
Ametryn	0.7		101 ± 18
Prometon	0.7	not analyzed	
Simazine	2.5	91 ± 20	92 ± 11
Trietazine	2.5	91	88 ± 8
Simetryn	2.5	not analyzed	
Terbuthylazine	4.0	106 ± 15	89 ± 7
Terbutryn	–	112 ± 36	98 ± 10
Cyanazine	40	not analyzed	

[a] ImmunoSystems Inc., Scarborough, Maine

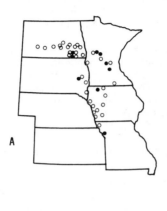

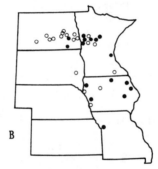

Figure 1. Collection sites. (A) Groundwater; (B) surface water; and (C) municipal water. Solid circles represent water containing ≥0.1 μg/L triazine by GC analysis.

standards and not analyzed. These three herbicides are not in
widespread use in the survey area. The other herbicides were
efficiently extracted from water prior to GC analysis (Table I).
Occasionally organic material in the extract interfered with one or
more analyte peaks (Figure 2), but the atrazine peak was always
resolved.
    On-site measurement of the EIA response with a portable
spectrophotometer is useful for documentation, although the response
to 1.0 µg/L of atrazine can be visually detected. The results with
a laboratory spectrophotometer and a portable hand-held spectrophotometer were compared. Standard curves recorded with these
instruments were nearly identical. The standard deviation for 18
replicates were similar with the two instruments (Table II).

Table II. Absorbance Values Obtained With the EIA Using
Distilled Water Spiked With Atrazine

			Atrazine Level	
		Control	0.1 µg/L	1.0 µg/L
		Beckman Model 24		
Average A	(n = 18)	0.643	0.566	0.262
Std. Dev.		0.074	0.057	0.022
		Hach Model DR100		
Average A	(n = 18)	0.44	0.40	0.21
Std. Dev.		0.04	0.03	0.01

EIA Performance. Triazines were found by GC in 36 of the 117 water
samples. Thirty-four of these samples contained atrazine in the
range of 0.1 to 12 µg/L. The other triazines found were simazine,
terbutryn, and trietazine.
    We began the survey with no criteria for the detection limit.
The EIA could detect 0.1 µg/L of atrazine in distilled water (Table
II). However, it was not known what effect organic material from
surface water would have in masking this level of detection. Also,
a large proportion of the solution in the competitive binding
portion of the assay is sample. This, and the variation of the
mineral and organic content of the water over the region of the
survey were expected to lead to large fluctuations in the non-
specific or background response. A related problem was the lack of
a standard sample matrix to prepare the response curve. Distilled
water was chosen for this purpose because it is supplied with the
EIA kits as control water.
    A significant background response was found but it was not
unusually large nor exceptionally varied (Figure 3). Application of
the conventional approach to determining the limit of detection was
adequate with the water sources tested (6): the standard deviation
of the mean response from distilled water (B_0 value; n = 20), was
multiplied by three and the product subtracted from 1.0 to give a
value of 0.7. This detection limit corresponded to an atrazine
level of 0.4 µg/L in distilled water. The criteria for the limit of
detection was adequate in that false-negative responses were not
found and the number of false positive responses were acceptable.
That is, eight samples found to have atrazine levels >0.4 µg/L by GC
also gave a positive response by the EIA. Of twenty-six samples

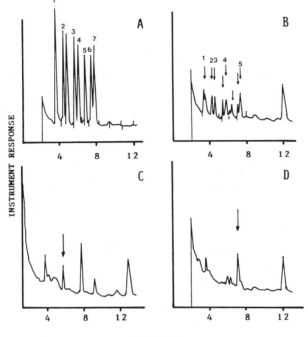

TIME (MIN)

Figure 2.   Gas chromatograms.
(A)   Water spiked with triazine standards (1.0 µg/L each) and
carried through the extraction;   (1) trietazine, (2) propazine,
(3) terbuthylazine, (4) atrazine, (5) prometryn, (6) terbutryn,
(7) simazine, and (8) ametryn; 12 Jul 88; 8×10^{-12} afs.
(B)   Water spiked with triazine standards (0.1 µg/L each) and
carried through the extraction; 4×10^{-12} afs.
(C)   Sample No. 66, 0.1 µg/L atrazine; 4×10^{-12} afs.
(D)   Sample No. 98, 0.2 µg/L simazine; 4×10^{-12} afs.

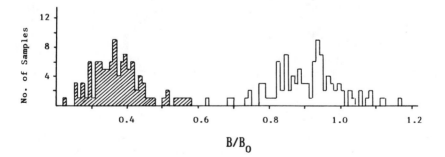

Figure 3.   Response of the EIA to samples containing <0.1 µg/L of
triazine, before (open) and after (solid) spiking with 1.0 µg/L
of atrazine.

containing 0.1–0.3 μg/L of atrazine by GC analysis, twelve gave B/B_o
responses <0.7 (false–positive). When used as a screening test,
false–positive responses of this nature will probably not be considered a drawback for use of the EIA. Samples containing <0.1 μg/L
by GC gave no false–positive responses. Twenty–five samples with
<0.1 μg/L of triazines all gave positive EIA responses when spiked
with 0.4 μg/L of atrazine.

Background levels of some water sources may require a multiplier either higher or lower than three in estimating the limit of
detection. A municipal water supply was monitored by GC and EIA
weekly for six months. No triazine herbicide was detected by GC in
this period. The nonspecific response of the EIA was lower than
that obtained in the six–state survey so that when a multiplier of
two was used to establish the limit of detection no false–negative
and only two false–positive responses were obtained.

Most of the samples collected in the six–state region gave B/B_o
values of <1.0 when the triazine levels were 0.1 μg/L by GC (Figure
3). The nonspecific response was greater with surface water than
with municipal or ground water (Table III). The nonspecific
response does not allow accurate quantitation near the limit of
detection. However, at low levels of analyte, the EIA appears to
serve as an excellent screening tool.

Table III. The EIA Response for Samples With
0.1 μg/L of Triazine Herbicide

Source	n	B/B_o	S.D.
Surface Water	19	0.83	0.07
Municipal Water	23	0.93	0.06
Ground Water	28	0.91	0.08

All water samples were found to give a strong EIA response if
1.0 μg/L atrazine was present. Figure 3 shows the EIA response of
samples found to have <0.1 μg/L of triazine by GC, both before and
after spiking to 1.0 μg/L of atrazine (100 μL of 0.1 μg/mL atrazine
to 10.0 mL of sample). The response to this level of atrazine could
be visually detected in most cases.

During this study, Bushway et al. (7), published a report on
the EIA used here with respect to measurement of atrazine in water
and soil. Their study found good correlation between HPLC and EIA
quantitation when water from a variety of sources was spiked with
atrazine between 1 and 80 μg/L. Our study focused on application of
the EIA as a screening method.

Bushway's group examined 10 ground water samples collected in
Maine, three of which had levels of triazine of 2–5 μg/L by the EIA.
Atrazine was not detected in these samples by HPLC.

Acknowledgments

We gratefully acknowledge the assistance of the Minnesota Department
of Agriculture, the U.S. Geological survey and the Nebraska
Department of Health. This investigation was project No. 447 of the
North Central Region Pesticide Impact Assessment Program. Any

opinions, findings, and conclusions or recommendations expressed in this publication are those of the authors and do not necessarily reflect the view of the U.S. Department of Agriculture.

Literature Cited

1. Maddy, K. T. Agric. Ecosyst. Environ. 1983, 9, 159–172.
2. Parsons, D. W.; Witt, J. M. Pesticides in Groundwater in the U.S.A.; A Report of a 1988 Survey of State Lead Agencies, Oregon State University Extension Service, June 15, 1988.
3. American Public Health Association, Standard Methods for the Examination of Water and Wastewater, 15th Ed., American Public Health Association, Washington, DC, 1980.
4. Muir, D. C.; Baker, B. E. J. Agric. Food Chem. 1976, 24, 122–125.
5. Roseboom, H.; Herbold, H. A. J. Chromatogr. 1980, 202, 431–438.
6. Schuurs, A. H. W. M.; Van Weemen, B. K. Clin. Chem. Acta 1977, 81, 1–40.
7. Bushway, R. J.; Perkins, B.; Savage, S. A.; Lekouse, S. J.; Ferguson, B. S. Bull. Environ. Contam. Toxicol. 1988, 40, 647–54.

RECEIVED August 30, 1990

Chapter 8

Immunoassay as a Screening Tool for Triazine Herbicides in Streams

Comparison with Gas Chromatographic–Mass Spectrometric Methods

D. A Goolsby[1], E. M. Thurman[2], M. L. Clark[3], and M. L. Pomes[2]

[1]U.S. Geological Survey, U.S. Department of the Interior, Denver, CO 80225

[2]U.S. Geological Survey, U.S. Department of the Interior, Lawrence, KS 66049

[3]U.S. Geological Survey, U.S. Department of the Interior, Iowa City, IA 53344

Immunoassay and GC/MS (gas chromatography/mass spectrometry) analysis for triazine herbicides were compared in order to evaluate the potential of immunoassay as a screening tool. Water samples were collected at 146 sites on streams in a 10-state region of the midwestern United States during 1989 before and shortly after application of pre-emergent herbicides that were used for weed control in the production of corn and soybeans. The sites were sampled a third time in fall 1989 during a low streamflow period. The two methods compared well with a rank correlation coefficient of 0.90 for immunoassay results less than 5 ug/L (micrograms per liter). The presence or absence of triazine herbicides can be determined visually at concentrations of about 0.5 ug/L, and the detection limit can be lowered to about 0.2 ug/L with a spectrophotometer. The median pre-application and fall concentrations of triazine herbicides in the streams were determined by immunoassay to be less than 0.2 ug/L and 0.3 ug/L, respectively, compared with a post-application median of 3.4 ug/L. Immunoassay appears to be a rapid, reliable, and low-cost analytical screening method for detecting triazine herbicides in water.

Large quantities of pre-emergent herbicides are used annually in the corn (Zea mays L.) and soybean (Glycine max L.) producing region of the midwestern United States. Approximately 1.5×10^8 kg (kilograms) are applied each year over a 10-state region (1). The herbicides are applied to cropland prior to or during planting of seed within about a 1-month period from late April through late May. Typical application rates for atrazine and alachlor, the

predominant herbicides used in the area, range from about 0.8 to
1.5 kg/hectare.
  Previous studies have shown that runoff from fields
immediately after application results in large concentrations of
herbicides in surface water (2, 3). This finding has been
documented in several small scale studies (4-10), but has not been
examined at a multi-state scale to determine if the herbicides
cause regional water-quality contamination. Large concentrations
of herbicides in drinking water, consumed over long periods of time
can have adverse effects on human health (11). Furthermore,
conventional water treatment practices do not effectively remove
herbicides, such as atrazine (11).
  In order to address these concerns, a reconnaissance was
conducted during 1989 on a large number of streams in 10 States
(Illinois, Indiana, Iowa, Kansas, Minnesota, Missouri, Nebrasks,
Ohio, South Dakota, and Wisconsin) in the midwestern United States.
Objectives of the reconnaissance were to: (1) Determine the
usefulness of immunoassay as a low-cost screening tool for triazine
herbicides prior to the analyses of water samples by more costly
GC/MS, and (2) determine if immunoassay is a viable method to
assess the geographic and seasonal distribution of herbicides in
streams in a large region of the midwestern United States. The
focus of this paper will be on the usefulness of immunoassay as a
screening tool. The seasonal and geographic distribution of
herbicides will be addressed elsewhere.

Experimental

Design of Reconnaissance. Sampling sites were selected at 150 U.S.
Geological Survey streamflow- gaging stations by a stratified
random procedure designed to ensure geographic distribution.
Drainage-basin areas represented by the sampling sites ranged from
less than 260 km^2 (square kilometers) to more than 1,800,000 km^2
for a site on the Mississippi River. The median drainage area was
1800 km^2. The aggregate drainage area of individual hydrologic
units sampled was about 518,000 km^2. The streams were sampled
three times during 1989: (1) In late March and April before
herbicides were applied to the fields (145 immunoassay analyses),
(2) in May and June during runoff from the first rainfall following
herbicide application (135 immunoassay analyses), and (3) in the
fall (October-November) during low-streamflow conditions when most
of the water in the streams was derived from ground-water sources
(146 immunoassay analyses).
  Vertically integrated water samples were collected at three to
five points across each stream. Samples were collected and
composited in glass containers and filtered through glass fiber
filters (GGF14230, Geotech, Inc., Denver, Colo.) having a pore
diameter of approximately 1 micrometer in order to remove
particulate material. The filtrate was collected in 125-mL
(milliliter) glass bottles that had been heated to 350°C (degrees
Celsius) to remove organic matter. Samples were refrigerated to
about 4°C until analyzed.

Immunoassay. All samples were analyzed for triazine herbicides using RES-I-MUNE immunoassay kits (ImmunoSystems Inc., Scarborough, Maine). The kits use polyclonal antibodies coated on the walls of polystyrene test tubes and an atrazine-enzyme conjugate prepared by covalently binding atrazine to horseradish peroxidase by a modified carbodiimide technique (12-13). Other reagents in the immunoassay kits include three standards (negative control, 0.1 ug/L atrazine solution and 1.0 ug/L atrazine solution), substrate, chromogen, and a "stop" solution of 2.5 N (normal) sulfuric acid.

   Instructions provided with the kit were followed. The kits were stored at 4 to 8°C and the contents were allowed to warm to about 23°C prior to making the analysis. The samples and standard solutions (160 microliters or 4 drops) were placed in separate test tubes, immediately followed by 4 drops of the enzyme conjugate. The test tube was gently mixed for 2-3 seconds, and then allowed to incubate for 5 minutes. Atrazine in the sample and the atrazine-enzyme conjugate compete for a limited number of antibody binding sites on the test tube. After the incubation period, the unreacted molecules were washed away by rinsing the test tube four times with reagent water. Substrate (4 drops) and chromogen (4 drops) were added, gently mixed in the tube, and allowed to incubate for 2 minutes. The presence of bound atrazine-enzyme conjugate in the test tube converts the substrate to a compound that causes the chromogen to turn blue. Because the number of antibody binding sites and the number of atrazine-enzyme conjugate molecules remain about constant, a sample containing a small concentration of atrazine allows many atrazine-enzyme conjugate molecules to be bound by the antibody. Therefore, a small concentration of atrazine in the water sample produces a dark-blue solution. Conversely, the more atrazine present in the water sample the lighter the color. Thus, concentration is inversely proportional to color. The reaction was arrested by adding 1 drop of the 2.5 N sulfuric acid solution, which halts the blue color development and causes the reaction solution to turn yellow. A visual comparison between the samples and the negative control then was made to determine if triazine herbicides were present or absent and the result was recorded. Because of the rapid reaction kinetics, only four to five samples, one standard, and a negative control could be analyzed in one run.

   After the visual comparison was made, 0.50 mL of deionized water was added to each tube to provide additional volume, and the solution was transferred into a 1 cm (centimeter) cuvette. The OD (optical density) of the samples, standards, and negative control were read using a Milton Roy Spectronic 401 single-beam spectrophotometer (Linden, New Jersey) that had been zeroed against a deionized water blank at 450 nm (nanometers). The OD of each standard and sample was normalized for color variations in the negative control by dividing the ODs by the OD of the negative control and multiplying by 100 to obtain percent NC (percent of negative control). A working curve was prepared for each of the three sampling periods by plotting the concentration of the standards as a function of the percent NC on semi-log paper for all measurements made on standards during the sampling period. A best fit line was drawn through the median percent NC for each standard (see fig. 1 for example). The concentration for each sample (in

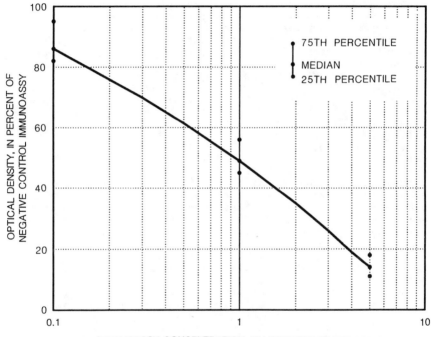

Figure 1. Immunoassay working curve for relation between triazine herbicide concentration and optical density for post-application sampling period.

ug/L) was obtained by reading the concentration corresponding to
the percent NC for that sample.

For quality-assurance purposes, about 8 percent of all samples
were submitted for immunoassay analysis as blind duplicates. In
addition, about 25 percent of the samples (mostly from the pre- and
post-application periods) were analyzed in a second laboratory by
immunoassay analysis. The procedure used by the second laboratory
was identical to that described above except that a differential
photometer (Artel, Inc., Windham, Maine) was used instead of the
Milton Roy spectrophotometer. These quality-assurance samples
provided a means to evaluate the variation in results within a
laboratory and between laboratories.

Thurman and others (14) have measured cross reactivity in the
immunoassay for 12 common triazine herbicides and metabolites
relative to atrazine at 1 ug/L concentrations for each herbicide.
The immunoassay response was expressed as ug/L of atrazine as shown
in figure 2. The response was greatest for atrazine (1.0 ug/L).
Ametryn's cross reactivity was equivalent to 0.6 ug/L of atrazine;
prometryn and propazine, 0.5 ug/L; and so forth (fig. 2). The
cross reactivity of each triazine is related to the extent of
binding that occurs between the antibody coated on the test tube
and the structure of the cross-reactive molecule. The binding was
strongest for compounds that have structures most closely
resembling the atrazine molecule, that is, a 4-ethylamino and a 6-
isopropylamino group. For example, a 1 ug/L solution of ametryn
(fig. 2) has a cross reactivity equivalent to 0.6 ug/L of atrazine.
Likewise a slight change on the amino group from ethyl to isopropyl
(propazine) elicits a response equivalent to 0.5 ug/L of atrazine.
Simazine, with 2 ethyl groups, has a cross reactivity equivalent to
0.2 ug/L of atrazine. Substitution of hydrogen for either the
ethyl (desethylatrazine) or the isopropyl (desisopropylatrazine)
lowers the cross reactivity to less than 0.1 ug/L of atrazine, or
essentially no response. Because the immunizing hapten (an
atrazine-like structure) was bound at the 2 position, it appears
that the relative response of the immunoassay is related to
antibody recognition and binding to the alkyl side chains on the
triazine ring.

Only hydroxyatrazine deviates from this pattern in cross
reactivity. Because it has the same alkyl structure as atrazine,
one would predict a cross reactivity similar to ametryn (about 0.6
ug/L as atrazine). However, hydroxyatrazine gave a response of
less than 0.1 ug/L as atrazine. A possible explanation is that the
hydroxyl group decreases the binding energy at the specific
antibody recognition site. The didealkylatrazine was nonreactive
(fig. 2). Neither alachlor nor metolachlor cross reacted with
atrazine, which is consistent with results of a previous study
(12).

The immunoassay also was examined for interference by
naturally occurring humic and fulvic acids (14), which account for
the majority of dissolved organic carbon in natural waters (15).
Atrazine was measured in water samples that contained from 5 to 100
mg/L of humic and fulvic acid from the Suwannee River (standard-
reference surface-water sample) (16) and humic and fulvic acids
from Biscayne aquifer near Miami, Florida (17). For all samples,
there was no difference between the immunoassay response in the
presence and absence of the humic material. These data indicate

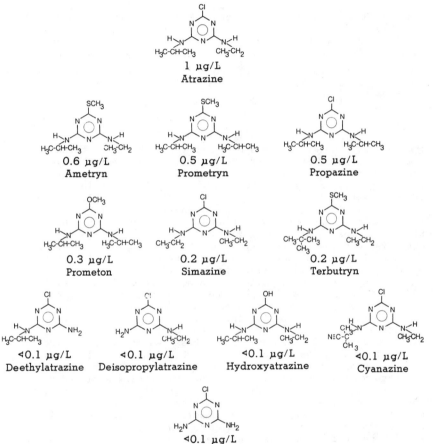

Figure 2. Chemical structure and immunoassay response of 1 ug/L (micrograms per liter) concentrations of triazines. Immunoassay response, expressed as ug/L of atrazine, shown for each herbicide. (Modified from Thurman and others, in press.)

that the immunoassay is not affected by cross reactivity with
natural dissolved organic matter.

GC/MS. About one-third of the pre-application samples and nearly
all post-application samples were analyzed by GC/MS for 11
herbicides and 2 metabolites of atrazine. Herbicides and
metabolites were isolated by solid- phase extraction and analyzed
by GC/MS ($\underline{14}$). GC/MS analyses of the eluates were performed on a
Hewlett Packard model 5890A gas chromatograph (Palo Alto, Calif.)
and a 5970A mass selective detector (MSD). Thirty-one ions were
selectively monitored, and the base-peak ion current was measured
for the quantification curve as a function of the response of the
mass 188 ion of d_{10}phenanthrene. Confirmation was based upon
presence of the molecular ion, two confirming ions (with area
counts $\pm$ 20 percent), and a retention time match of $\pm$ 0.2 percent
relative to d_{10}phenanthrene.

Results

Immunoassay. Working curves prepared for each of the three
sampling periods were linear to slightly curvilinear throughout the
range 0.1 to 5 ug/L (fig. 1). Concentration is inversely related
to OD expressed as percent NC. Because of the variation measured
in the negative control against the deionized water blank, the
detection limit for the method was established to be about 0.2 ug/L
for atrazine. Any value less than this was reported as <0.2 ug/L.
Similarly, the upper reporting limit for the procedure was
established to be about 5 ug/L. The ODs expressed as a percent of
the negative control obtained for all standards used for each
sampling period are summarized in Table I.
    During the three sampling periods, blind duplicates of 28
samples were submitted for immunoassay analyses by a single
laboratory as a quality-assurance measure. Results for the samples
are summarized in table II. The maximum difference between
duplicates was 0.5 ug/L and 86 percent of the samples had absolute
differences of 0.14 ug/L or less. These results indicate that
within-laboratory variation of the method is relatively small.
    Variation between laboratories was much larger than the within-
laboratory variation (Table II). For 107 samples analyzed by two
laboratories, the largest differences were >5 ug/L. However, for
most samples, the difference was much smaller. The median
difference was 0.0 ug/L, and the absolute difference was 0.1 ug/L
or less for 50 percent of the samples. More than 80 percent of the
samples had differences smaller than 0.6 ug/L. As expected, the
largest differences occurred in samples with largest herbicide
concentrations. Of the 13 samples with differences greater than 1
ug/L, 11 samples had atrazine concentrations determined by GC/MS to
be greater than 1 ug/L and 8 had atrazine concentrations larger
than 2 ug/L. However, neither laboratory recorded consistently
higher concentrations.
    The results show that within-laboratory variation is small and
indicate that within-laboratory precision can be very good.
Between-laboratory variation was much larger than within-laboratory
variation, but there was no apparent bias between the laboratories
(Table II). About 20 percent of the between-laboratory differences

Table I. Optical density of immunoassay standards expressed as a percent of the negative control

[ug/L, micrograms per liter; % NC, percent of negative control; IQR, interquartile range--75th percentile minus 25th percentile; (1), unitless]

Sampling round	Concentration of standard ug/L	Number of measurements	Optical density, % of NC					
			Mean	Standard deviation	Standard error of mean	Median	IQR	Coefficient of variation (1)
1	0.1	9	89.6	2.2	0.7	89	4	0.025
2	0.1	6	87.5	7.3	3.0	86	13	0.083
3	0.1	5	98.2	2.3	1.0	98	3.5	0.023
1	1.0	16	49.9	7.4	1.9	50	10	0.148
2	1.0	12	50.2	7.1	2.0	49	11	0.141
3	1.0	18	44.3	6.0	1.4	44.5	8	0.135
2	5.0	6	13.8	3.4	1.4	14	5	0.246
3	5.0	5	15.0	3.1	1.4	15	6	0.207

were larger than 0.6 ug/L. The reasons for these differences were
not determined.

Table II. Comparison of immunoassay results obtained on blind
duplicate samples by a single laboratory (within lab) and results
on samples analyzed by two different laboratories (between labs)

[Results shown are differences between duplicate analyses
within a single laboratory and between two laboratories;
differences are in micrograms per liter]

	Within laboratory	Between laboratory
Number of samples	28	107
Mean difference	0.01	-0.02
Standard deviation of differen	0.15	1.1
Maximum difference	0.5	>5.0
Minimum difference	-0.3	<-5.0
Percentile differences		
10th	-0.12	-0.4
25th	0.0	-0.1
50th (median	0.0	0.0
75th	0.08	0.0
90th	0.14	0.6

Immunoassay by Visual Comparison. During the course of this study,
421 immunoassay analyses were made by visual comparison as
described previously in this paper. Immunoassay analyses on these
same samples were also made with the aid of a spectrophotometer; in
addition, 180 of the samples were analyzed by GC/MS. A comparison
of the visual analysis with spectrophotometric immunoassay results
and GS-MS results is given in Table III. The results indicate a
visual detection limit for immunoassay of about 0.5 ug/L. For the
173 samples in which the visual analysis results were negative, 98
percent of the spectrophotometric analyses were less than 0.5 ug/L.
Also, for 45 of these samples that were analyzed by GC/MS, 98
percent contained less than 0.5 ug/L of atrazine.

Table III. Comparison of visual immunoassay results for presence
or absence of triazine herbicides with immunoassay results
obtained from spectrophotometric analysis and with
atrazine results obtained from GC/MS analyses

Immunoassay results by visual analysis	Immunoassay results by spectophotometric analysis	Atrazine results by GC/MS analysis
Negative	N = 173	N = 45
(herbicides not	% <0.2 ug/L = 91	% <0.2 ug/L = 66
detected)	% <0.5 ug/L = 98	% <0.5 ug/L = 98
Positive	N = 248	N = 135
(herbicides	% >0.2 ug/L = 88	% >0.2 ug/L = 95
detected)	% >0.5 ug/L = 60	% >0.5 ug/L = 82

[GS-MS, gas chromatography-mass spectrometry; N, number of samples;
%, percent of samples; ug/L, micrograms per liter; =, equals]

Visual analysis results were positive for triazines in 248 samples, 135 of which were analyzed by GC/MS. In 88 percent of the 248 samples, triazines were determined to be present in concentrations larger than 0.2 ug/L by immunoassay using spectrophotometric analysis. Also, atrazine concentration in 95 percent of the samples analyzed by more sensitive GC/MS analysis, was determined to be larger than 0.2 ug/L These results indicate that even without a spectrophotometer, the presence or absence of atrazine at concentrations of about 0.5 ug/L can be determined with reasonable reliability by a visual comparison against solutions of known atrazine concentration.

Comparison of Immunoassay with GC/MS Analysis. The relation between atrazine concentration determined by GC/MS analysis and triazine concentration determined from immunoassay analysis on 127 samples is shown in figure 3. Samples with immunoassay results larger than 5 ug/L are not plotted. Although the two determinations are highly correlated (rank correlation coefficient is 0.90; p<0.0001) the relation is not linear over the 0.2 ug/L to 5 ug/L range of the immunoassay results. Linear and multiple-linear regression models were fitted to the data to enable prediction of atrazine concentrations from the immunoassay data. The best model obtained (R^2 = 0.86) is as follows:

$$\text{Atrazine (ug/L)} = -0.21 + 2(I) - 0.73(I)^2 + 0.15(I)^3; \qquad (1)$$
where: I = Triazine immunoassay result in ug/L as atrazine.

As may be expected, the residuals from this model become larger as the immunoassay concentration becomes larger. The difference between the measured and predicted concentrations was less than 0.15 ug/L for 50 percent of the samples, and less than 0.6 ug/L for 80 percent of the samples. Only 10 percent of the samples had differences larger than 0.8 ug/L.

The spectrophotometric immunoassay analysis produced no false positives. Every sample in which triazines were detected by immunoassay (0.2 ug/L or larger) was shown by GC/MS to contain atrazine at concentrations of 0.18 ug/L or larger. Furthermore, the immunoassay produced no false negatives for samples with atrazine concentrations of 0.5 ug/L or larger.   For the 64 samples with atrazine concentrations determined by GC/MS to be larger than 0.5 ug/L, all immunoassay analyses were larger than the 0.2 ug/L detection limit. These results indicate that immunoassay has good reliability in predicting the presence or absence of atrazine, and immunoassay can provide a semiquantitative estimate of concentration in the range from about 0.2 ug/L to 5 ug/L if the response is primarily due to atrazine. The cost of the immunoassay is about $15 per sample compared to more than $200 for a GC/MS analysis in a commercial laboratory. Also, immunoassay results can be obtained in 15 minutes or less. GC/MS analyses typically require shipment of samples to a laboratory and results are not available for days or often weeks after sample collection.

Seasonal Distribution. Immunoassay results from the three sampling periods were analyzed to determine the seasonal distribution of triazine herbicides in midwestern United States streams during 1989. Concentrations of triazine herbicides generally were small

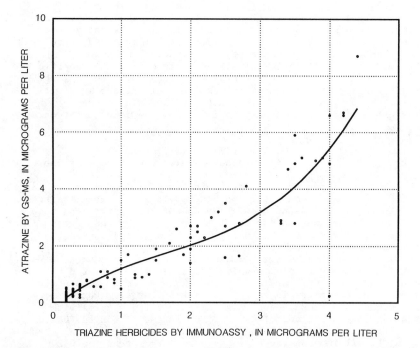

Figure 3.   Relation between triazine herbicide concentration
determined by immunoassay and atrazine concentration determined
by GC/MS.   Solid line shows fit for multiple regression model.

in the spring before application to cropland (Table IV). About 75 percent of the samples had triazine concentrations of 0.2 ug/L or less. Two samples had a triazine concentration larger than 3 ug/L. However, GC/MS analyses were not made on these two samples and it is not known if the atrazine concentration in these samples exceeded 3 ug/L, which is the U.S. Environmental Protection Agency's Health Advisory Level (HAL) for atrazine (11).

During the first rainfall runoff following herbicide application, large triazine herbicide concentrations were detected (Table IV). More than 85 percent of the samples had detectable concentrations of triazine herbicides, and more than 50 percent of the samples had triazine concentrations larger than 3 ug/L. Of the 69 samples with immunoassay results of 3 ug/L or larger, all but five had atrazine concentrations (by GC/MS) that equalled or exceeded the atrazine HAL of 3 ug/L. Four samples with immunoassay results less than 3 ug/L had more than 3 ug/L of atrazine (by GC/MS). Concentrations of triazines exceeded the upper limit of the immunoassay method (5 ug/L) in more than 40 percent of the samples. The large concentrations measured during this sampling period probably persist for a relatively short period of time. Large increases in herbicide concentrations in streams following application have been reported previously by numerous investigators (4-10).

Table IV.   Summary of immunoassay results for pre-application, post-application, and fall low-flow sampling periods

[ug/L, micrograms per liter; %, percent; >, greater than; <, less than]

	Pre-application	Post-application	Fall low-flow
Number of samples analyzed	145	135	146
		Concentration ug/L	
10th percentile	<0.2	<0.2	<0.2
25th percentile	<0.2	0.6	<0.2
50th percentile (median)	<0.2	3.4	0.3
75th percentile	0.2	>5	0.5
90th percentile	0.7	>5	1.1
		Percent	
Percent of samples in indicated range			
<0.3 ug/L	67	15	49
0.3-3 ug/L	32	32	50
>3 ug/L	1	53	1

Immunoassay results for samples collected during the fall low-streamflow period were similar to the pre-application results (Table IV). Ninety percent of the samples had triazine concentrations of 1.1 ug/L or less and only one sample had triazine concentrations larger than 3 ug/L. This sample was determined by GC/MS to contain 3.1 ug/L of atrazine. However, about 60 percent of the fall low-flow samples contained detectable levels of triazines by immunoassay (0.2 ug/L or more) compared to 45 percent of the pre-application samples. Additional analysis will be made

on these data using streamflow hydrographs in an attempt to
determine if the source of the triazine herbicides during the fall
low-flow period is overland flow or discharge from groundwater
systems.

Conclusions

Immunoassay appears to be a rapid, reliable, and inexpensive
screening tool for triazine herbicides in water. The presence or
absence of triazine herbicides at concentrations of about 0.5 ug/L
can be determined by visually comparing water samples with standard
solutions containing atrazine. When a spectrophotometer is used to
quantify immunoassay results, triazine concentrations as small as
0.2 ug/L can be detected. Within-laboratory and between-laboratory
reproducibility of results was reasonably good, especially at
concentrations less than 1 ug/L. There was an excellent but
slightly non-linear correlation between atrazine concentrations
determined by GC/MS and triazine immunoassay results. A multiple-
linear regression model was developed in which the immunoassay
results explained 86 percent of the variation in atrazine
concentrations. No false positive identifications of atrazine were
made by immunoassay, and no false negative identifications were
made on samples containing 0.5 ug/L or more of atrazine. The cost
for immunoassay analysis is about $15 per sample compared with more
than $200 for a GC/MS analysis in a commercial laboratory.
Immunoassay appears to be an excellent screening and quality-
assurance tool for use prior to GC/MS analysis of water samples.
It also appears to be good tool for visually determining the
presence or absence of triazine herbicides, and with a
spectrophotometer, a semiquantitative estimate of concentration can
be made if the immunoassay response is primarily from atrazine.
However, immunoassay has some cross reactivity with triazines other
than atrazine and can produce a positive response in the absence of
atrazine.
        The immunoassay results were very useful in determining the
occurrence and seasonal distribution of triazine herbicides in the
corn and soybean belt of the midwestern United States. Major
differences in concentrations existed between pre-application and
post-application sampling periods. About 85 percent of the samples
from the post-application period had detectable concentrations of
triazines. The immunoassay indicates that more than 50 percent of
the samples had triazine concentrations larger than 3 ug/L, and
nearly all of these also had atrazine concentrations (determined by
GC/MS) that exceeded the U.S. Environmental Protection Agency HAL
for atrazine of 3 ug/L. Forty percent of the post-application
samples exceeded the upper limit (5 ug/L) of the immunoassay
method.

Literature Cited

1.   Gianessi, L.P.; Puffer, C.M. "Use of Selected Pesticides for
      Agricultural Crop Production in the United States, 1982-85,"
      NTIS, Springfield, Va., 1988, 490
2.   Leonard, R.A. in "Environmental Chemistry of Herbicides," CRC
      Press, Boca Raton, 1988, 45-87.

3.   Nash, R.G. in "Environmental Chemistry of Herbicides," CRC Press, Boca Raton, 1988, 131-169.
4.   Frank, R.; Sirons, G.J. Sci. Total Environ., 1979, 12, 223-229.
5.   Frank, R.; Braun, H.E.; Holdrinth, M.; Sirons, G.J.; Ripley, B.D. J. Environ. Qual., 1982, 11, 497-505.
6.   Glotfelty, D.E.; Taylor, A.W.; Isensee, A.R.; Jersey, J.; and Glen. S. J. Environ. Qual., 1984, 13, 115-121.
7.   Baker, D.B. "Sediment, Nutrient, and Pesticide Transport in Selected Lower Great Lakes Tributaries," 1988, EPA-905/4-88-001, U.S. Government Printing Office, 244.
8.   Squillace, P.J.; Engberg, R.A. "Surface-water quality of the Cedar River basin, Iowa-Minnesota, with emphasis on the occurrence and transport of herbicides," 1988, U.S. Geol. Survey WRIR 88-4060, 81.
9.   Snow, D.D.; Spalding, R.F. "Soluble pesticide levels in the Platte River basin of Nebraska," in Agricultural Impacts on Ground Water Conference, Des Moines, IA., NWWA, 1988, 211-233.
10. Spalding, R.F.; Snow, D.D. "Stream Levels of Agrichemicals During a Spring Discharge Event," Chemosphere, in press.
11. U.S. Environmental Protection Agency, "Drinking Water Health Advisory: Pesticides," CRC Press, Boca Raton, 1989, 43-67.
12. Bushway, R.J.; Perkins, B.; Savage, S.A.; Lekousi, S.L.; Ferguson, B.S. Bull. Environ. Contam. Toxico., 1988, 40, 647-654.
13. Bushway, R.J.; Perkins, B.; Savage, S.A.; Lekousi, S.L.; Ferguson, B.S. Bull. Environ. Contam. Toxico., 1989, 40, 647-654.
14. Thurman, E.M.; Meyer, M.D.; Pomes, M.L. "Comparison of an Enzyme-Linked Immunoassay and Gas Chromatography/Mass Spectrometry for the Determination of Triazine Herbicides in Water," in press.
15. Thurman, E.M. "Organic Geochemistry of Natural Waters," Martinus Nijhoff/DRW Junk publishers, Boston, 1985, 496.
16. Malcolm, R.L.; MaCarthy P. Environ. Sci. and Technol., 1986, 20, 904-911.
17. Thurman, E.M. in "Humic Substances in Soil and Water: Geochemistry and Analysis," 1985, John Wiley and Sons, Inc., New York, 87-103.

RECEIVED August 30, 1990

Chapter 9

Development of Immunoassays for Thiocarbamate Herbicides

Shirley J. Gee[1], Robert O. Harrison[1,3], Marvin H. Goodrow[1],
Adolf L. Braun[2], and Bruce D. Hammock[1]

[1]Departments of Entomology and Environmental Toxicology, University of
California, Davis, CA 95616
[2]Environmental Monitoring and Pest Management, California Department
of Food and Agriculture, Sacramento, CA 95814

Immunoassays for the thiocarbamate herbicides,
molinate, thiobencarb and EPTC (Eptam) are described.
Using hapten synthesis strategies similar to those
reported earlier for molinate, several haptens were
synthesized for EPTC and thiobencarb. Rabbits were
immunized with conjugates of two haptens for each target
compound. Lower titer antibodies were produced against
EPTC haptens, resulting in less sensitive assays (I_{50}
for EPTC = 35 uM; 6.6 ppm). Cross reactivity with
related thiocarbamates was 9-50%. The antibodies raised
against thiobencarb haptens were of higher titer and of
similar sensitivity (I_{50} for thiobencarb = 0.3 uM; 158
ppb) to the molinate assay. With thiobencarb, three
assays were characterized using different combinations
of antibodies and antigens. Antibodies against an
azophenyl hapten of thiobencarb used in a homologous
assay showed very high specificity for thiobencarb
(cross reactivity by other thiocarbamates was below
0.1%). In the other two assays related thiocarbamates
cross reacted less than 2%. This assay has been applied
to the analysis of split samples from a field study to
evaluate assay performance and to compare to gas
chromatographic analysis.

The thiocarbamate herbicides, molinate and thiobencarb (Figure 1),
are used as pre-emergence herbicides in rice culture in California.
Between early May and the end of June each year, about 900,000 kg of
these materials are applied in California alone (1). Molinate has
been implicated in fish kills in drainage canals (2) and thiobencarb
imparts an off taste to the drinking water that can be tasted by
some people at very low concentrations. Because of these problems,
the California Department of Food and Agriculture (CDFA), in
conjunction with the California Department of Fish and Game, the
local Water Quality Control board and manufacturers of these
thiocarbamates, has a program to monitor drainage canals and river
water for these two compounds. In collaboration with CDFA, our
laboratory was asked to develop immunoassays for molinate and

[3]Current address: ImmunoSystems, Inc., 4 Washington Ave., Scarborough, ME 04074

0097–6156/91/0451–0100$06.00/0
© 1991 American Chemical Society

thiobencarb. CDFA's primary goal was to be able to analyze the
samples from this monitoring program by a quicker more cost
effective method than the currently used gas liquid chromatography
(GC) method. A second, longer term goal of CDFA was to develop an
understanding of immunoassays and determine the potential
contribution of this analytical method to the Environmental
Monitoring and Pest Management Branch's analytical program
(Stoddard, ACS Symposium Series, in press).

The history, principles and justification for development and
use of immunoassays are discussed elsewhere in this volume
(Vanderlaan et al.) and in reviews from this laboratory (3-8,
Hammock et al. ACS Symposium Series, in press) and others (9-10).
The former also include examples and references to many of the
assays which we have developed. The thiocarbamates, in particular,
were of interest to us because they present a challenge for assay
development. For example, molinate is a low molecular weight,
relatively volatile, and somewhat hydrolytically unstable compound.
Thus the very properties that make molinate easy to analyze by GC
make it a difficult candidate for immunoassay. Some of the
information presented has been published, but is condensed here to
serve as background to explain the unified development strategy for
the thiocarbamate class. It also provides a basis for comparison of
previously obtained data to those newly reported here.

Molinate Hapten Synthesis, Conjugate Preparation and Antibody Screening

Details on strategies for thiocarbamate hapten design can be found
in Gee et al. (11) and Harrison et al. (this volume). Haptens were
synthesized by a thio replacement reaction in which the parent
compound was oxidized to the sulfone using 3-chloroperoxybenzoic
acid. The sulfone was then displaced with the appropriate thiol to
yield either a carboxylic acid hapten or a nitrophenyl hapten with
varying aliphatic chain spacers (Figure 2). The nitrophenyl haptens
were reduced to aminophenyl haptens using dodecacarbonyltriiron and
then coupled to carrier proteins by diazotization. Carboxylic acid
haptens were coupled to carrier proteins by the mixed anhydride
method using isobutylchloroformate and tributylamine (11).

Rabbits were immunized with conjugates of the carboxylic acid
(Ia) and aminophenyl (Ic) haptens. The resulting polyclonal
antibodies were screened for spacer recognition and target
specificity. Antibodies raised against hapten Ia bound strongly to
the coating antigen having a homologous hapten. This binding could
not be inhibited by the target analyte, molinate. However this
binding could be inhibited by molinate when the coating antigen
contained a heterologous hapten, i.e. antigens containing haptens
Ib, Ic or Id. Antibodies raised against hapten Ic also bound
strongly to the homologous coating antigen and could not be competed
off by molinate, presumably due to strong linker recognition
(Harrison et al., this volume). In addition, molinate only slightly
inhibited the binding of this antibody to the heterologous haptens.
Thus, the most useful antibodies were raised against hapten Ia.
These antibodies were used in an indirect competition enzyme linked
immunosorbent assay (ELISA) (11-13). Details of the synthesis (11)
and assay development (13) have been reported previously.

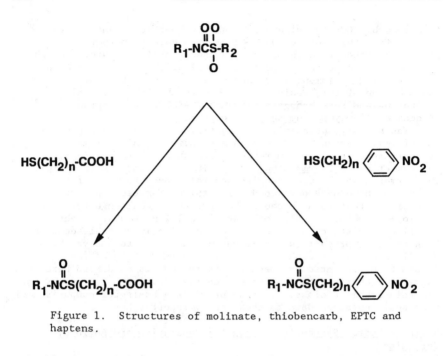

Figure 1.   Structures of molinate, thiobencarb, EPTC and haptens.

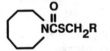

1

molinate

II

EPTC

III

thiobencarb

I, II: R = −CH$_3$

III: R = ⟨benzene⟩−Cl

Ia, IIa, IIIa: R = −CH$_2$CH$_2$COOH

Ib, IIb, IIIb: R = −(CH$_2$)$_5$COOH

Ic, IIc, IIIc: R = −CH$_2$⟨benzene⟩−NH$_2$

Id, IIId: R = ⟨benzene⟩−NH$_2$

Figure 2.   Synthetic routes for the thiocarbamate haptens.

Molinate Field Study Validation

This assay was validated in a field study which examined assay performance and compared data obtained on the same samples by GC analysis. These data have been published ($\underline{12}$), and to our knowledge, is the first study of its kind which addresses procedural error and data handling for real world samples in immunoassay for pesticides. Thus, it seems appropriate to reiterate some of the salient points here.

Samples obtained from a treated rice field were analyzed by both GC and ELISA. Samples for GC analysis were extracted with toluene and the extracts analyzed. Samples for ELISA analysis were buffered, diluted and analyzed without further workup.

The limit of reliable measurement of the ELISA was 21 ppb. The limit of reliable measurement is a calculated concentration that is approximately the mean plus two standard deviations, multiplied by two, of a negative control sample ($\underline{12},\underline{14}$). This measurement is probably more meaningful than a conventional limit of detection because it provides a conservative and statistically well supported estimate of the operating characteristics of the assay, obtained under realistic conditions. The I_{50}, which is the amount of molinate needed to inhibit the assay by 50%, was 106 ± 32 ppb. The working range was approximately 35 to 500 ppb.

We used this study to assess the importance of a number of potential sources of error in the ELISA procedure. The variability of the baseline ELISA signal (interwell coefficient of variation at the absorbance for the zero dose control) was 4%, of which 0.3% was reader error (instrument imprecision plus inaccuracy; $\underline{15}$). The between well variability is akin to the variability in baseline signal in GC analysis. Pipetting error was measured gravimetrically to be less than 1%. Two procedural variables were tested for their effect on reproducibility. Shaking the plate before reading decreased the average coefficient of variation for quadruplicate wells by almost two fold. Similar improvements in readings were obtained by Kemp et al. ($\underline{16}$). Reading in dual wavelength mode (405-650 nm) accounted for a small but reproducible decrease in the coefficient of variation.

A nested analysis of variance (ANOVA) was conducted for two control samples analyzed in 37 assays. Variability among replicate wells (within plates) accounted for more than 90% of the total variability for both controls. Among day and among plate (within day) variability constituted 5-10% and less than 1% of the total, respectively. This interwell variability is a characteristic of microplate ELISAs which is not widely documented, but it is known to those who work in this field. This variability is due to several compounded factors, including intrinsic variability in the binding characteristics of the plates, pipetting error, thermal variations across the plate and interwell variability of washing. The relative contributions of these factors have not been studied adequately.

The ELISA data obtained for field samples and spiked samples compared favorably to data obtained by GC. The correlation coefficient (r) exceeded 0.90 in each of two separate comparisons. Other issues addressed in this study were development of quality assurance criteria, analysis of four parameter standard curves and interpretation of resulting data, pipetting techniques, and study

design. Complete details are given in the report by Harrison et al.
(12).

Assay Development for EPTC and Thiobencarb

The thiocarbamates, thiobencarb and EPTC, were also targets for
immunoassay development. Using a synthetic strategy similar to that
used for molinate, several haptens were synthesized (Figure 1) and
the resulting antibodies were evaluated. Rabbits were immunized
with mercaptopropanoic acid haptens IIa, IIIa, coupled to keyhole
limpet hemocyanin (KLH) via mixed anhydrides, and p-aminophenyl
haptens IIc and IIId, coupled to KLH by diazotization. The
resulting antibodies were tested in both homologous and heterologous
assay systems. In general, antibody titers in homologous assays
were similar for each antigen. Combinations of coating antigens and
antibodies resulting in titers of less than 1000 from a checkerboard
titration were not screened further. The coating antigen and
antibody combinations that passed this first screen were then tested
for the ability of the target analyte to inhibit binding.

Antibodies directed against the EPTC hapten IIa could only be
inhibited by EPTC when used in a heterologous system. The amount of
EPTC needed to inhibit one of these indirect ELISAs by 50% (I_{50}) was
35 uM (6.6 ppm) under optimized conditions. Other structurally
related thiocarbamates such as vernolate, pebulate, butylate and
cycloate cross reacted from 9-50% (Table 1) relative to EPTC. There
was high cross reactivity with EPTC haptens IIa and IIc, as expected
(I_{50}s were 0.4 and 0.5 uM, respectively). Antibodies against hapten
IIc were not inhibited by levels of EPTC up to 2 ppm, regardless of
the coating antigen used. Apparently binding of antibodies to the
linker moiety of EPTC conjugate IIc-BSA was stronger than to the
N,N-dipropyl substituent of the thiocarbamate in homologous assays.
Similar results were obtained with molinate (11). In heterologous
assays, the antibody recognition of the N,N-dipropyl substituent of
the thiocarbamate was so poor as to make the assay unusable.
Further discussion of this and related problems in hapten design can
be found in Harrison et al. (this volume).

Antibodies made against thiobencarb hapten IIIa could not be
competed off in a homologous assay, probably due to poor recognition
of the N,N-diethyl substituent on the thiocarbamate. Thiobencarb
effectively inhibited the binding of this antibody to the
heterologous antigen. After optimization, these assays had good
sensitivity for thiobencarb (I_{50} = 0.7 uM) and cross reactivity with
structurally related compounds of less than 2% (Table 2) relative to
thiobencarb. The maximum absorbance of this assay was very small,
but could be improved using signal amplifying systems. Antibodies
against thiobencarb hapten IIId gave the best assays with both
homologous and heterologous coating antigens due to the close
structural similarity of immunizing hapten, coating hapten and
target compound. This result implies that the aromatic ring is
important to antibody binding and is certainly important to antibody
specificity. The I_{50}s with this antibody in heterologous and
homologous assays were 0.6 and 0.3 uM, respectively (Table 2).
Cross reactivities of other thiocarbamates in the heterologous assay
were less than 2% and strikingly, in the homologous assay was less
than 0.07%, denoting a particularly specific antibody.

Table 1. Relative Cross Reactivity of Some Thiocarbamates in the EPTC Assay

Inhibitor	I_{50} (uM)	
	Assay I[a]	Assay II[b]
II (EPTC)	35 (6.6 ppm)	67 (12.6 ppm)
IIa	0.5	0.9
IIb	-	-
IIc	0.4	0.6
pebulate	400	800
cycloate	200	400
butylate	c	c
vernam	75	130
thiobencarb (III)	c	c
molinate (I)	c	c

Mean I_{50} values for 2 experiments; for each experiment standard curves for each inhibitor were prepared using quadruplicate wells at each of 10 concentrations. Dashes indicate compound was not tested.
[a]Assay I: Coating antigen: IIc-BSA; Immunizing antigen: IIa-KLH
[b]Assay II: Coating antigen: IIc-CONA; Immunizing antigen: IIa-KLH
[c]The I_{50} is greater than 5 X 10^{-4}M, the highest concentration tested.
Abbreviations: BSA = bovine serum albumin; KLH = keyhole limpet hemocyanin; CONA = conalbumin

Table 2. Relative Cross Reactivity of Some Thiocarbamates in Three Thiobencarb Assays

Inhibitor	I_{50} (uM)		
	Assay I[a] (heterologous)	Assay II[b] (heterologous)	Assay III[c] (homologous)
III (thiobencarb)	0.60 ± 0.03 (0.3 ppm)	0.70 ± 0.21 (0.4 ppm)	0.3 (0.2 ppm)
IIIa	421 ± 162	1.91 ± 0.14	350
IIIb	13.2 ± 3.9	1.28 ± 0.34	9
IIIc	0.90 ± 0.33	0.50 ± 0.15	1.1
IIId	0.22 ± 0.07	0.42 ± 0.22	0.1
EPTC (II)	25 ± 4.3 (4.7 ppm)	54 ± 24.5 (10.2 ppm)	d
vernolate	54 ± 1	62 ± 35.2	500
pebulate	97 ± 0.5	198 ± 74	2500
cycloate	361 ± 0.1	198[e]	d
molinate	462 ± 71 (87 ppm)	d	500
butylate	d	d	d

Mean I_{50} values for 3 experiments ± SD unless otherwise indicated; for each experiment, standard curves for each inhibitor were prepared using quadruplicate wells at each of 10 concentrations.
[a]Assay I: Coating antigen: IIIb-OA; Immunizing antigen: IIId-KLH.
[b]Assay II: Coating antigen: IIIc-THY; Immunizing antigen: IIIa-KLH.
[c]Assay III: Coating antigen: IIId-OA; Immunizing antigen: IIId-KLH, n = 2
[d]The I_{50} is greater than 5 X 10^{-4}M, the highest concentration tested.
[e] N = 1

Conclusions
We have shown that the thiol replacement reaction chemistry for the
thiocarbamates can be applied to other members of that class for
development of antigens and successful antibody production. The use
of this reaction to synthesize haptens with different spacers and
coupling chemistries was crucial to the production of useful assays
for molinate and EPTC because of strong spacer recognition. Assays
for the three compounds had different sensitivities. The EPTC assay
was 100 fold less sensitive compared to the molinate assay, likely
due to the lack of the rigid structure found in molinate. The EPTC
assay also lacked sensitivity compared to the thiobencarb assay
since EPTC lacks structures capable of pi stacking or dipole-dipole
interactions (Harrison et al., this volume). The assays described
here demonstrate that antibodies can be made against haptens that
are relatively hydrolytically unstable. The molinate assay has
clearly been demonstrated to be useful for quantitative analysis of
environmental samples. The data analyzed therein provides
guidelines for the optimization, validation, quality control and
assurance of other assays.

Acknowledgments
This work was supported in part by a grant from the California
Department of Food and Agriculture, NIEHS Superfund grant PHS
ES04699-01 and Environmental Protection Agency Cooperative Agreement
CR 814709-01-0. B.D.H. is a Burroughs Wellcome Scholar in
Toxicology.

Literature Cited
1. California Department of Food and Agriculture; California
    Department of Food and Agriculture Pesticide Use Report, Annual
    Report; California Department of Food and Agriculture:
    Sacramento, CA, 1985.
2. Finlayson, B.J.; Lew, T.L. Rice Herbicide Concentrations in
    Sacramento River and Associated Agricultural Drains,
    Administrative Report #83-7; California Department of Food and
    Agriculture: Rancho Cordova, CA; 1983.
3. Harrison, R.O.; Gee, S.J.; Hammock, B.D. In Biotechnology in
    Crop Protection: ACS Symposium Series No. 379; Hedin, P.A.,
    Menn, J.J., Hollingworth, R.M., Eds.; American Chemical
    Society: Washington, DC, 1988; pp 316-330.
4. Jung, F.; Gee, S.J.; Harrison, R.O.; Goodrow, M.H.; Karu, A.E.;
    Braun, A.L.; Li, Q.X.; Hammock, B.D. Pest. Sci. 1989, 26, 303-
    317.
5. Hammock, B.D.; Mumma, R.O. In Pesticide Analytical
    Methodology; ACS Symposium Series No. 136; Zweig, G., Ed.;
    American Chemical Society: Washington, DC, 1980; pp 321-352.
6. Cheung, P.Y.K.; Gee, S.J.; Hammock, B.D. In The Impact of
    Chemistry on Biotechnology: Multidisciplinary Discussions; ACS
    Symposium Series No. 362; Phillips, M.P.; Shoemaker, S.P.;
    Middlekauff, R.D.; Ottenbrite, R.M., Ed.; American Chemical
    Society: Washington, DC, 1988; pp 217-229.
7. Hammock, B.D.; Gee, S.G., Cheung, P.Y.K.; Miyamoto, T.;
    Goodrow, M.H.; Seiber, J.N. In Pesticide Science and
    Biotechnology; Greenhalgh, R.; Roberts, T.R. Eds.; Blackwell
    Scientific Publications: Oxford, 1987; pp 309-316.

8. Van Emon, J.; Seiber, J.N.; Hammock, B.D. In Analytical Methods for Pesticides and Plant Growth Regulators: Advanced Analytical Techniques, Vol. XVII; Sherma, J., Ed.; Academic Press: New York, 1989, pp 217-263.

9. Mumma, R.O.; Brady, J.F. In Pesticide Science and Biotechnology; Greenhalgh, R.; Roberts, T.R. Eds.; Blackwell Scientific Publications: Oxford, 1987; pp 341-348.

10. Vanderlaan, M; Watkins, B.E.; Stanker, L. Environ. Sci. Technol. 1988, 22, 247-254.

11. Gee, S.J.; Miyamoto, T.; Goodrow, M.H.; Buster, D.; Hammock, B.D. J. Agric. Food Chem. 1988, 36, 863-870.

12. Harrison, R.O.; Braun, A.L., Gee, S.J.; O'Brien, D.J.; Hammock, B.D. Food Agric. Immunol. 1989, 1, 37-51.

13. Li, Q.X.; Gee, S.J., McChesney, M.M., Hammock, B.D., Seiber, J.N. Anal. Chem. 1989, 61, 819-823.

14. Wernimont, G.T.; Spendley, W. Use of Statistics to Develop and Evaluate Analytical Methods; Association of Official Analytical Chemists: Arlington, VA, 1985, pp 76-78.

15. Harrison, R.O.; Hammock, B.D. J. Assoc. Offic. Anal. Chem. 1988, 71, 981-987.

16. Kemp, M.; Husby, S.; Jensenius, J.C. Clin. Chem. 1985, 31, 1090-1091.

RECEIVED August 30, 1990

Chapter 10

Analysis of Heptachlor and Related Cyclodiene Insecticides in Food Products

Larry H. Stanker[1], Bruce Watkins[1], Martin Vanderlaan[1], Richard Ellis[2], and Jess Rajan[2]

[1]Biomedical Sciences Division, Lawrence Livermore National Laboratory, University of California, P.O. Box 5507, L-452, Livermore, CA 94550
[2]Food Safety and Inspection Service, U.S. Department of Agriculture, Washington, DC 20250

A rapid, competition immunoassay has been developed that detects chlorinated cyclodiene insecticides in meat, fish and milk products. The assay, a competition enzyme-linked immunosorbent assay, employs a monoclonal antibody that recognized all of the cyclodiene insecticides it was tested against. A simple method to extract heptachlor and related cyclodiene insecticides that is compatable with the immunoassay is described. The sensitivity of the immunoassay is suffcient to detect cyclodienes at the tolerance level in beef. We anticipate initial application of this assay as a screening aid.

Cyclodiene insecticides are chlorinated hydrocarbons, often referred to as organochlorides and include compounds such as heptachlor, heptachlor epoxide, chlordane, and dieldrin. In addition to having long environmental half-lives, cyclodienes are lipophillic and thus in biological systems tend to accumulate in adipose tissue. Their mode of action, toxicology, carcinogenic potential, and metabolism has been extensively reviewed (1,2). The use of heptachlor and related cyclodiene insecticides on agricultural products in the United States was phased out in the 1970's (endosulfan represents an exception and is currently used on a variety of crops). Nevertheless, numerous incidents of heptachlor contamination of food products have occured, the most recent a case in Arkansas involving heptachlor contamination of chickens. Contamination of the human food supply, specifically by heptachlor, heptachlor epoxide, aldrin, and dieldrin, have been documented by the World Health Organization (3), the U. S. Department of Agriculture Food Safety and Inspection Service (USDA-FSIS) (4) and the United States Food and Drug Administration (USFDA)(5). These studies indicate that main source route of exposure is via the food supply through meat and dairy products.

Traditional analysis for cyclodiene insecticides involves their extraction from rendered adipose fat followed by gas or thin layer chromatography and detection by electron capture or mass spectrometry (6-7). These methods are time consuming, require sophisticated equipment and highly trained personnel. Because of the persistence of cyclodiene insecti-

cides in the environment and in the food supply, a rapid, inexpensive detection method is desirable.   Immunoassays have these characteristics and represent potential alternatives to traditional chemical methods
Development of immunoassays for residue analysis of small molecules has been well documented in the literature (8-9, 14) and by the articles in this volume.   Recently, Dreher and Podratzki (10) reported the development of an immunoassay for endosulfan and its metabolites using a rabbit polyclonal antiserum.   This assay however, did not readily detect other related cyclodiene insecticides.   We report here the development of a monoclonal antibody that detects all nine of the cyclodiene insecticides tested plus toxaphene, and the incorporation of this antibody to an immunoassay for detecting these compounds in   meat, fish, and dairy products at, or below, the tolerance levels.

Materials and Methods

Reagents.   The following chemicals were obtained from Chem Serve (West Chester, PA) as analytical standards: aldrin, chlordane, dieldrin, endosulfan (mixed isomers), α-endosulfan, β-endosulfan, endosufansulfate, endrin, heptachlor, heptachlor epoxide, toxaphene, lindane (γBHC), and kepone.

Hapten Synthesis.   A hapten suitable for conjugation to a carrier protein was synthesized as outlined in Figure 1.   Basically, The Diels-Alder addition of hexachlorocyclopentadiene and cyclopentadiene gave 4,5,6,7,8,8-hexachloro-3a,4,7,7a-tetrahydro-4,7-methanoindene that was oxidized with selenium dioxide to 1-exo-hydroxy-4,5,6,7,8,8-hexachloro-3a,4,7,7a-tetrahydro-4,7-methanoindene as described (11).

Synthesis of Immuogen.   1-Exo-hydroxy-4,5,6,7,8,8-hexachloro-3a,4,7,7a-tetrahydro-4,7-methanoindene was treated with succinic anhydride in pyridine.   The resulting solution was evaporated in vacuo, dissolved in 5% aqueous sodium bicarbonate, washed with chloroform, and the aqueous solution was acidified with concentrated hydrochloric acid to give the unpurified hemisuccinate.   Ths reaction product was conjugated to bovine serum albumin (BSA) (the heptachlor-BSA conjugate, Hept-BSA) and keyhole limpet hemocyanin (KLH) (the heptachlor-KLH conjugate, Hept-KLH) by a N-hydroxysuccinimide procedure (12).

Monoclonal Antibody Production.   Six-month old BALB/cBkl mice (Bantin and Kingman Laboratories, Fremont, CA) were injected intraperitoneally (IP) with 100 µg of the heptachlor-KLH conjugate mixed 1:1 with complete Freund's adjuvant.   Mice received a single IP injection every other week.   A total of three injections were administered.   Four days prior to fusion, each mouse was given an intrasplenic injection of 100 µg of the heptachlor-BSA conjugate in sterile saline.   The spleen was removed and the splenocytes fused with SP2/0 myeloma cells and grown under conditions described by Stanker et al. (13).
A direct-binding ELISA, described by Stanker et al. (13) and modified as described below was used to screen culture fluid from the growing hybridomas for antibodies to heptachlor.   In this procedure, 96-well Nunc round bottom microtiter plates were coated with the heptachlor-BSA (hept-BSA) conjugate as follows.   The heptachlor-BSA conjugate was dissolved in distilled water, 100 µL (100 ng/well), of this solution was added to each well and then allowed to evaporate to dryness at 37° C.   Immediately before use, remaining reactive sites on the plastic microtiter plates are blocked by a 1 h

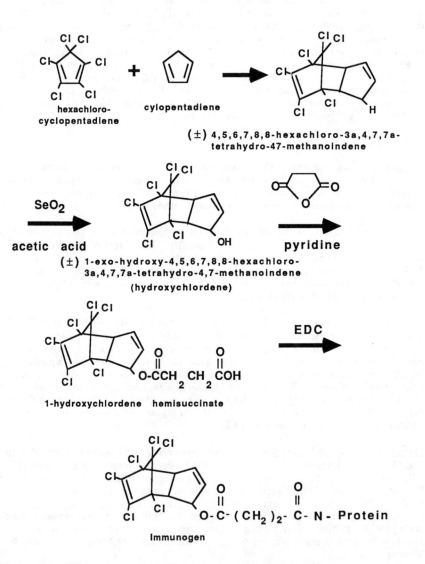

Figure 1.   Schematic outline of hapten synthesis and linkage to a carrier molecule.

incubation at room temperature with dilution buffer (DB) (0.005 M sodium-phosphate, 0.075 N NaCl, 0.001% Tween-20, 1% calf serum, pH 7.2). The plates were then incubated for 1 h at 37° C. with the hybridoma supernatants. The plates were carefully washed with a solution of 0.05% Tween-20 in water, peroxidase conjugated goat anti-mouse antiserum (Sigma, St Louis, MO) diluted 1:500 in DB was added to each well, and the plates incubated at 37° C for 1-h. Next, the plates were again washed with the 0.05% tween-water solution and the substrate, 2,2 azino-di-3-ethylbenzthiazoline sulfonic acid (ABTS), added. Absorbance was measured at 405 nm and transferred to a Macintosh (Apple Computer, Inc., Cupertino, CA) computer for subsequent analysis.

Hybridoma cells from wells showing a positive response in the ELISA screen were expanded and subcloned twice by limiting dilution to insure their monoclonal origin. Ascites fluid was prepared in irradiated mice (13), and the monoclonal antibodies were purified by affinity chromatography on Protein G (Pharmica, Uppsala, Sweden) as described by the manufactuer. Isotype determination was done by ELISA using mouse heavy- and light-chain specific antisera (Southern Biotechnology Assoc., Birmingham, AL).

Competition Enzyme-Linked Immunosorbent Assay. A competition enzyme-linked immunosorbent assay (cELISA) was developed to quantify the amount of heptachlor in solution and to evaluate the ability of the antibodies to distinguish among various cyclodiene insecticides and related chemicals. Microtiter plates were coated with 0.25 ng/well hept-BSA (100 µl/well of a 2.5 ng/mL solution of hept-BSA in distilled water was allowed to evaporate onto the bottoms of the wells at 37° C). The plates were then blocked with DB as described above. Competitors (analytical standards dissolved in methanol) were added to the DB such that the resulting solution was 50% methanol. An aliquot (200 µL) of this competitor-dilution buffer solution was added to an antigen-coated well. Then, the amount of competitor was serially diluted down the microtiter plate, so each well contained 100 µL of competitor in a 50% methanol-dilution buffer solution. Next an equal volume of dilution DB 100 ng of anti-heptachlor monoclonal antibody was added to each well. Thus, each well contained 200 µl of a 25% methanol solution in DB, antibody and competitor. Plates were incubated for 1 h at 37° C and then processed as described above.

In each experiment microtiter wells containing all components except competitor were prepared and the activity in these wells was taken to represent 100% activity. The test wells, each containing different amounts of competitor, were normalized to the 100% activity wells. Percent inhibition then was calculated by subtracting the normalized percent activity from 100.

Extraction of Heptachlor from Beef Fat. Beef fat was obtained from local supermarkets. The fat was chilled to 4° C and ground with a food processor. Fat samples were heated at 110° C until they liquified (ie., rendered). The rendered fat samples were then spiked with various concentrations of heptachlor dissolved in hexane. In some experiments, ground fat samples (50 g) were spiked with various concentrations of heptachlor in hexane (a 100 µL spike resulting in a 100, 300 and 1000 part-per-billion (PPB)), and stored at 4° C until used (the spiked fat samples were stored at least 24-h before use). Spiked fat samples were rendered as above. Heptachlor was removed from the fat as follows. Rendered fat, 0.2 g, was dissolved in 2 mL of hexane, applied to a deactivated florisil column using a Baker vacuum manifold (Baker Chemicals, Phillipsburg, NJ). The heptachlor was then eluted with 20 mL of hexane using the vacuum manifold. The heptachlor containing hexane wash (20 mL) was then dryed at 30 °C under a gentle stream of N_2 gas, the heptachlor dissolved in 200 μL of methanol, and used in a cELISA.

Column Chromatography. Florisil (Baker Chemicals, Phillipsburg, NJ) was activated by heating at 650° C for 24-h. The activated florisil was then de-activated with water (7% water was found to give optimum results). The de-activated resin was stored at room temperature in sealed containers, and used within 2 weeks. Activated florisil was stored at 130° C.

Deactivated florisil (5 grams) was added to a disposable, polycarbonate, syringe barrel fitted with a glass-wool plug. The resin was then washed with 20 mL of hexane immediately prior to sample application.

Radiolabeled Heptachlor. Radiolabeled heptachlor, ring-labeled with ^{14}C, having a specific activity of 5.2 mCi/mmol and a purity greater than 98% was obtained from Sigma Chemical Co. (St Louis, MO).

Results

Hapten synthesis. Heptachlor is a small organic molecule, and thus, it must be conjugated to a carrier protein in order to render it immunogenic. Synthesis of the hapten (-Exo-hydroxy-4,5,6,7,8,8-hexachloro-3a,4,7,7a-te-trahydro-4,7-methanoindene) and its conjugation to a carrier protein (both KLH and BSA were used as carrier proteins) is shown in Figure 1. The KLH conjugate served as immunogen whereas the BSA conjugate served as antigen for subsequent direct binding ELISA's and cELISA's.

Antibody Production. Spleen cells from a BALB/c mouse immunized with Hept-KLH were fused with SP 2/0 myeloma cells and cultured in Thirty, 96-well microculture dishes. Hybridomas were observed in a typical fusion in greater than 90% of the wells at the time cultures were screened. The hybridomas were screened for anti-heptachlor antibody production using the direct binding ELISA in which the Hept-BSA conjugate served as antigen. In a typical screen of 30, 96-well microtiter plates, 50 wells were observed to contain hybridomas that appeared to be secreting antibody that recognized the Hept-BSA but not BSA itself. The cells from these wells were expanded and tested against Hept-BSA, Hept-KLH, BSA, and KLH. Antibody that recognized both hapten conjugates, but did not bind either of the un-conjugated carrier proteins, was observed in approximately 10% of the original positives (i.e., 5 wells). These were next evaluated for their ability to recognize "free" unconjugated heptachlor in a competition ELISA. In most cases no competition was observed. In only 2 cases, representing the results from 5 independent fusion experiments, were antibodies observed that recognized the nonconjugated compounds. These were subcoloned twice and are referred to as monoclonal antibody LLNL-Hept-1 and LLNL-Hept-2. Results from competition ELISA experiments suggested that the concentration of heptachlor resulting in a 50% inhibition of the control (i.e. the I_{50} value) for LLNL-Hept-2 was approximately 10-fold lower than the I_{50} for LLNL-Hept-1 (I_{50} for LLNL-Hept-1 = 37 ng and 3.8 ng for LLNL-Hept-2). A 5-fold difference in sensitivity was observed when the competitor was heptachlor epoxide. Thus, the more sensitive LLNL-Hept-2 antibody was chosen for use in assay development. Isotype analysis indicated that LLNL-Hept-1 is an IgG1 antibody with a kappa light chain and that LLNL-Hept-2 is of isotype IgG2a, kappa light chain.

Antibody Characterization. Typical c-ELISA's for LLNL-Hept-2 using heptachlor as competitor (15 samples run over a 6-month interval) are presented in Figure 2. These data indicate that the average I_{50} for heptachlor occured when 3.0 ng of analyte was added to the reaction. A standard deviation of ± 7% was observed causing the I_{50} values to range between 1.9 and 4.5 ng/well. Similar variations were observed with other competitors. The LLNL-Hept-2 antibody used in the above experiments, and all subsequent experiments, was purified from ascites fluid by affinity chromatography on Sepharose Protein G (see methods section).

Next, LLNL-Hept-2 was characterized for its ability to recognize related cyclodiene insecticides. Representative inhibition curves for LLNL-Hept-2, using, heptachlor epoxide, chlordane, aldrin and endrin, in addition to the results obtained with heptachlor, are shown in Figure 2. The degree of cross reactivity of LLNL-Hept-2 to different chemicals was calculated by comparing the I_{50} values obtained with each chemical to the I_{50} obtained with heptachlor (the heptachlor value was arbitrarily assigned a value of 100%). These cross-reactivity results are summarized in Table I.

Immunoassay Development in Meat. The results presented above clearly indicate that the LLNL-Hept-2 antibody recognized all of the cyclodiene insecticides tested with roughly equal affinity. The next problem that needed to be solved was analyzing for pesticide residues in adipose tissue, the site of accumulation in animals. The heptachlor had to be extracted from the fat sample in a manner that is chemically compatible with an immunoassay. In addition, the cleanup method should be rapid and simple to take full advantage of an immunochemical detection method.

Fat samples were spiked or rendered and spiked with heptachlor at the levels indicated in the figures. The rendered samples (0.5 g of fat) were then dissolved in hexane (2 mL), passed over a solid-phase column prepared in a disposable syringe barrel, and eluted with 20 mL of solvent. The solvent was collected, evaporated under a stream of N_2 at 30° C, the residue resuspended in 110 µL of methanol, and 50 µL (the equivalent of 0.2 g of starting material) analyzed in the cELISA.

Different resins, including acid alumina, silica gel, C-18 reverse phase, and florisil, as well as different solvents, were evaluated for their ability to clean-up the heptachlor-spiked samples for an immunoassay. The data obtained using acid alumina and C-18 reverse phase resin as the solid support and elution with 20 mL of the solvent indicated are summarized in Table II. It may be noticed that many of the resin/elution solvent combinations were able to give >95% recoveries of the heptachlor (based on ^{14}C-heptachlor recovery), but when non-spiked beef fat samples were analyzed in the cELISA inhibition values of from 18 to 75% were observed. The inhibition, observed in nonspiked samples represents unknown interfering compounds that coeluted with heptachlor. In the case of chromatography on an acid alumina column followed by elution with 20 mL of 75% acetonitrile in water, only 18% inhibition in nonspiked samples was observed. Material from identical experiments using nonspiked and spiked beef fat were collected as 3 mL fractions. Each fraction was then analyzed

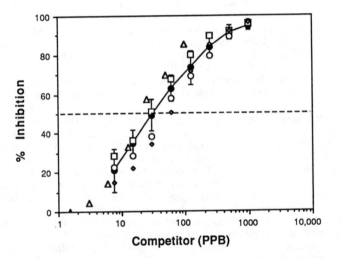

Figure 2. Competition ELISA results using Mab LLNL-Hept-2 with various cyclodiene insecticides as competitors: heptachlor (solid circles), heptachlor epoxide (open circles), aldrin (open squares), chlordane (open diamond), and endrin (open triangle). Bars represent ± 1 standard deviation for heptachlor.

Table I: Cross reactivity of LLNL-Hept-2.

Compound	% Cross Reactivity
heptachlor	100
heptachlor epoxide	100
chlordane	75
aldrin	100
endrin	150
dieldrin	300
endosulfan (mix)	173
α-endosulfan	43
ß-endosulfan	250
endosulfansulfate	182
toxaphene	100
lindane (BHC)	7.3
kepone	5
hexa-Cl-cylopentidiene	1.6
hexa-Cl-butadiene	0.4
1,2,4-triCl-Bz	0
hexa-Cl-ethane	0
2-Cl-naphthalene	0
1,2-diCl-Bz	0
1,3-diCl-Bz	0
1,4-diCl-Bz	0
2,5-diCl-nitrobenzene	0
2,4,6-triCl-phenol	0
2-Cl-phenol	0
2,4-diCl-phenol	0
2,4-diMe-phenol	0
2-nitro-phenol	0
4-nitro-phenol	0
2,4-dinitro-phenol	0.2
4,6-dinitro-cresol	0.14
penta-Cl-phenol	0
phenol	0
4-Cl-3-Me-phenol	0
2,4,5-triCl-phenol	0
4,5-diCl-catecol	0
2,4-diCl-6-nitrophenol	0
2,2,2-triCl-ethanol	0.4
DDT	0
Chlorobenzene	0
2,4,5-triCl-phenoxyacetic acid	0
2,4-D	0

A value of 0 indicates that the cross reactivity was less than 0.6%. A value greater than 100% indicates a higher relative affinity.

Table II.　Recovery efficiency of chlorinated
cyclodienes from beef fat

Expt. #	Resin	Elution Solvent	% Recovery	%Inhibition of 0 Spike
1	AA[1]	MeOH[2]	100	50
2	AA	50% MeOH/H_2O	0	
		75% MeOH/H_2O	0	
		100% MeOH/H_2O	100	nd[3]
3	AA + silica gel	100% MeOH	100	88
4	AA	Cyclohexane	100	61
5	AA	50% MeOH (acid)	0	
		100% MeOH	100	67
6	AA	50% MeOH (basic)	0	
		100%	100	75
7	AA	100% ACN[4]	100	nd.
8	AA	25% ACN/H_2O + 75%ACN/H_2O	100	39
9	AA	75% ACN/H_2O	100	18

1.　AA = acid alumina
2.　MeOH = methanol
3.　nd = not done
4.　ACN = acetonitrile
5.　10 mL was with 25% followed by a 10 mL wash with 5=75% ACN

for ^{14}C-heptachlor and percent of starting mass in the spiked samples and for spurious inhibition of the cELISA in the unspiked samples. The results of these experiments are summarized in Figure 3. These data clearly indicate that the majority of the heptachlor was recovered in the initial 2 fractions and that 6-8% of the fat (% of the starting mass) also was recovered in fractions 1 and 2. These same fractions from the nonspiked sample were then evaluated in a cELISA to determine if the material responsible for the background inhibition could be isolated. Background inhibition of the cELISA was substantially lower in fraction 1 than in fraction 2 (Figure 2, panel B). Fractions 1 and 2 were then spiked with heptachlor and analyzed in the cELISA. No increased inhibition was observed, suggesting that a sufficient amount of fat remained to allow the heptachlor spike to partition with the fat and thus be sequestered from the antibody. The use of acid alumina was thus abandoned.

The results obtained using florisil with hexane as the elution solvent are summarized in Figure 4. Triplicate experiments with radiolabeled heptachlor suggested that 93.6% ± 6.7 of the ^{14}C-labeled heptachlor was recovered in the 20 mL wash with only a few % of the starting mass in this fraction. The average inhibition observed when nonspiked beef-fat samples were analyzed using this method in the cELISA was 25, ±6%. However, in contrast to the acid alumina results, these samples when spiked with heptachlor showed an increased inhibition in the cELISA proportional to the spike.

Typical cELISA results obtained with rendered beef-fat samples spiked with heptachlor at 10, 1, 0.5 and 0.1 parts-per-million (PPM) and then cleaned-up on a florisil column with hexane as the elution solvent are shown in Figure 5. A linear response was obtained in these experiments and the 0.1 PPM sample resulted in an inhibition value of approximately 50% (twice the background value, arrow in Figure 5) The results obtained when fish fillets and whole heavy cream were spiked and analyzed are shown in Figure 6. In each case the arrow indicated the level of inhibition observed in nonspiked samples (the matrix effect).

Next we evaluated the efficiency of our assay to detect other cyclodiene insecticides contaminating beef fat. Results from these experiments are summarized in Table III. Since radiolabeled compounds for other cyclodienes were not available to us and the antibody recognized heptachlor, and the other cyclodienes with rough equivalency we normalized the cELISA results to that observed with heptachlor. Our results show that only heptachlor, aldrin and chlordane were readily detected. Dieldrin, endosulfan, endrin and heptachlor epoxide were not detected over the background. To determine if we were not extracting the compounds or if they were being retained on the column, we chromatographed analytical standards in hexane. The results (Table III, % recovered from column) were identical with those observed when meats were spiked and evaluated. Thus those cyclodienes not detected in the cELISA were being retained on the column.

Discussion

The use of chlorinated, cyclodiene insecticides in agricultgure production by and large has been phased out with the exception of endosulfan. Nevertheless, studies by the World Health Organization (3) by the USDA-FSIS (4), and by the US-FDA (5) indicate that these pesticides still are found commonly as contaminants in foods, particularly in meat and dairy products.

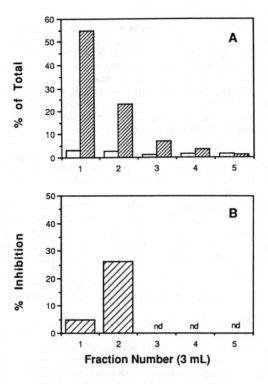

Figure 3. Chromatography of heptachlor spiked beef fat on an acid alumina column. Panel A. Percent recovery of ^{14}C-heptachlor (hatched bars) and % recovery of fat (% of starting mass) (open bars). Panel B. cELISA results for fractions 1 and 2 from nonspiked beef samples. nd = not done.

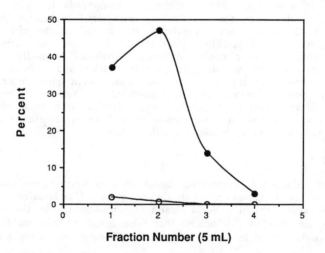

Figure 4. Recovery of ^{14}C-heptachlor (solid circles) and fat (open circles) following clean-up on a florisil column using hexane as solvent.

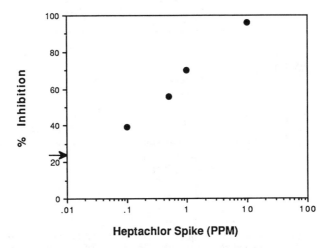

Figure 5. cELISA data obtained with heptachlor spiked beef-fat. Samples were spiked and cleaned-up using a 10 g florisil column and hexane as solvent. Arrow indicates level of background inhibition in nonspiked samples.

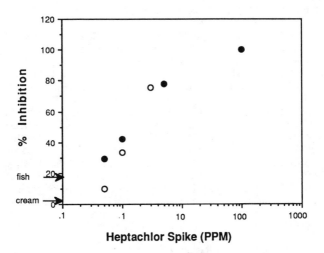

Figure 6. cELISA of heptachlor spiked fish (solid circles) and heavy cream (open circles) following clean-up on florisil with hexane as solvent. Arrows indicate level of inhibition in nonspiked samples.

Table III.   Recovery efficiency of cyclodienes from beef fat.

Compound	% Recovered from Column	% Recovered in Beef Fat
Aldrin	97	93
Chlordane	69	67
Dieldrin	nd	nd
Endosulfan	nd	nd
Endrin	nd	nd
Heptachlor	100	100
Hept-Epoxide	nd	nd

nd = below background level of cELISA

Their presence in these products is not surprising since as a group the cyclodienes are apolar, lipophilic compounds.   Acceptable daily intake for cyclodienes has been established by the World Health Organization and tolerance limits in meats and poultry set by the US-EPA.   Development of an immunoassay for these compounds would greatly facilitate improved monitoring for cyclodienes in foods and environmental samples.   We have isolated a monoclonal antibody, LLNL-Hept-2, that detects cyclodiene insecticides and have used this antibody to develop an immunoassay that is able to detect heptachlor (and related chlorinated cyclodienes) in meat, fish and dairy products.

The ability of LLNL-Hept-2 to recognize different chlorinated cyclodienes was studied using a cELISA.   The results of these studies indicate that LLNL-Hept-2 binds all of the cyclodiene insecticides evaluated with roughly equal relative affinity.   Since the norborene structure is held in common by all of the cyclodiene insecticides tested, these data suggest that this portion of the molecule is recognized by the antibody.   By design, this portion of the immunogen was most distal to the region used for linkage of the hapten to the carrier molecule.   Thus, the antibody appears to bind the least altered portion of the hapten.   Recognition of this region of the molecule by LLNL-Hept-2, agrees with results by us (14-15) and others (16) suggesting that specific recognition is often to those regions distal to the linkage chemistry.

The cELISA can easily detect heptachlor contamination in the range of 0.1 PPM to 10 PPM, with minimal sample clean-up.   Only 5 grams of material in the case of beef fat and fish or 1 mL of dairy cream was necessary for analysis.   The large amount of florisil used was necessary to remove as much fat as possible since excess fat appears to interfere with the immunoassay.   Commercial solid-phase florisil columns did not give satisfactory results.   Failure of these columns may be related to the small amount of florisil in commercial columns being insufficient to remove the fat, to improper activation and deactivation of the resin, or a combination of these.   In our experiments we observed an absolute necessity to activate the florisil at 650 °C followed by deactivation with water. Failure to deactivate resulted in large amounts (e.g., the entire fat sample) co-chromatographing with the heptachlor.

A cELISA using LLNL-Hept-2, capable of detection heptachlor and other related chlorinated cyclodiene insecticides in spiked meat, fish and dairy samples was developed. Our cleanup method did not remove all of the substance(s) that interfered with the cELISA. Such substances were observed in the hexane extracted material from all three matrixes investigated. Our data suggests that the beef fat has the highest level of interfering substances and heavy dairy cream the lowest level. The exact nature of these substances are not known.

The failure to detect an increase in inhibition following addition of a heptachlor spike to the extracts derived from the acid alumina cleanup method suggest that in addition to interfering substances, the cleanup method must be able to remove most of the fat from the sample or the cyclodienes, because of their lipophilic nature, simply will partition with the fat when the sample is suspended in an aqueous buffer system that is compatible with an immunoassay. We have observed similar results with immunoassays to other lipophilic pesticides (pyrethroids) (14). Chromatography on florisil appears to remove a sufficient amount of fat and of the interfering compound(s) from the sample to allow for detection of heptachlor using the cELISA. However, comparison of the sensitivity of the cELISA for cyclodiene analytical standards versus meat, fish or dairy extracts clearly indicates that a loss of sensitivity was realized in "real" samples. This loss of sensitivity is due in part, if not solely, to the coextraction of interfering compounds or the loss of some heptachlor to the small amount of fat that the method failed to separate.

Our results clearly point out the critical role of sample cleanup in development of an immunoassay for lipophilic compounds. The methods should be simple, rapid, and capable of being exported into the field if the full potential of immunoassays are to be realized. The sample cleanup method we describe here meets these criteria for heptachlor, aldrin and chlordane. However, recovery experiments clearly indicate that the more polar epoxides of these insecticides are retained on the column. Thus, while the antibody binds all cyclodiene insecticides with roughly equal affinity, the cleanup method will not equally present the cyclodienes to the assay. Attempts at using mixed solvent elution systems such as ether/hexane remove the more polar insecticides but also remove more of the starting fat and interfering compound(s). Thus, further efforts in the area of sample cleanup are needed.

Since LLNL-Hept-2 binds all of the chlorinated cyclodiene insecticides tested, a positive response in the cELISA represents an integrated value and does not identify a specific insecticide. Uncertainty about the exact composition of the contamination is a result of cross-reactivity with related chemicals and occurs even with monoclonal antibodies. However, monoclonal antibodies are invariant reagents that need to be extensively characterized only once to determine the range of compounds detected and their relative sensitivities. In contrast, polyclonal antibodies need to be characterized, and the assay optimized following each cycle of antibody production.

We anticipate that the immunoassay described here will be most useful as a screening tool, allowing for a rapid, inexpensive, easily automatable test for cyclodiene contamination. A positive response can then be confirmed using a more traditional chemical method if the exact nature of the contamination must be ascertained. Thus, the power of this method is to eliminate the vast majority of samples that are not contaminated and there by allow resources to be concentrated on those few samples that screen positive.

Further studies of spiked and incurred tissues using both the cELISA and a more traditional chemical assay for chlorinated cyclodiene insecticides are needed to demonstrate the reliability of the cELISA. Such studies also will permit an evaluation of the level of false negatives and/or false positives detected by the immunoassay. This latter point is of key importance for defining a useful screening method.

Acknowledgments

Funding for this work was provided by the U.S. Department of Agriculture, Food Safety Inspection Services under interagency agreement number 13-37-7-046, and the work was performed under the auspices of the U.S. Department of Energy by the Lawrence Livermore National Laboratory under contract number W-7405-ENG-48.

Literature Cited

1. Matsumura, F. 1985. Toxicology of Insecticides; Plenum Press: New York, 1985; 2nd edition, 598p.

2. Ware, G. (Ed.) Reviews of Environmental Contamination and Toxicology; Springer-Verlag: New York, 1988; Vol 104, p131.

3. Global Pollution and Health: Results of Health-related Environmental Monitoring; GEMS: Global Environment Monitoring System, United Nations Environment Programme, World health Organization, 1211 Geneva 27, Switzerland: 1987,22p.

4. Domestic Residue Data Book National Residue Program 1987, United States Department of Agriculture, Food Safety and Inspection Service,U.S. Government Print Office, Washington, DC, 1987.

5. Gunderson, E.L. J. Assoc. Off. Anal. Chem. 1988, 71,1200-09.

6. Chemistry Laboratory Guide Book, U. S. Department of Agriculture, Food Safety and Inspection Service, U.S. Government Print Office: Washington, DC, 1986.

7. Corneliussen, P.E.; McCuly, K.A.; McMohon, B.; Newsome, W.H. In Official Methods of Analyhsis of the Association of Official Analytical Chemists; Williams, S., Ed.; Assoc. Official Analytical Chem., Inc.: Arlington, Virginia, 14th edition, 1984; pp 533-562.

8   Hammock, B.D.; Gee, S.J.; Cheung, P.Y.K.;Miyamoto, T.; Goodrow, M.H.; Van Emon, J.; Seiber, J.N.   Utility of Immunoassay in Pesticide Trace Anlaysis. In Proc. Int. Congr. Pesticide. Chem. 6th Mtg.,Greenhalgh, R; Roberts, T. Eds.; Blackwell: Oxford, UK., 1987; pp 309-316.

9.   Vanderlaan, M.; Watkins, B.E.; Stanker, L.H.   Environmental Sciences and Technology, 1988b, 22, 247-254.

10. Dreher, R.M. and Podratzki, B. 1988. J. Agric. Food Chem. 1988, 36, 1072-1075.

11. Büchel, K.H.; Ginsberg, A.E.; Fischer, R. Chem. Ber. 1966, 99, 405.

12. Lauer, R.C.; Solomon, P.H.; Nakanishi, K.; Erlanger, B.F.   Experimentia 1974, 30, 558.

13. Stanker, L.H.; Branscomb, E.; Vanderlaan, M.; Jensen R.H.   J. Immunol. 1986, 136, 615-622.

14. Stanker, L.H.; Bigbee, C.; Van Emon, J.; Watkins, B.; Jensen, R.H.; Morris, C.; Vanderlan, M. J. Argi. Food Chem. 1989, 37; 34-839.

15. Vanderlaan, M.; Watkins, B.E.; Hwang, M.; Knize, M.G.; Felton, J.S. Carcinogenesis 1988a, 9, 153-160.

16. Hammock, B.D.; Mumma, R.O.   In Recent Advances in Pesticides: Analytical Methodology; Harvey, J. Jr.S.; Zweig, G., Eds.; ACS Symposium Series No.136; American Chemical Society: Washington DC, 1980; pp 321 52.

RECEIVED September 21, 1990

Chapter 11

Testing Cereal Products and Samples by Immunoassay

Tests for Organophosphate, Carbamate, and Pyrethroid Grain Protectants

John H. Skerritt[1], Lisa G. Robson[1], David P. McAdam[2], and Amanda S. Hill[1]

[1]Wheat Research Unit, Division of Plant Industry, Commonwealth Scientific and Industrial Research Organisation, North Ryde, New South Wales 2113, Australia
[2]Wheat Research Unit, Division of Plant Industry, Commonwealth Scientific and Industrial Research Organisation, Canberra, Australian Capital Territory 2601, Australia

Organophosphates, synthetic pyrethroids and/or carbaryl are applied to stored grain and grain storage facilities to minimize insect infestation. However, local industry deregulation coupled with increasingly stringent export residue tolerances and consumer demand for "chemical-free" foods have markedly increased demand for pesticide residue testing. Current instrumental testing requires skilled analysts and expensive capital equipment, and is too slow and costly for large-scale market, silo or field-based testing. We aim to develop rapid, simple and inexpensive tests for detection of key pesticides by modern immunological and enzyme-based methods. Monoclonal and polyclonal antibodies have been prepared and test methods developed for fenitrothion, the major organophosphate pesticide used on grain. Some antibodies were specific for fenitrothion and sensitive to 0.1 - 0.5ng - other antibodies bound to closely related organophosphates as well. A simple cholinesterase inhibition test for general screening for organophosphates and carbamates is being adapted for use in field situations or high-throughput laboratories. Finally, antibodies to certain pyrethroids (phenothrin, permethrin) are being assessed for use in simple test kits.

A variety of pesticides (grain protectants) is applied to wheat and barley, usually upon receival of grain at silos (termed elevators in the US) to prevent infestation with insects during storage. Some compounds may also be applied at grain export terminals or by farmers storing grain on their farms. While considerable progress has been made in chemical-free storage of grain under controlled atmospheres (inert gases) or by grain

0097–6156/91/0451–0124$06.00/0
© 1991 American Chemical Society

chilling, these approaches will remain economical in only a
minority of silos for some time to come.

Compounds currently used for treatment of grain or grain
storage facilities are shown in Table I.   Those indicated in
parentheses are either not used widely, are only of experimental
status or are not currently recommended(1).   In the late 1980's
the most commonly-used compounds in Australia were fenitrothion
and bioresmethrin.   Clearly, grain analysis differs from, say,
analysis of organochlorine contaminants in meat or dioxins in
water, in that the compounds under analysis in grain are
deliberately applied by the industry to this produce.   Modern
grain protectants are however, designed to decompose appreciably
before the grain is processed (eg. milled or malted) for
consumption by humans.

Table I.   Pesticides Used for Treatment of Stored Grain

ORGANOPHOSPHATES	Fenitrothion	(Malathion)
	Chlorpyrifos-methyl	(Azinphos-methyl)
	Pirimiphos-methyl	
CARBAMATE	Carbaryl	
PYRETHROIDS	Bioresmethrin	(Permethrin)
	Deltamethrin	((1R)-phenothrin)
		(fenvalerate)
PYRETHROID SYNERGIST	Piperonyl butoxide	
INSECT GROWTH REGULATOR	Methoprene	
OTHER	Phosphine	
	Methyl bromide	

One aim of our work is to enable the decentralization of
residue testing in grain and cereal products by the development of
simpler, faster and less expensive test methods.   This should
enable the potential of load-by-load screening of export shipments
for central grain terminals and grain receivals from farmers by
silos (elevators) and mills.   Both semi-quantitative screening
methods (to check that residue levels in grain are below legally-
set maximum residue limits, MRLs) and quantitative tests for use
in regional or mill laboratories are envisaged.   In addition,
tests are needed to ensure that grain has been adequately and
evenly treated with protectants so that subsequent infestation
will not occur.   Both of these aspects will become increasingly
important as more grain is treated on-farm by relatively
inexperienced operators.   Central laboratories and conventional
instrumental analytical techniques will, of course, maintain their
importance, both to confirm results obtained in field tests by
newer methods and in some cases to obtain quantitative results for
samples found to be positive in "field" screening tests.

Apart from the analysis of anticholinesterase nerve gases such

as soman(2,3), most work to date has been done on immunoassays for
malathion(4,5) and parathion/paraoxon(6-11).   Immunoassays have
also been developed which specifically detect organophosphate
pesticides with diethyl-thiophosphate ester groups, but not the
dimethyl-ester groups(12).   Because of the need for low toxicity
in grain protectants, all organophosphates approved by the Codex
Alimentarius Commission contain the O,O-dimethyl group(1).   Thus
work on the compounds of greatest importance to our industry has
not been done.   In the case of other compounds such as malathion,
tests are of insufficient sensitivity, speed, or robustness.

Experimental Methods

For both organophosphates and pyrethroids we have concentrated on
the use of polar water-miscible extractants(13).   These provide
as good extraction of these compounds as immiscible extractants
such as methylene chloride, and have the advantage that the grain
extract can be added directly to the antibody incubation mixture.
In extraction studies either 40g grain was added to 80mL solvent
(whole grain) or 10g ground grain added to 20mL solvent (ground
grain).   Acetonitrile and dichloromethane solutions were firstly
evaporated and the residue reconstituted in ethanol before gas
chromatography.
        Both polyclonal and monoclonal antibodies to fenitrothion were
raised using several methods of conjugating fenitrothion to
carrier proteins such as chicken IgG, ovalbumin and hemocyanin.
The pesticide was coupled to protein in several ways; through the
aromatic nitro-group by reduction and diazotization of the
resulting amino group and by coupling through a 6-carbon spacer
arm based on the acyl chloride of ethyl adipate.   In addition,
coupling of fenitrothion through the thio-phosphate group was
performed by reaction of the methyl ester of thiophosphoryl
chloride with a protected primary amino acid and 3-methyl-4-
nitrophenoxide (or other organophosphate substituents).   The
protecting group was hydrolyzed and reacted with N-
hydroxysuccinimide to yield an activated moiety, reactive both
with the carrier proteins and with marker enzymes.   Since all of
the organophosphates used in Australia (and in many other
countries) are dialkoxyphosphorothioates, this strategy was also
used for the other compounds of interest.
        Two formats were developed for the ELISA assays (Table II),
and also for those with permethrin/phenothrin-binding monoclonal
antibodies (see below).
        Analysis of fenitrothion by technique (A) is based on the
competition between fenitrothion in the test sample and enzyme-
labeled fenitrothion, for binding to fenitrothion-specific
antibody on the solid phase.   There is a decrease in the binding
of the enzyme-labeled fenitrothion as the concentration of
fenitrothion increases in the test sample; this is manifest in
decreasing absorbance of the colored enzyme product.   The
concentration of fenitrothion in an unknown sample may be
determined by comparing the degree of inhibition of antibody
binding caused by the addition of the sample extract, with that
resulting from the addition of known amounts of fenitrothion.
        For the results shown in Figure 1, polyclonal IgG antibodies

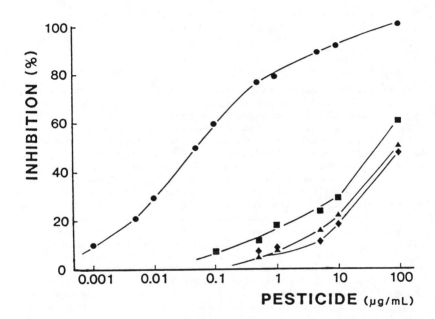

Figure 1. Reaction of an antibody to fenitrothion with organophosphates
used on stored grain in Australia: fenitrothion (●), chlorpyrifos-
methyl (■), pirimiphos-methyl (◆), azinphos-methyl (▲).

Table II. Organophosphate and Pyrethroid ELISA Methods

(A)   Tube precoated with PESTICIDE-SPECIFIC ANTIBODY
1.   Add buffer, pesticide-containing grain extract then enzyme-
        labeled pesticide.
2.   Mix. Incubate 2-30 minutes.
3.   Tip out unbound components. Wash tube.
4.   Add enzyme substrate. Incubate 2-15 minutes.
5.   Stop color development. Read absorbance.
        Compare with controls and pesticide standards

OR

(B)   Tube precoated with PESTICIDE-PROTEIN CONJUGATE
1.   Add buffer, pesticide-containing grain extract then enzyme-
        labeled antibody.
2-5.   As for (A)

(to fenitrothion amine coupled to chicken serum IgG through ethyl
adipyl chloride) were used; similar but not identical
specificities were obtained with monoclonal antibodies prepared to
this immunogen and to related immunogens. The antibody was coated
(1hr, 20°) at 10μg/mL in 50mM sodium carbonate buffer, pH 9.6.
Non-specific binding of immunoreactants was minimized by treatment
with 10mg/mL bovine serum albumin in 50mM sodium phosphate, pH
7.2 - 0.9% sodium chloride (PBS).   Antibody-coated solid phases
may be prepared (for incorporation in kits) and stored many months
before use.   The grain or food or other sample to be analyzed was
simply extracted in neat ethanol.   The extract is diluted five-
fold in 10mg/mL bovine serum albumin in 0.05% (v/v) Tween 20-PBS,
applied to the solid phase and enzyme-labeled fenitrothion is
immediately added.   Fenitrothion standards are prepared in
ethanol and dilutions assayed simultaneously.
    The second enzyme-immunoassay format (B) involved competition
for fenitrothion-specific antibody between pesticide present in a
sample extract and solid-phase-bound fenitrothion.   The antibody
may either be enzyme-labeled by direct covalent association or a
second labeled anti-mouse antibody used to detect binding of the
first antibody.
    The physiological basis of action of organophosphates and
carbamates is inhibition of acetylcholinesterase at the insect
neuromuscular junction.   Accordingly, for almost 30 years
attempts have been made to analyse these pesticides using
cholinesterase inhibition assays(14).   Many of these assays have
involved an initial separation of the compound by thin-layer
chromatography and treatment with crude cholinesterase sources
such as horse serum.   We have assessed and modified a method
using Taiwanese-made reagents (Tai Da Chemicals, Taipei), most
notably a high-affinity bee-head cholinesterase enzyme.   A
laboratory format using ELISA microwell plates, and a field test-
tube format have been developed by us from these reagents. Also we
have attempted to make the steps in the method similar to those in
ELISA assays, so that the assays may be performed together using

the same grain extract. In these assays, the methanol grain extract and buffered enzyme are added to the test tube, incubated 5 or 10 minutes, then a mixture of acetylthiocholine and Ellman's reagent (dithionitrobenzoate,DTNB) is added. After 2 minutes, stopping solution is added. The action of cholinesterase on acetylthiocholine releases a thiol which reacts spontaneously with DTNB to yield a yellow color. Thus, like immunoassays for pesticides, presence of the pesticide compound results in inhibition of color development.

Results and Discussion

Extraction of organophosphates. Where rapid results are required, the results indicate alcohols can be used together with high-frequency homogenization or blending of ground grain or flour (Table III), or where batches of grain samples are to be analyzed together, whole or ground grain can be sat overnight in solvent with occasional mixing. The rapid extraction methods do not give complete extraction, but rather give reproducible extraction.

Table III. Extraction of Fenitrothion (% available)

Solvent	Whole grain shaking		Ground grain homogenization (Ultraturrax, 2/3 maximal speed)	
Extraction time	5 min	>12 hr	30 sec	2 min
ethanol	61	93	63	96
50% ethanol	23	39	nt	nt
methanol	nt	nt	27	78
acetone	78	100	72	78
acetonitrile	38	95	68	82
dichloromethane	nt	nt	61	73

Analysis by glc, total available = 4.1ppm, nt = not tested
Mean standard deviations of determinations were ± 6%.

Immunoassays for organophosphates. High-titer antibodies to fenitrothion (coupled through the phosphate group) were obtained: of the other conjugates, those through the C6 spacer arm produced the best antibody responses. Horseradish peroxidase conjugates prepared in this manner were both enzymically active and able to bind to the antibodies.

Fenitrothion was readily detectable in wheats dosed with differing amounts of fenitrothion used in commercial grain treatment practice (Table IV). For analysis of extracts of grain, no clean-up of the extracts was necessary, if a fenitrothion standard curve prepared in undosed grain extract was used for calibration.

Table   IV.   Analysis of Fenitrothion in Wheat by ELISA

Original application (ppm)	Actual amount[a] (ppm)	Extraction method (ELISA results, ppm)		
		Overnight		Rapid
		Whole grain	Ground grain	Ground grain
3	2.1	2	2	2
9	6.3	5	6	6
14	9.8	9	13	9

[a] Analysis by gas-liquid chromatography.   Actual values are lower than original application due to losses on grain during storage. Data are means of 3 experiments.

The assay was sensitive to 0.5ng fenitrothion, with 50% inhibition obtained at 7ng fenitrothion.   This sensitivity is sufficient for analysis of fenitrothion both at low levels of addition and at levels near or in excess of specified maximum residue levels.   The test did not determine 4-nitro-3-methylphenol, a major (inactive) breakdown product of fenitrothion, produced during cooking(1).   The assay was specific for fenitrothion and very closely related compounds. 0,0-dimethyl-phosphorothioate esters with 4-substituted phenyl moieties were most active, and of these, the most active were substituted at the 3-position on the phenyl moiety.   0,0-diethyl-phosphorothioates had little or no activity.   Other common organophosphates such as dichlorvos, chlorpyrifos-ethyl, pirimiphos-methyl, azinphos-methyl, dimethoate and malathion were detected extremely weakly or not at all (Figure 1).   Carbamates (eg. carbaryl), organochlorines (eg. dieldrin) and synthetic pyrethroids (eg. bioresmethrin) were not detected.   Parathion-methyl was detected weakly.   Thus for pesticides encountered in Australian cereal foods, this assay is functionally specific for fenitrothion.
  The assay with a polyclonal antibody (P7) was sensitive to 0.1ng fenitrothion, with 50% inhibition obtained at 100ng fenitrothion.   The assay with a monoclonal antibody (F3) was sensitive to 100ng fenitrothion.   These sensitivities are sufficient for monitoring samples at levels equal to or in excess of specified legal maximum residue levels, when simple ethanol extracts of the food, grain or produce sample are used. Sensitivity can be increased by use of volatile organic solvents for extraction of the sample, evaporation of solvent and redissolving the residue in a smaller volume of organic solvent.
  The specificity of the assay differed from that obtained with the first assay format (Table V).   The assay with antibody P7 detected fenitrothion, amino-fenitrothion, methyl-parathion, parathion, fenthion and dicapthon with similar potency to

Table V. Reaction of antibodies with fenitrothion and related organophospahtes

Compound	Relative potency[a]		
Assay format	A	B	
Antibody	P1	P7	F3

1. Dimethyl-o-phosphorothioates $((CH_3O)_2$ PS-Ox$)$

a) x = substituted phenyl

fenitrothion	1.0	1.0	1.0
reduced fenitrothion	2.0	2.0	-
methyl-parathion	0.04	1.4	2.0
fenthion	0.6	1.0	-
cythioate	0.2	0.1	-
dicapthon	0.01	1.1	0.15
temephos	-	-	-

b) x = other

chlorphyifos-methyl	0.001	0.1	-
pirimiphos-methyl	0.0005	-	-

2. Diethyl-o-phosphorothioates $(C_2H_5O)_2$ PS-Ox$)$
x = substituted phenyl

parathion	0.0005	0.1	-
fensulfothion	0.01	-	-
dichlofention	-	-	-
chlorpyrifos-ethyl	-	-	-
pirimiphos-ethyl	-	-	-

3. Other organophosphates
a)

dichlorvos	-	-	-
paraoxon	-	-	-

b) Dimethyl phosphorodithioates $(CH_3O)_2$ PS-Sx$)$

azinphos-methyl	0.0005	-	-
dimethoate	0.0005	-	-
malathion	0.0008	-	-

[a] Potency relative to fenitrothion determined as relative concentration yielding 50% inhibition of antibody binding. Assay format 1: - = <0.0005 relative cross reaction, Assay format 2: - = fenitrothion IC_{50} > 200ppm

fenitrothion. Related (oxo)-phosphates such as paraoxon and dichlorvos were not detected. Dimethyl-O-phosphorothioate pesticides with other than phenyl-substitutions were detected very weakly. Thus, 0,0-dimethyl-phosphorothioates with 4-nitrophenyl or 4-thiomethylphenyl groups were detected, but not compounds with larger groups substituted in the 4-position. The assay using antibody F3 was even more specific, detecting fenitrothion, methyl-parathion and dicapthon only. Other pesticides, including a carbamate (carbaryl), an organochlorine (dieldrin) and a pyrethroid (bioresmethrin) were not detected by either antibody.   Thus, these assays detect fenitrothion and related organophosphorus pesticides.

Cholinesterase inhibition assays.   With the compounds used for treatment of Australian grain, the cholinesterase inhibition assay was specific for carbaryl and dichlorvos.   Each of the other major organophosphates that are widely used are actually thiophosphates, and do not potently inhibit the enzyme unless they are first oxidised, either *in vivo* by enzymes in the target insect or chemically by the analyst.   Therefore, we have developed this assay either for carbaryl-specific analysis or for broad-specificity analysis of carbaryl and all of the organophosphates in grain (Table VI).   In the latter case, the methanol extract of grain was treated with bromine water for three minutes before exposure to the enzyme.   These assays were quite sensitive, detecting down to 10ng/mL carbaryl, 1ng/mL dichlorvos and after bromine treatment, 0.2ng/mL of compounds such as chlorpyrifos-methyl in the final incubation mixture.   The current assay required only 10 to 15 minutes and both microwell (ELISA-like) and both tube and cuvet assays have been developed (Figure 2).   Further modifications can reduce the assay time further or convert the assay to a spot-test, yielding an insoluble enzyme product.

Table VI.   Analysis of pesticides used for treatment of stored grain using a cholinesterase inhibition assay

Compound	Pesticide solutions not oxidised IC_{50}[b] (ppm)	Pesticide solutions oxidised[a] IC_{50} (ppm)
Carbaryl	0.8	0.9
Dichlorvos	0.07	0.08
Fenitrothion	100	1
Chlorpyrifos-methyl	5	0.01
Pirimiphos-methyl	100	0.3
Malathion	10	0.4
Bioresmethrin	>> 1000	>> 1000

[a] treatment with bromine water (3 min); 10 min enzyme incubation
[b] concentration in methanol solution yielding 50% inhibition of color development.   Final concentration is 16-fold lower.

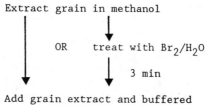

Extract grain in methanol

OR   treat with Br_2/H_2O

3 min

Add grain extract and buffered
enzyme to test tube

5 min or 10 min

Add acetylthiocholine and DTNB

2 min

Add stop solution. Read color versus a
pesticide-free control and pesticide standards

Figure 2. Cholinesterase Inhibition Assay - Method

Synthetic pyrethroids.   Apart from widespread use as
household insecticides, and in treatment of fruit, vegetables,
tobacco, and cotton, synthetic pyrethroids are used for
treatment of stock and wool.   Some such as (1R)-phenothrin
also have public health applications such as treatment of head
lice.   In different parts of the world, permethrin,
bioresmethrin, deltamethrin and natural pyrethrins are used,
and (1R)-phenothrin has been commercially evaluated.   Few
ELISA assays for pyrethroids have been developed to date, and
those that have, are for other compounds, such as S-
bioallethrin(15), cypermethrin(16) and tetramethrin(17).   We
have used monoclonal antibodies generated by collaborators at
the University of California(18) in our initial studies.
These antibodies were raised to a 3-phenoxybenzyl-2,2-
dimethylcyclopropane-1,3-dicarboxylate hapten coupled to bovine
serum albumin, and thus bind mainly to permethrin and (1R)-
phenothrin.   These collaborators had developed a two-step
competitive ELISA for analysis of permethrin in meat.   This
assay used the phenoxybenzyl half of the pyrethroid molecule
immobilized on the ELISA solid phase through conjugation with
bovine serum albumin; pyrethroid was quantified using a three
hour assay by sequential incubation with monoclonal antibody,
enzyme-labeled anti-mouse antibody followed by enzyme
substrate.   In addition, some clean-up of pyrethroid-
containing meat extracts was required.   We have simplified
this assay in several ways, for adaption to the analysis of
grain and flour.   Firstly, the antibody has been directly
conjugated to the enzyme horseradish peroxidase (HRP) which
removes a separate incubation and washing step from the assay.
Secondly, an alternate assay format has been developed, using
microwell-bound monoclonal antibody and a HRP-labeled
phenoxybenzoic acid derivative (Table VII).   The results
obtained were similar to those of Stanker and coworkers(18),
except that using our assay format the difference in potency
between (1R)-phenothrin and permethrin, and deltamethrin and
cypermethrin was greater.   This assay format may have
advantages in sensitivity and in stability for long-term
storage of test kits and in reproducibility of results obtained
with such kits.   Thirdly, we have developed a means of
assaying permethrin in grain and flour without the need for
clean-up of grain extracts.   In standard ELISA diluents,
permethrin is approximately 10-fold less potent in a methanol
or acetonitrile extract of ground wheatgrain than in pure
solvent.
     Components in the wheat extract itself did not inhibit
antibody binding, but decreased the inhibition potency of the
pyrethroids.   Some of the results obtained from initial
experiments with permethrin-treated wheat and flour samples are
shown in Table VIII. Both assay formats using antibody
immobilized on the microwell and antigen on the microwell gave
similar results.   In virtually all cases, values obtained were
slight underestimates, due to incomplete extraction of the
pyrethroid from the grain - in these experiments the grain was
either gently shaken overnight or homogenized for two minutes
in methanol or acetonitrile.   Although such extraction

techniques do not provide complete extraction, they are reproducible and as such can form the basis of a useful screening method for residue levels. Use of a constant correction factor (for incomplete extraction) results in quite accurate estimates. In addition a simple way to overcome a sample matrix effect was found, by inclusion of certain additives to the ELISA assay diluent used for the grain extract, enabling quantitative results to be obtained with wheat samples.

Table VII. Specificity of a Pyrethroid Monoclonal Antibody

Compound	Solid phase antigen		Solid phase antibody
Extractant	Acetonitrile	Methanol	Methanol
(1R)-Phenothrin	20	10	30
Permethrin	30	20	20
Deltamethrin	2000	1500	2000
Fenvalerate	~5000	~5000	~5000
Cypermethrin	~5000	~5000	~5000
Bioresmethrin	-	-	-
Tetramethrin	-	-	-
Bio(d-trans)allethrin	-	-	-
S-bioallethrin	-	-	-
Piperonyl butoxide	-	-	-
Fenitrothion	-	-	-

Data shown are concentrations yielding 50% inhibition of antibody binding in ng/mL (ppb) - = no inhibition at 2000ppb

Table VIII. Analysis of Permethrin by ELISA

Matrix	Actual (ppm)[a]	Determined (ppm)[b]			
Assay format		Antibody on plate		Antigen on plate	
Extraction method		overnight	rapid	overnight	rapid
Wheat	0.68	0.6	0.5	0.5	0.6
	0.73	n.t.	n.t.	0.5	0.5
	0.91	0.9	0.8	0.9	0.8
	2.6	1.6	1.5	1.9	1.7
Flour	0.26	0.5	0.3	0.3	0.3
	0.74	0.7	0.6	0.6	0.4

[a] by gas-liquid chromatography with electron capture detection
[b] using methanol extracts of ground grain or flour. Standard errors of the mean were typically 10-15%.

Summary and Conclusions

We have developed useful immunoassays for members of the major
classes of pesticides used for treatment of stored grain -
organophosphates and synthetic pyrethroids. While work to
date has mainly involved the organophosphate, fenitrothion, and
the pyrethroids, permethrin and (1R)-phenothrin, work on the
development of assays for other important compounds in these
groups (chlorpyrifos-methyl, pirimiphos-methyl and
bioresmethrin) is ongoing. While these immunoassay tests have
quite narrow specificity, enabling the identity of the residue
to be established, broad screening tests are sometimes
preferred. Immunoassay is not as amenable to such broad
screening as other methods(19), and, for organophosphate and
carbamate residues, we have modified an existing cholinesterase
inhibition method to make it suitable for use with the broad
range of these pesticides.
     Both the ELISA and enzyme-immunoassay tests seem applicable
to wheat and flour with little need for sample clean-up.
While the assays are in theory adaptable to other matrices such
as vegetables, wool and citrus fruits, the effect of the
particular matrix must be studied for each assay. For
example, clean-up is essential for the pyrethroid antibodies
when meat or fat is analyzed(17).
     Current work is aimed at developing prototype antibody kits
for use by the grain industry. Considerable stability testing
and collaborative trialling is, however, required before this
can be done. Assay validation of pesticide ELISAs has also
been reviewed elsewhere(20). While quantitative ELISA kits
will be useful in regional laboratories, we also aim to produce
simple "tube" and dipstick-type tests for use in field
situations, such as at grain elevators or in produce markets.
Much of our experience in this area comes from earlier
development of the first ELISA test (now commercially
available) that is designed to be used in the home for food
analysis, in this case, for gluten. We aim to produce
something similar to our rapid gluten test, and we will
describe this test in detail. This test allows sufferers of
gluten intolerance to inexpensively test foods for gluten
before they consume them. It is suitable for use with all
types of foods, raw, cooked or processed. The principle of
the test is similar to the now-commonplace home pregnancy
tests. The test is designed for use by untrained individuals
without need for any facilities or equipment except running tap
water, provides rapid (5 minute), results and is suitable for
home use by celiacs or in-process quality control in food and
wheat starch manufacture. For the test, we selected
monoclonal antibodies to certain heat-stable proteins. The
antibodies chosen also cross-react appropriately with prolamins
from different cereals such as rye and barley but not maize and
rice, yet showed little difference in binding to different
varieties of wheat, barley and rye(21,22). Foods are coarsely
ground, and gluten extracted by shaking for 60 seconds in the
extractant provided, and a drop of sample is added to the
coated tube and mixed for 90 seconds. The antibody solution

is added, the tube shaken then unbound components tipped out
and the tube washed under tap water, and finally substrate is
then added.   Presence of gluten in the food results in the
development of a blue color within seconds of addition of color
developer (enzyme substrate).   A pre-formed colored standard
corresponding to results obtained with "just acceptable"
amounts of gluten is proved in the kit for comparison.   The
rapid test kit was successfully trialed by over 50 celiacs,
dieticians and food industry laboratories early in 1989.
    A field test for grain protectants would be extremely
valuable.   It would enable grain managers to ensure that
protectant levels fell within a "desirable" range such as very
low or less than Codex "Maximum Residue Levels".   It would
also enable consumers such as flour millers and lot-farmers to
test residue levels.   These tests will be a valuable tool in
the rational and safe use of these pesticides.

Acknowledgments

The authors are grateful to Dr J. Desmarchelier and Mr N.
Jancic for advice and assistance and to Mrs B. Arneman for
typing.   The provision of some pyrethroid-specific antibodies
by Drs L Stanker and M Vanderlaan (University of California) is
greatly appreciated.   The work was partly supported by the
Wheat Research Council of Australia.

Literature Cited

1.   Snelson, J. T. Grain Protectants; Australian Centre for
      International Agricultural Research Monograph No 3,
      Department of Primary Industries and Energy, Canberra,
      Australia, 1987, p448.
2.   Hunter, K. W.; Lenz, D. E.; Brimfield, A. A.; Naylor, J.
      A. FEBS Lett. 1982, 149, 147-152.
3.   Buenafe, A. C.; Rittenburg, M. B. Mol. Immunol. 1987, 24,
      401-407.
4.   Haas, J. H.; Guardia, E. J. Proc. Soc. Exp. Biol. Med.
      1968, 129, 546-562.
5.   Centeno, E. R.; Johnson, W. J.; Sehon, A. H. Int. Arch.
      Allergy 1970, 37, 1-13.
6.   Ercegovich, C. D.; Vallejo, R. P.; Gettig, R. R.; Wood,
      L.; Bogus, E. R.; Mumma, R. O. J. Agr. Food Chem. 1981,
      29, 559-563.
7.   Brimfield, A. A.; Lenz, D. E.; Graham, C.; Hunter, K. W.
      J. Agric. Food Chem. 1985, 33, 1237-1242.
8.   Lober, M.; Krantz, S.; Herrmann, I. Acta Biol. Med. Ger.
      1982, 41, 487-496.
9.   Vallejo, R. D.; Bogus, E. R.; Mumma, R. O. J. Agric. Food
      Chem. 1982, 30, 572-580.
10.   Ngeh-Ngwainbi, J.; Foley, P. H.; Tran, S. S.; Guilbault,
      G. J. Amer. Chem. Soc. 1986, 108, 5444.
11.   Heldman, E.; Balan, A.; Horowitz, O.; Ben-Zion, S.;
      Torten, M. FEBBS Lett. 1985, 180, 243-248.
12.   Suedi, J.; Heeschen, W. Kiel. Milchwirtsch. Forschungsber
      1988, 40, 179-204.

13. Sharp, G. J.; Bryan, J. G.; Dilli, S.; Haddad, P. R.;
    Desmarchelier, J. M. Analyst 1988, 113, 1493-1507.
14. Mendoza, C.E. Residue Rev. 1972, 43, 105-142.
15. Wing, K. D.; Hammock, B. D.; Wustner, D. A. J. Agric. Food
    Chem. 1978, 26, 1328-1333.
16. Roberts, T. R. Trends Anal. Chem. 1985, 4, 3-7.
17. Demoute, J-P.; Tower, G.; Mouren, M. French Patent 2 593
    503, 1986.
18. Stanker, L. H.; Bigbee, C.; Van Emon, J.; Watkins, B.;
    Jensen, R. H.; Morris, C.; Vanderlaan, M. J. Agr. Food
    Chem. 1989, 37, 834-839.
19. Harrison, R. O.; Gee, S. J.; Hammock, B. D. In
    Biotechnology in Crop Protection; Hedin, P., Menn, J. J.,
    Hollingworth, R. M., Ed.; ACS Symposium Series No 379,
    American Chemical Society: Washington, DC, 1988; pp316-
    330.
20. Jung, F.; Gee, S. J.; Harrison, R. O.; Goodman, M. H.;
    Karu, A. E.; Braun, A. L.; Li, Q.X.; Hammock, B. D.
    Pestic. Sci. 1989, 26, 303-317.
21. Hill, A.S.; Skerritt, J.H. Food Agr. Immunol. 1990, in
    press.
22. Skerritt, J.H.; Hill, A.S. J. Agr. Food Chem. 1990, in
    press.

RECEIVED August 30, 1990

IMMUNOASSAYS
FOR NATURAL TOXINS

Chapter 12

Current Immunochemical Methods for Mycotoxin Analysis

Fun S. Chu

Food Research Institute and Department of Food Microbiology and Toxicology, University of Wisconsin, Madison, WI 53706

Mycotoxins are small molecular weight secondary fungal metabolites and are not immunogenic. However, mycotoxins can be conjugated to a protein or a polypeptide carrier and subsequently be used for immunization. Using this approach, antibodies against a number of mycotoxins have been developed. With the availability of these antibodies, simple and rapid radioimmunoassays (RIA) and enzyme-linked immunosorbent assays (ELISA) for the determination of these mycotoxins in grains and biological fluids have been developed. The present review will cover recent progress on this rapidly developing research area with emphasis on: (1) the approaches used for the preparation of immunogens and antibody production; (2) diversity of specificity of antibodies obtained from various immunogens and (3) different immunochemical approaches that are currently used in mycotoxin analysis.

Mycotoxins are small molecular weight secondary fungal metabolites and are not immunogenic. However, mycotoxins can be conjugated to a protein or a polypeptide carrier and subsequently be used for immunization. Using this approach, antibodies against a number of mycotoxins have been developed. With the availability of these antibodies, both polyclonal and monoclonal types, several types of immunoassays have been developed for mycotoxin analysis. The rapid progress on the immunoassay of mycotoxins research area can be seen from the large volume of literature published in the last few years as well as from various talks on the application of this methodology that are covered in this symposium. Such progress together with an overall recognition of the potential wide application of this powerful technique have not only led to more rapid and accurate protocols that allow detection of small amounts of mycotoxins present in foods, feeds and biological fluids, but have also generated considerable interest in the preparation of

0097–6156/91/0451–0140$06.00/0

commercial kits for mycotoxin analysis. Due to limited time for
the talk and limited space for the manuscript, I will confine my
discussion to the following three areas: (1) the approaches used
for the preparation of immunogens for antibody production; (2)
diversity of specificity of antibodies obtained from various
immunogens and its role in the immunoassay and (3) different
immunochemical approaches that are currently used in mycotoxin
analysis. Most references on the immunoassay of mycotoxins before
1985 are also omitted from this article; for a more extensive
discussion and complete references on the immunoassay of
mycotoxins, several recent reviews should be consulted (1-13). The
application of immunoassays for mycotoxins can been seen from
several articles in this book.

Approaches used for the preparation of immunogens:

Since mycotoxins are not antigenic, they must be conjugated to a
protein or polypeptide carrier before immunization. The approaches
that have been used for conjugation of mycotoxins to a protein
carrier and methods for the preparation of mycotoxin derivatives
before conjugation are summarized in Table I.
    Whereas some mycotoxins have a reactive group which can be
directly conjugated to a protein carrier for antibody production,
most mycotoxins do not; thus, a reactive group must first be
introduced into the molecule before coupling to the protein (1-7).
Additional reactive groups can be introduced into different
analogues of a specific mycotoxin, so that the reactive group would
be located at different positions in the related parent molecule.
For mycotoxins containing a carbonyl group, a carboxymethyl oxime
(CMO) could be prepared (1-6). Approaches for the preparation of
various aflatoxin B_1 and T-2 toxin derivatives are shown in figures
1 and 2. For mycotoxins containing a hydroxyl side chain,
acylation of the mycotoxin with bifunctional anhydrides such as
succinic acid and glutaric acid anhydride to form the corresponding
hemisuccinate (HS) and hemiglutarate (HG) are generally used.
However, the stability of the HS and HG should be tested before
conjugation to the proteins. Since the ability of a hapten to
elicit antibody in animals depends both on the amount of the
haptenic group conjugated to the protein as well the orientation of
the hapten on the protein molecule, the presence of a spacer alkyl
chain, such as glutarate, between protein and mycotoxins has an
additional advantage in improving antibody production both in
quantity and in quality. We have also found that immuogenicity of
mycotoxin-protein conjugates improved when additional spacer groups
were introduced into bovine serum albumin (BSA) with ethylene
diamine (14). An alternative method for mycotoxin containing an
primary alcohol is by oxidizing the OH group to a carbonyl group
for subsequent derivatization and conjugation (15).
    Among different methods that have been used in conjugation of
mycotoxin to protein carriers, the water soluble carbodiimide (WSC)
method is most commonly used. Alternatively, mycotoxin or its
derivative is first activated with N-OH-succinimide (NHS) in the
presence of carbodiimide and then conjugated to protein directly

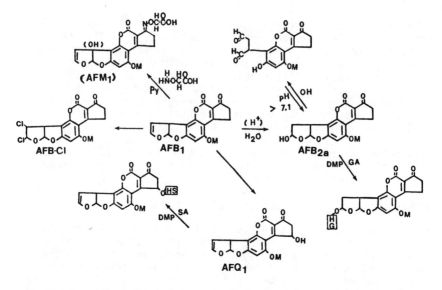

Figure 1. Approaches used for the preparation of various aflatoxin B1 derivatives. Abbreviations are given in Table I.

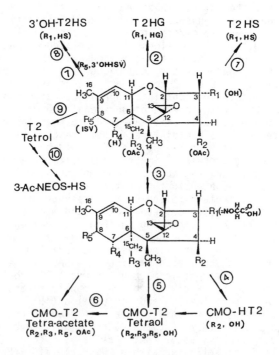

Figure 2. Approaches used for the preparation of various T-2 toxin derivatives. Abbreviations are given in Table I except ISV and NEOS which represent isovaleroxy and neosolaniol, respectively.

(7-8, 13). The mixed anhydride method has also been used for conjugation of some mycotoxins to proteins. For mycotoxins containing an active aldehyde group, reductive alkylation was used. Cross-linking agents such as glutaraldehyde have been used for mycotoxins containing an amino group such as fusarochromanone (Yu et al., 1989; unpublished).

Table I.   Approaches used for the conjugation of mycotoxins to protein (a)

Derivatives prepared	Mycotoxins used	Methods of conjugation
CMO	AFB, AFM, DAS, T-2, ZEARA	WSC, MA, NHS
HA	AFB2a, AFG2a, AFQ2a, ST,	RA
HG	AFB2a, DAS, T-2	WSC, MA, NHS
HS	AFQ, AFB3, DAS, DON, Ac-DON, DOVE, T-2, HT-2, acetyl-NIV, Roridin, 3'-OH-T-2, T-2-tetra-acetate	WSC, MA, NHS
None	PR toxin	RA
None	AFB2a, KJ	TA
None	OTA, Rubratoxin B	WSC, MA
None	AFB-Cl, AFB-Br, SA	Direct
None	AFB m-Cl-PBA	Direct
None	TDP-1 (Fusarochromanone)	WSC, GA

(a) Abbreviation used: CMO, carboxymethyl oxime; DAS, diacetoxyscirpenol; DON, deoxynivalenol; DOVE, deoxyverrucarol; NIV, nivalenol; HA, hemiacetal; HG, hemiglutarate; HS, Hemisuccinate; GA, glutarladehyde; MA, mixed anhydride; NHS, n-hydroxylsuccinimde; RA, reductive alkylation; TAB, tetrazobenzidine; WSC, water soluble carbodiimide; AFB, AFM, AFB2a, AFG2a, AFQ2a, AFB3, AFQ, AFB-Cl and AFB-Br are aflatoxins Bl, Ml,B2a, G2a, Q2a, B3, Ql, chloride and bromide, respectively; KJ, kojic acid; OTA, ochratoxin A; ST, sterigmatocystin; ZEARA, zearalenone; m-Cl-PBA, m-chloroperoxybenzoic acid; SA, secalonic acid.

Using different mycotoxin protein conjugates as immunogens, specific antibodies, both polyclonal and monoclonal antibodies, against most mycotoxins have been obtained (1-13). Table II summarizes the different antibodies which have been made against various mycotoxins and mycotoxin metabolites.

Diversity of antibody specificity:

The specificity, generally expressed as the cross-reactivity, of an antibody is determined primarily by the type of mycotoxin or its metabolites that has been used in the antibody production as well

as by the site in the mycotoxin molecule where linkage was made
when it was conjugated to the protein carrier. Thus, the degree of
cross-reactivity of these antibodies with their respective
structural analogs vary considerably (1-9). The accuracy of
immunoassay of mycotoxins for naturally-contaminated samples
generally is affected by both the specificity of the antibody to be
used and by the possible presence of structurally related analogs
of the mycotoxin in the sample that may react with the antibody.
For example, if the antibodies used in an immunoassay for aflatoxin
(AF) are 100% cross-reactive with aflatoxin B1 (AFB1), but only 33%
with AFG1, the presence of AFG1 in the sample would result in a
lower "apparent" total AF level by the immunoassay method (16).
Thus, it is very important to know the specificity of the antibody
to be used in the immunoassays. The specificity of antibody
against various mycotoxins has been documented in a number of
reviews (1-6, 17). For the purpose of discussion, the diversity of
antibody specificity for aflatoxins (AFs) and triochothecenes
(TCTCs) are presented here.

Table II. Antibodies Against Mycotoxins

Mycotoxins	Type	References
Aflatoxins		
B1, B2, M1	M, P	5, 13, 17-25
G1, B1&G1	M, P	13, 26, 27
Aflatoxin metabolites		
B2a, Q1, M1, Ro	M, P	5, 9, 17, 22, 25, 28
DNA-Adducts	M, P	17, 29-32
Kojic Acid	P	5
Ochratoxin A	M, P	5, 13, 33-36
Patulin	P	37
PR-toxin	P	38
Rubratoxin B	P	5
Secalonic acid	P	39
Sterigmatocystin	P	5, 40
Trichothecenes		
DAS, DON, DOVE	M, P	5, 15, 41-44
AcDON, NIV, Roridin A	M, P	45-48
T-2 toxin	M, P	5, 49-54
T-2 toxin metabolites		
HT-2, T-2-tetraacetate	M, P	5, 55-58
3'-OH-T-2, de-ep-T-2	M, P	51, 59-61
Zearalenone	M, P	5, 63

Abbreviation used: M, monoclonal antibody; P, polyclonal
antibodies; other abbreviations are same as Table I.

Specificity of antibodies against aflatoxin and related
metabolites: The relative cross-reactivity of different types of
antibodies against aflatoxin with some selected AF analogs are
illustrated in Fig. 3. The importance of the site of conjugation
to the antibody specificity is well demonstrated from these

examples. When rabbits were immunized with conjugates prepared by linking the CMO of AFB1 and AFM1, the antibodies generally recognize the dihydrofuran ring of the molecule as well as structural variation in the dihydrofuran portion of aflatoxin molecules. Thus, antibodies highly specific against AFB1 and AFM1 were produced in this manner. These two type of antibodies are currently used for most of the immunoassays for both aflatoxins. Whereas antibodies for AFB1 cannot recognize AFM1, antibodies for AFM1 bind AFB1 effectively. This property has been used as an approach for the immunoassay of AFM1 so that radioactive AFB1 or enzyme-linked AFB1 could be used as the marker in the assays (5, 8, 9, 63, 64). From these studies, it is apparent that the side chain hydroxyl group in AFM1 plays an important role in eliciting antibodies that can recognize AFM1.

When conjugates prepared through the dihydrofuran portion of the AF molecule were used in immunization, the antibodies had a specificity directed toward the cyclopentane ring. Antibodies with similar specificity and a wider range of cross-reactivities, were obtained from rabbits when conjugates prepared by coupling AFB2a (designated as B2a antibody; 5, 17), AFG2a, AFB diol (designated as AFB-diol antibody; 5, 17) and AFB chloride (designated AFB-Cl antibody; 38) to protein were used for immunization. These antibodies also have been shown to cross-react with AFB-DNA adducts and other related compounds. Since the majority of the structure of AFB1 and AFG1 is identical to AFB3, we conjugated the AFB3-HS to protein which was then used in the immunization. We found that the antibodies elicited showed good reaction with both AFB1 and AFG1. Thus, these type of antibodies are very useful to for immunoassays of both aflatoxins (27; Schubring, Zhang, and Chu, Abstracts of Annual meeting-1989 of American Society for Microbiology. New Orleans, May, 1989).

A number of monoclonal antibodies against AFs have been obtained in the last few years (17-25, 29-32). In general, the specificity of different monoclonal antibodies for aflatoxins is also depend on the type of conjugates used in the immunization. When the same immunogens are used for the preparation of either polyclonal or monoclonal antibody, most of the high affinity monoclonal antibodies have almost the same specificity as that observed for the polyclonal antibodies (Fig. 3). Although clones that elicited antibodies which have different cross-reactivities have been found, these clones generally have lower affinities. For example, Kawamura et al., (25) have recently found some of the monoclonal antibodies for AFB1 had higher cross reactivity with aflatoxicol (AFRo), one of the aflatoxin metabolites. Nevertheless, the affinity of these antibodies to aflatoxicol was still lower than the affinity of the best monoclonal antibody to AFB1. Two types of AFB-DNA adducts have been prepared (17, 29-32). The monoclonal antibodies prepared against AFB-guanine modified DNA and AFB-ring opened modified DNA showed virtually no cross-reactivity with aflatoxins B1, B2a, diol and AFB-guanine. They have a high degree of recognition for AFB-DNA adducts regardless of whether they are AFB-DNA or AFG-DNA (17).

Specificity of antibodies against trichothecene mycotoxins: The relative cross-reactivity of different types of antibodies against T-2 toxin and other trichothecenes with selected TCTCs are illustrated in Figs. 4 and 5. It is apparent that these antibodies are very specific to the toxins that were originally used in the conjugation. Since T-2 antibody cross-reacted with HT-2 to some extent (5, 55, 56), it is possible to use this antiserum for monitoring HT-2; however, it is useless for the analysis of other important metabolites of T-2 toxin. Likewise, it is also possible to use the HT-2 and 3'-OH-T-2 antibodies to measure T-2 toxin (56, 60). The antibodies for DAS (15, 41-44) are highly specific for DAS itself, and cannot be used for monitoring DAS metabolites including either 4-mono-acetyl-DAS (4-MAS) or 15-mono-acetyl-DAS (15-MAS) as well as scirpentriol. Antibodies against DOVE (5), triacetyl-DON (45), and 3-acetyl-DON (46), however, can only be used for measuring DOVE, triacetyl-DON and 3-acetyl-DON. Thus, DON must be first converted to triacetyl-DON before analysis of DON can be made (8). Antibodies against T-2 tetraol tetra-acetate (55) were also obtained, with specificity to the tetra-acetate of T-2 toxin but not to the tetraol. Antibodies against macrocylic TCTC roridin A has recently obtained (47, 48). Such antibodies had highest cross-reaction with roridin family TCTCs. As more data accumulated, we found that the side chain of TCTC groups play an important role in eliciting antibody against this group of mycotoxins. Consequently, an immunogen which was prepared by conjugating 3-acetyl-neosolaniol-HS (3-Ac-NEOS-HS) to protein at the C-8 position instead of using conjugates that have been coupled to the protein through the C-3 position of TCTCs, was used (58,59). We found that antibodies, both polyclonal and monoclonal type, obtained from animals after immunization with this immunogen showed shown good cross-reactivities with most of the group A TCTCs (58, 59). The affinity of these antibodies to the acetylated group A trichothecenes is almost 10 times higher than to the non-acetylated group A trichothecenes. This property has become very useful for the development of an ELISA method for the detection of less than 10 ppb of group A trichothecene in the urine and serum (65). The epoxide group in the TCTC appears not to be an important epitope for the production of antibody because we found that two types of antibodies raised against T-2 toxin, i.e. anti-T-2-HS-BSA and anti-3-Ac-NEOS-HS-BSA, showed good cross-reactivity with deepoxy T-2 toxin (66).

The specificity of monoclonal antibodies against several types of TCTCs are also shown in Figs. 4 and 5. Several monoclonal antibodies were found to have specificities different from the polyclonal antibodies when the same type of immunogen was used. For example, immunization with T-2-HS-BSA conjugates resulted in monoclonal antibodies from different laboratories that each have different specificities. Some monoclonal antibodies have high specificity against HT-2 toxin (49), the others have specificity against 3'-OH-T-2 toxin (51). Since TCTCs have different side chains, the diversity of antibody specificity for those monoclonal antibodies might be due to the hydrolysis of either the immunogen in the immunization process or the hydrolysis of the T-2 toxin

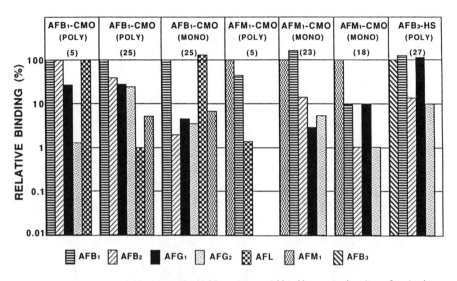

Figure 3. Specificity of different antibodies against selected aflatoxins. Reference numbers are given in the parenthesis for each immunogen used. Abbreviations are given in Table I except AFL which represents aflatoxicol.

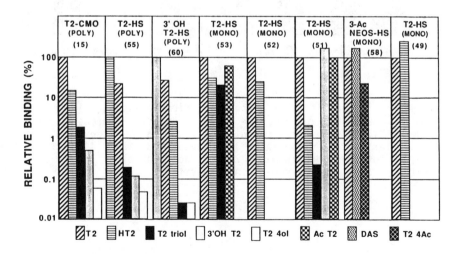

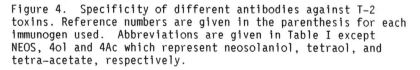

Figure 4. Specificity of different antibodies against T-2 toxins. Reference numbers are given in the parenthesis for each immunogen used. Abbreviations are given in Table I except NEOS, 4ol and 4Ac which represent neosolaniol, tetraol, and tetra-acetate, respectively.

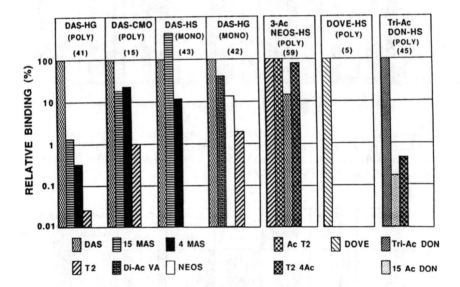

Figure 5. Specificity of different antibodies against T-2 toxin metabolites and selected trichothecenes. Reference numbers are given in the parenthesis for each immunogen used. Abbreviations are given in Table I except MAS, NEOS, 4Ac and VA which represent monoacetoxyscirpenol, neosolaniol, tetra-acetate, and verrucarin A respectively.

conjugate in the microtiter plate during selection of the specific clones.

Specificity of antibodies against other ochratoxin, sterigmatocystin and zearalenone: The antibody for ochratoxin A (OTA) was most specific for OTA and OTC, and was not specific for ochratoxins B, alpha and other coumarins (5, 34). Thus, they can not be used for monitoring OT-alpha which has been reported to be a major ochratoxin metabolite in animals. These results indicate that phenylalanine and chloride ion in OTA play an important role in eliciting OTA antibody. The antibodies obtained from pigs which were immunized with the CMO derivative of zearalenone conjugated to BSA were most specific for zearalenone with good cross-reactivity with zearalanone, and low melting point zearalenol at the same degree as with zearalenone (5, 62). Thus, these antibodies could be used for the determination of both zearalenone in agricultural commodities and for monitoring zearalenone metabolites in serum (62, 67). The antibody against ST has some cross-reactivity with dehydro-ST (16 times less than ST), but had no cross-reaction with aflatoxins and 6-O-methyl-ST (5). Similar to the antibodies against AFB2a, antibodies against ST also cross-react with ST-DNA adducts effectively (Olson and Chu, unpublished observation).

Immunochemical approaches for mycotoxin analysis:

Although a number of immunochemical methods have been used for the analysis of small molecular weight biological substances, only radioimmunoassay (RIA), enyzme-linked immunosorbent assay (ELISA) and immuno-affinity assay (IAA), have been developed for the analysis of mycotoxins. Recent developments have led to several quick screening tests and more than 10 types of commercial kits have become available in the last few years (8, 10, 13). In most cases, sample after extraction from the solid matrix and diluted in buffer can be directly used in the assay. Since the application of immunoassay for several mycotoxins are covered by other speakers, I will only briefly highlight some of the recent progress on these methods.

Progress on the RIA of mycotoxins: Since RIA involves the competition between the binding of antibody with a radioactive marker and the toxin in the sample, the specific activity of radioactive ligands plays an essential role in the sensitivity of RIA. Whereas tritated mycotoxins are generally used in the RIA with a sensitivity between 0.25-0.5 ng/assay (2-5 ppb), recent developments have led to the use of more sensitive labeled markers such as 125I-AFB1, 125I-AFM1 (68-71) and 125I-OA (72, 73). C-14 labeled has also been for OTA analysis (36). RIA is now more simple to perform using a magnetic gel which antibody is conjugated to (Wei and Chu, unpublished observations) as well as other solid phases (6-9). The sensitivity of RIA is improved by a more efficient simple cleanup treatment (5, 8, 36, 68, 74). However, because of the need for radioactive toxin, RIA can only be used in a laboratory which has facilities for determining radioactivities.

Competitive enzyme-linked immunosorbent assays (ELISA): Two types
of ELISA have been used for the analysis of mycotoxins and both
types are heterogenous competitive assays. One type, i.e. direct
ELISA, involves the use of a mycotoxin-enzyme conjugate and the
other system, i.e. indirect ELISA, involves the use of a
protein-mycotoxin conjugate and a secondary antibody to which an
enzyme has been conjugated. Although horseradish peroxidase (HRP)
is most commonly used as the enzyme for conjugation, other enzymes
such as alkaline phosphatase and beta-galactosidase, also have been
used (5, 9, 13).

Direct Competitive ELISA: In this assay, specific antibodies are
first coated to a solid phase, most commonly to a microtiter plate
although other solid-phases have been used (1-13). The sample
solution or standard toxin is generally incubated simultaneously
with enzyme conjugate or incubated separately in two steps. After
washing, the amount of enzyme bound to the plate is then determined
by incubation with a substrate solution (5, 12, 13). The resulting
color, which is inversely proportional to the mycotoxin
concentration present in the sample, is then measured
instrumentally or by visual comparison with the standards. In
general, direct ELISA is approximately 10-100 times more sensitive
than RIA when purified mycotoxin is used. As low as 2.5-5 pg of
pure mycotoxin can be measured. However, the sample matrix
sometimes causes interference with the assay. This problem is
generally overcome by dilution of the sample to a range which does
not affect the assay (75) or by using sample extract in the
preparation of the standard curve (6-8, 12, 13). Since a clean-up
step is generally not necessary, many samples can be analyzed
within a relatively short period. The sensitivity of the direct
ELISA for the analysis of mycotoxins in foods and feeds is in the
range of 0.05 to 50 ppb (6-8, 12, 13). Like RIA, the sensitivity
of ELISA can be improved when a cleanup treatment is included
(6-9). For example, when milk or urine samples were subjected to a
simple cleanup treatment with a C-18 reversed-phase Sep-Pak
cartridge (76) or through an affinity column (9, 22, 36, 77), the
sensitivity of ELISA for AFM in milk and urine reached the 10-25
ppt level. Use better antibody and toxin-enzyme conjugates, the
time required to run the ELISA has improved considerably. Thus,
the whole ELISA can be completed within one hour. By shortening
the incubation time, and also using of removable microtiter wells,
this method could be used as a screening test for mycotoxins (8-13,
75, 77, 78). A protocol for screening AF in cotton seed and mixed
feed has been approved as the interim official first action by the
AOAC (13, 80).

Indirect competitive ELISA (or double antibody ELISA): In the
indirect ELISA, instead of using a mycotoxin-enzyme conjugate, a
mycotoxin-protein (or polypeptide) conjugate is first prepared and
then coated to the microplate (1-13). The plate is then incubated
with specific antibody against the mycotoxin in the presence or
absence of the homologous mycotoxin. The amount of antibody bound
to the plate coated with mycotoxin-protein conjugate is then
determined by reaction with a second antibody-enzyme complex such

as goat anti-rabbit IgG-enzyme complex or goat anti-mouse-IgG
(IgM)-enzyme complex which is generally commercially available, and
by subsequent reaction with the substrate. Thus, toxin in the
samples and toxin in the solid-phase competes for the same binding
site with the specific antibody in the solution. The sensitivity of
the indirect ELISA is comparable to or slightly better than the
direct ELISA with the advantage that much less antibody (100 times
less) is required. In addition, it is not necessary to prepare a
toxin-enzyme conjugate. The disadvantage of the method is that an
additional incubation step is necessary for the assay and thus, it
requires more analytical time (2-3 hours). Modifications on the
indirect ELISA have been made for the analysis of several
mycotoxins. One approach involved the conjugation of antibody,
especially monoclonal antibody, to an enzyme and then use directly
in the assay (81, 82). The second approach involved pre-mixing the
antibody with the second antibody-enzyme conjugate. Such
modifications have shortened the assay time without sacrificing the
sensitivity. Because only small amounts of antibody are needed for
the indirect ELISA, this method has been used extensively for
monitoring the antibody titers of hybridoma culture fluids for the
screening of monoclonal antibody producing cells (1-13) in addition
to toxin analysis.

Immunoaffinity assay: The immunoaffinity assay which involves the
use of an antibody column that traps the mycotoxins has been used
for AFB1, AFM1 and OTA (8, 9, 13, 36, 83,84). The toxin can be then
eluted from the column for subsequent analysis or adsorbed in a
solid-phase to which the fluorescence is then read directly. Thus,
the affinity column serves as a specific cleanup and concentration
tool for the analysis. Recent advances in improvement of
instrumentation of fluorescence detection and post-column
derivatization have led to a wider application of this method for
AF detection. An AOAC collaborative study showing good result has
been completed (85). In such an assay, AF extracted from the
sample is first diluted with buffer at pH 7.0 and then subjected to
a disposable affinity column containing anti-AF antibody Sepharose
gel. After washing, AF is removed from the column with methanol,
subjected to treatment with iodine solution, and the fluorescence
determined. Nevertheless, this method cannot be used for
mycotoxins, such as TCTCs, which do not have high fluorescence or a
chromophore.

New developments: Several new approaches for the analysis of
mycotoxins have been made in recent years. One method called
the"hit and run" assay (86) was developed for T-2 toxin. In this
assay, an affinity column which was prepared by conjugating T-2
toxin to Sepharose gel, was equilibrated with fluorescein
isothiocyanate (FITC)-labelled Fab fractions of IgG (anti-T-2
toxin). Samples containing T-2 toxin were injected into the
column. The FITC-Fab which eluted together with the samples
containing T-2 toxin was then determined in a standard flow-through
fluorometer. The detection limit for T-2 toxin was found to be
between 25 and 50 ng/mL. Although this assay is rapid and the
column could be used many times, it was very selective and required

specific monoclonal antibodies (Chu, unpublished observations). The
sensitivity was also not very high. Another approach that may lead
to the development of a biosensor is a homogenous immunoassay for
T-2 toxin which involves the use of liposomes and complement (87).
Unfortunately, the system is not very sensitive. Antibody against
T-2 toxin has been conjugated to fiber optics which could be used
in the future for the development of a biosensor (88).

Quick screening tests: By shortening the incubation time and
adjusting the antibody and enzyme concentrations in the
direct/modified indirect competitive microtiter plate ELISA assay
system, it is possible to obtain a positive or negative test at
certain toxin levels (such as 20 ppb) as a quick screening test for
AF. This approach has been adapted by the AOAC as first action for
screening AF at 20 ppb in several commodities (8, 9, 13, 78, 80).
Several other types of immunoscreening tests which have similar
sensitivity to those of ELISA, were developed for mycotoxin
analysis in recent years (8, 13, 78, 79, 83, 89, 90). Rather than
coating the antibody onto the microtiter ELISA plate, the new
approach immobilizes the antibody on a paper disk or other membrane
which is mounted either in a plastic card (card screen test) or in
a plastic cup (8). The reaction is carried out on the wetted
membrane disk. The principle of the reaction in this assay format
is similar to that of direct ELISA. In practice, sample extracts
(a few drops) could be first applied to the test spot on the
plastic card or in the cup, followed by reaction with
aflatoxin-enzyme conjugate, washing, and finally reaction with the
substrate. The absence of color (or decrease in color), generally
blue, at the spot indicates the sample contains mycotoxins as
comparing with a negative control spot where a bright blue color
appears. The reaction is generally very rapid and takes less than
10 min. to complete. Thus, the rapid screening test kits permit
monitoring of mycotoxins semi-quantitatively in the field. Good
correlation between the quick screening methods with conventional
methods such TLC and HPLC methods has been found. Collaborative
studies for some of these new screening tests are currently being
carried out (80).

Complimentarily of chemical analyses and immunochemical methods:

Immunoaffinity chromatography as a cleanup tool: The application
of affinity column chromatography for the immunoassay of mycotoxins
began as early as 1977 when Sun and Chu immobilized antibody to
Sepharose gel, which was then used in a RIA (91). Subsequent
studies led to use of the affinity column as a tool for
pre-concentration and cleanup of samples of mycotoxins before RIA
or HPLC analysis (30, 92). With the availability of different
monoclonal antibodies against AFB and AFM, affinity chromatography
has gained wide application as a cleanup tool for AF analysis in
recent years. Because the first step in affinity chromatography
involves the binding of toxin with the column at neutral pH, this
method is ideal for biological fluids. Thus, this method is used
more frequently for the analysis of AF metabolites, especially AFM
in biological fluids including urine and milk (8, 9, 13). More

recently, a robotic system involving the use of an affinity column and HPLC was developed for routine analysis of AFM in milk (93).

Immunochromatography: A new approach involving the use of ELISA as a post column monitoring system for HPLC for the analysis of different group A TCTCs was developed. Various group A trichothecenes were first separated on a C-18 reversed-phase column. Individual fractions eluted from the column were analyzed by ELISA using generic antibodies against group A trichothecenes. This approach not only can identify each individual group A trichothecene, but also can determine their concentration quantitatively. As low as 2 ng of T-2 toxin and related trichothecenes as well as their metabolites can be monitored by this method. Combination of HPLC and ELISA technology proved to be an efficient, sensitive and specific method for the analysis of trichothecenes (94) and other mycotoxins.

Concluding Remarks:

Different approaches used in the preparation of immunogens have led to production of various antibodies, including both monoclonal and polyclonal antibodies, with diverse specificity against mycotoxins. With availability of such antibodies, good and reliable immunoassay protocols for mycotoxins, in different commodities have been established. These methods are very simple, sensitive and specific. Several immunoassay kits for aflatoxin and other mycotoxins are currently available. It is imperative for those who perform the immunoassay to know the specificity of the antibody used in each assay. Depending on the purpose of research or routine analysis, selection of appropriate antibody plays an important role in determining the success of a typical assay. Antibodies highly specific to certain mycotoxins as well as antibodies which cross-react with a group of mycotoxin (generic type) are important, and both should be considered for immunoassays. Immunochemical methods, including affinity chromatography as a clean-up tools and immunoassay as a post-column HPLC monitoring system, have been used in conjunction with other analytical methodology. Immunohistochemical methods have also been used for localize mycotoxins in animal tissues (4, 8). Such innovative approaches have led to a new dimension in the immunochemical method for mycotoxins. This type of research should be continued and extended in addition to the development of new antibodies for other important mycotoxins and mycotoxin metabolites, especially generic type antibodies and improvement in the existing immunoassay protocols.

Acknowledgments

This work was supported by the College of Agricultural and Life Sciences, the University of Wisconsin, Madison, Wisconsin. Part of the work described in this contribution was supported by a Public Health Service grant (CA 15064) from the National Cancer Institute, a contract (DAMD17-86-C-6173) from the U.S. Army Medical Research and Development Command of the Department of Defense, and an USDA

North Central Regional Project (NC-129). The author thank Dr. R. D. Wei, Miss Racheal Lee and Ms. Susan Hefle for their help in the preparation of the manuscript.

Literature cited

1. Chu, F. S. Proc. Int. Symp. Mycotoxins. Cairo, Egypt (Ed. Naguib et al; National Information Documentation Center, Dokki, Cairo, Egypt), 1983; p. 177.
2. Chu, F. S. J. Food Prot. 1984, 47, 562-569.
3. Chu, F. S. 1984, In Toxigenic Fungi-Their Toxins and Health Hazard; Kurata, H. and Ueno, Y., Eds.; Kodansha (Tokyo) and Elsevier: Amsterdam-Oxford-N.Y.& Tokyo, 1984; p. 234.
4. Chu, F. S. In Diagnosis of Mycotoxicoses; Richard, J. L.; Thurston, J. R. Eds.; Martinus Nijhoff Publishers: Dordrecht/Boston/Lancester, 1986; p.163.
5. Chu, F. S. In Modern Methods in the Analysis and Structural Elucidation of Mycotoxins; Cole, R. J., Ed.; Academic Press: N.Y., 1986, p 207.
6. Chu, F. S. In Mycotoxins and Phycotoxins; Steyn, P. S.; Vleggaar, R., Eds.; Elsesvier: Amsterdam, 1986, p. 277.
7. Chu, F. S. Proc. International Workshop on Aflatoxin Contamination of Groundnut, 6-9 Oct.; 1987 at ICRISAT Center, Hyderabad, India, 1988.
8. Chu, F. S. Vet. and Human Toxicol. 1989, (In press).
9. Fremy, J. M.; Chu, F. S. In Mycotoxin in Dairy Products; van Egmond, H. P. Ed.; Elsevier Sci.: London, 1989, p.97.
10. Kukal, L.; Kas, J. Trends in Anal. Chem. 1989, 8, 112-116.
11. Martlbauer, E.; Terplan, G. Proc. 2nd World Congree of Foodborne Infections & Intoxications, 1986, 2, 920-923.
12. Morgan, M. R. A. J. Assoc. Pub. Anal. 1985, 23, 59-63.
13. Pestka, J. J. J. Assoc. Off. Anal. Chem. 1988, 71, 1075-1081.
14. Chu, F. S.; Lau, H. P.; Fan, T. S.; Zhang, G. S. J. Immunol. Methods 1982, 55, 73-78.
15. Zhang, G. S.; Schubring, S. L.; Chu, F. S. Appl. Environ. Microbiol. 1986, 51, 132-137.
16. Chu, F. S.; Lee, R. C.; Trucksess, Mary W.; Park, D. L. J. Assoc. Off. Anal. Chem. 1988, 71, 126-129.
17. Garner, C.; Ryder, R.; Montesano, R. Cancer Research 1985, 45, 922-928.
18. Woychik, N.A.; Hinsdill, R.D.; Chu, F. S. Appl. Environ. Microbiol. 1984, 48, 1096-1099.
19. Blankford, M.B.; Doerr, J. A. Hybridoma 1986, 5, 57-.
20. Candlish, A. A. G.; Stimson, W. H.; Smith, J. E. Lett Appl. Microbiol. 1985, 1, 57-61.
21. Candlish, A. A. G.; Stimson, W. H.; Smith, J. E. Food Microbiol. 1987, 4, 147-153.
22. Kaveri, S.V.; Fremy, J.M.; Lapeyre, C.; Strosberg, A.D. Lett. Appl. Microbiol. 1987, 4, 71-75.
23. Dixon-Holland, D. E.; Pestkan, J. J.; Bridigare, B. A.; Casale, W. L.; Warner, R. L.; Ram, B. P.; and Hart, L. P. J. Food Prot. 1988, 51, 201-204.
24. Hastings K. L.; Tulis, J.J.; Dean J. H. J. Agric. Food. Chem. 1988, 36, 404-408.

25. Kawamura, O.; Nagayama, S.; Sato, S.; Ohtani, K.; Ueno, I.; Ueno, Y. Mycotoxin Res. 1988, 4, 75-27.
26. Chu, F. S.; Steinert. B. W.; Guar, P. K. J. Food Safety 1985, 7, 161-170.
27. Zhang, G. S.; Chu, F. S. Experientia 1989, 45, 182-184.
28. Fan, T. S.; Zhang, G. S.; Chu, F. S. Appl. Environ. Microbiol. 1984, 47, 526-532.
29. Hertzog, P. J.; Lindsay Smith, J. R.; Garner, R. C. Carcinogenesis 1982, 38, 825-828.
30. Groopman, J. D.; Trudel, L. J.; Donahue, P. R. Marshak-Rothstein, A.; Wogan, G. N. Proc. Natl. Acad. Sci. (USA) 1984, 81, 7728-7731.
31. Groopman, J. D.; Kemsler, T. W. Pharmacol and Therapy 1987, 34, 321-334.
32. Garner, R. C.; Dvorackova, I.; Tursi, F. Int. Arch. Occup. Environ. Health. 1988, 60, 145-150.
33. Candlish, A.A.G.; Stimson, W. H.; Smith, J. E. Lett. Appl. Microbiol. 1986, 3, 9-11.
34. Candlish, A. A.; Stimson, W. H.; Smith, J. E. J. J. Assoc. Off. Anal. Chem. 1988, 71, 961-964.
35. Morgan, M. R. A.; McNerney, R.; Chan, H. W. S.; Anderson, P. H. J. Sci. Food Agric. 1986, 37, 475-480.
36. Rousseau, D. M.; Candlish, A.G.; Slegers, G. A.; Van Peteghem, C. H.; Stimson, W. H.; Smith, J. E. Appl. Environ. Microbiol. 1987, 53, 514-518.
37. Mehl, M.; Strke, R.; Jacobi, H. D. Schleinitz, K.D.; Wasicki, P. Pharmazie 1986, 41, 147-148.
38. Wei, R. D.; Chu, F. S. J. Food Protect. 1988, 51, 463-466.
39. Neucere, J. N. Mycopathologia 1986, 93, 39-43.
40. Morgan, M. R. A.; Kang, A. S.; Chan, H. W. S. J. Sci. Food Agric. 1986, 37, 837-880.
41. Chu, F. S.; Liang, M. Y.; Zhang, G. S. Appl. Environ. Microbiol. 1984, 48, 777-780.
42. Pauly, J. U.; Bitter-Suermann, D.; Dose, K. Biol. Chem. Hoppe-Seyler 1988, 369, 487-492.
43. Hack, R.; Klaffer, U.; Terplan, G. Lett. Appl. Microbiol. 1989, 8, 71-75.
44. Casale, W. L.; Pestka, J. J.; and Hart, P. J. Agric. Food Chem. 1988, 36, 663-668.
45. Zhang, G. S.; Li, S. W.; Chu, F. S. J. Food Prot. 1986, 49, 336-339.
46. Kemp, H. A.; Mills, E. N. C.; Morgan, M. R. A. J. Sci. Food Agric. 1986, 37, 888-894.
47. Hack, R.; Martlbauer, E.; Terplan, G. Appl. Environ. Microbiol. 1988, 54, 2328-2330.
48. Martlbauer, E.; Gareis, M.; Terplan, G. Appl. Environ. Microbiol. 1988, 54, 225-230.
49. Hunter, K. W.; Jr.; Brimfield, A. A.; Miller, M.; Finkelman, F.D.; and Chu, F. S. Appl. Environ. Microbiol. 1985, 49, 168-172.
50. Ohtahni, K.; Kawamura, O.; Kakii, H.; Chiba, J.; Ueno, Y. Proc. Jpn. Assoc. Mycotoxicol. 1985, 22, 31-32.
51. Gendloff, E. H.; Pestka, J. J.; Dixon, D. E.; and Hart, L. P. Phytopathology 1987, 77, 57-59.

52. Goodbrand, I. A.; Stimson, W. H.; and Smith, J. E. Lett. Appl. Microbiol. 1987, 5, 97-99.
53. Hack, R.; Martlbauer, E.; and Terplan, G. J. Vet. Med. 1987, B34, 538-544.
54. Chiba, J.; Kawamura, O.; Kajii, H.; Ohtani, K.; Nagayama, S.; Ueno, Y. Food Addti. Contaim. 1988, 5, 629-639.
55. Fan, T. S. L.; Zhang, G. S.; Chu, F. S. Appl. Environ. Microbiol. 1987, 53, 17-21.
56. Fan, T. S. L.; Xu, Y. C.; Chu, F. S. J. Assoc. Off. Anal. Chem. 1987, 70, 657-661.
57. Fan, T. S.; Zhang, G.S.; Chu, F. S. J. Food Prot. 1984, 47, 964-968.
58. Fan, T. S. L.; Schubring, S. L.; Wei, R. D.; Chu, F. S. Appl. Environ. Microbiol. 1988, 54, 2959-2963.
59. Wei, R. D.; Chu, F. S. Anal. Biochem. 1987, 160, 399-408.
60. Wei, R. D.; Bischoff, W. B.; Chu, F. S. J. Food Prot. 1986, 49, 267-271.
61. Wei, R. D.; Swanson, S.; Chu, F. S. Mycotoxin Res. 1988, 4, 15-19.
62. Liu, M. T.; Ram, B. P.; Hart, P.; Pestka, J. J. Appl. Environ. Microbiol. 1985, 50, 332-336,
63. Harder, W. O. and Chu, F. S. Experientia 1979, 35, 1104-1104.
64. Martlbauer, E.; Terplan, G. Arch. fur Lebensmittelhyg. 1985, 36, 53-55.
65. Lee, R. C., Wei, R. D. and Chu, F. S. J. Assoc. Off. Anal. Chem. 1989, 72, 345-348.
66. Wei, Ru-dong, Swanson, S.; Chu, F. S. Mycotoxin Res. 1988, 4, 15-19.
67. Thouvent, D.; Morfin, R. F. Appl. Environ. Microbiol. 1983, 45, 16-23.
68. Fukal, L.; Prosek, J.; Rauch, P.; Sova, Z.; Kas, J.; J. Radioanal. Nucl. Chem. Articles 1987, 109, 383-391.
69. Rauch, P.; Fukal, L.; Prosek, J.; Bresina, P.; Kas, J. J. Radioanal. Nucl. Chem. Lett. 1987, 117, 163-169.
70. Fukal, L. Melkarske Listy. 1988, 14, 196.36-196.38.
71. Fukal, L.; Reisnerova, H.; Rauch, P. Sci. des Aliments. 1988, 8, 397-403.
72. Ruprich, J; Veres, K; Schmiedova, D. Vet. Med. Praha. 1988, 33, 101-108.
73. Ruprich, J.; Piskac, A.; Mala, J. Vet. Med. Praha. 1988, 33, 165-173.
74. Qian, G.S.; Parvin, Y.; Yang, G. C. Anal. Chem. 1984, 56, 2079-2080.
75. Chu, F. S.; Fan, T. S. L.; Zhang, G. S.; Xu, Y. C.; Faust, S.; and McMahon, P. L. J. Assoc. Off. Anal. Chem. 1987, 70, 854-857.
76. Hu, W. J.; Woychik, N.; Chu, F.S. J. Food Prot. 1983, 47, 126-127.
77. Mortimer, D.N.; Gilbert, J.; Sheperd, M. J. J. Chromatogr. 1987, 407, 393-398.
78. Park, D. L.; Miller, B. M.; Nesheim, S.; Trucksess, M. W.; Vekich, A.; Bidigare, B.; McVey, J. L.; Brown, L. H. J. Assoc. Off. Anal. Chem. 1989, 72, 638-643.

79. Trucksess, M. W.; Stack, M. E.; Nesheim, S.; Pohland, A. E.; Park, D. L. J. Assoc. Off. Anal. Chem. 1989, 72, 957-962.
80. Scott, P. M. J. Assoc. Off. Anal. Chem. 1989, 72, 75-80.
81. Itoh, Y.; Nishimura, M.; Hifumi, E.; Uda, T.; Ohtani, K.; Kawamura, O.; Nagayama, S.; Ueno, Y. Proc. Jpn. Assoc. Mycotoxicol. 1986, 24, 25-29.
82. Sidwell, W. J.; Chan, H. W. S.; Morgan, M. R. A. Food & Agric. Immunol. 1989, 1, 111-118.
83. Groopman, J. D.; Donahue, K. F. J. Assoc. Off. Anal. Chem., 1988, 71, 861-867.
84. Candlis, A. A.; Haynes, C. A.; Stimson, W. H. Intl. J. Food Sci. Techn. 1988, 23, 479-485.
85. Trucksees, M.; Stack, M. E.; Nesheim, S.; Hansen, T. J.; Donahue, K. F. Abst. The 103rd AOAC Annual International Meeting, St. Louis, Mo. Sept. 25-28, 1989.
86. Warden, B. A.; Allam, K.; Sentissi, A.; Cecchini, D. J.; and Giese, R. W. Anal. Biochem. 1987, 162, 363-369.
87. Ligler, F. S.; Bredehorst, R.; Talebian, A; Shriver, L. C.; Hammer, C. F.; Sheridan, J. P.; Vogel, C.; Gaber, B. P. Anal. Biochem. 1987, 163, 369-375.
88. Williamson, M. L.; Atha, D. H.; Reeder, D. L.; Sundaram, P. V. Anal. Lett. 1989, 22, 803-816.
89. Dorner, J. W.; Cole, R. J. J. Assoc. Off. Anal. Chem. 1989, 72, 962-964.
90. Patey, A. L.; Sharman, M.; Wood, R.; Gilbert, J. J. Assoc. Off. Anal. Chem. 1989, 72, 965-969.
91. Sun, P.; Chu, F. S. J. Food Safety 1977, 1, 67-75.
92. Wu, S.; Yang, G.; Sun, T. Clin. J. Oncol. 1983, 5, 81-84
93. Carman, A.S.; Kuan, S. S.; Ware, G. M.; Francis, O. J.; Miller, K. V. Abst. The 103rd AOAC Annual International Meeting, St. Louis, Mo. Sept. 25-28, 1989.
94. Chu, F. S.; Lee, R. C. Food & Agric. Immunol. 1989, 1, 127-126.

RECEIVED August 30, 1990

Chapter 13

Aflatoxin Immunoassays for Peanut Grading

R. J. Cole, J. W. Dorner, and F. E. Dowell

National Peanut Research Laboratory, Agricultural Research Service, U.S. Department of Agriculture, 1011 Forrester Drive, SE, Dawson, GA 31742

Recent studies conducted to determine the accuracy, speed, and expense of Enzyme-Linked Immunosorbent Assay (ELISA) screening methods for aflatoxin are reviewed in light of proposals that such a method be implemented as part of the official grading system for farmers stock peanuts. The studies have shown several of the ELISA methods to be reliable with regard to screening accuracy. This combined with the speed of the analyses and the relatively low costs could make a practical change in the way incoming loads of farmers stock peanuts are evaluated for aflatoxin contamination.

The U.S. peanut industry has developed a system for aflatoxin management during the production and processing of peanuts. This system includes diversion of aflatoxin-suspect lots at the initial point of sale of farmers stock peanuts, analysis of shelled peanuts prior to sale, and removal of aflatoxin-contaminated kernels during the shelling, blanching, and manufacturing processes (1). Removal of contaminated kernels during these processes is achieved by screening out small kernels and removal of damaged kernels by electronic color sorting and hand-picking.

Sale of farmers stock peanuts starts at the grading point where a value is placed on the peanuts based on several grade factors. The official U.S. peanut grading system currently relies on an indirect visual method for detecting loads of peanuts suspected of being contaminated with aflatoxin and diverting them from edible channels to oil production. This procedure involves a visual examination of loose-shelled kernels from approximately 1800 g of farmers stock peanuts and of kernels from a 500 g subsample of pods for the presence of the aflatoxin-producing fungi. If any kernel shows evidence of *Aspergillus flavus* growth, the entire lot of approximately 4000 kg of farmers stock peanuts is diverted to oil stock without further evaluation. The refined oil is free of any aflatoxin, which resides primarily in the meal, and residual aflatoxin contained in the crude oil is degraded during the refining process. The meal cannot be used for feed purposes and must be destroyed, exported, or used for fertilizer purposes only.

The U.S. peanut industry is currently considering the feasibility of implementing a new approach for aflatoxin management at the grading point. This involves different sampling and detection methods for more effective aflatoxin control. This paper discusses the feasibility of using Enzyme-Linked Immunosorbent Assay (ELISA) screening methods as part of the grading of farmers stock peanuts. In order to implement use of a direct analytical method for detecting aflatoxin-contaminated lots at peanut grading points, the method should meet several criteria. First, it should be relatively inexpensive, because of limited available resources. Second, the method should be relatively accurate, particularly around the current FDA action level of 20 ppb. Third, it must be relatively simple to conduct. The peanut grading season lasts only 2-3 months; therefore, employees must be trained each year. In addition, employees may work 10-15 hours per day, which dictates that the analytical method be simple in order to minimize physical and mental fatigue. Finally, and, perhaps, most important criterion is speed, since any analytical method cannot impede the flow and sale of peanuts through the grading process, particularly during peak processing periods. Since the peanut harvest season can be very short and grading points are extremely variable in size (capacities for processing range from 20-200 samples daily), each test needs to be completed within five minutes.

Recently, the effectiveness of the official visual method was compared with a rapid immunoassay (EZ-Screen Quick Card [Environmental Diagnostics, Inc., Burlington, NC]) in determining the presence or absence of aflatoxin in farmers stock grade samples (2). The 152 samples used for comparison were official grade samples obtained for analysis after the grading inspectors had completed their inspection. The comparative analyses were conducted on common methanol-water (80:20;2 ml/g) extracts. The ELISA test was conducted according to recommendations of the suppliers. HPLC analyses were done according to the method of Dorner and Cole (3).

The results showed 41% of the grade samples determined to be contaminated by visual inspection contained less than 20 ppb aflatoxin when analyzed by both ELISA and HPLC methods; 18.7% of peanuts determined to be uncontaminated by visual inspection actually contained aflatoxin with a range of 26-2542 ppb. The results of ELISA and HPLC agreed in 98.6% of the composite sample analyses with the detection of 20 ppb or greater. However, the ELISA screening method failed to give positive tests 12 of 13 times when the aflatoxin content was between 20-43 ppb in the component samples. Samples were analyzed at the rate of one every two minutes when duplicate analyses of ten samples (20 analyses) were performed.

It was concluded that the direct ELISA method was considerably more effective (97.6% effective) than the visual method (51% effective) in identifying farmers stock grade samples that were contaminated with aflatoxin at levels >20 ppb.

This prompted a subsequent study comparing duplicate analyses of common extracts using the EZ-Screen Quick Card and Afla-10 Cup Test (International Diagnostics Systems Corp., St. Joseph, MI) with HPLC (4). However, in this case a large number of samples in the critical range of 0-50 ppb range were selected from an unrelated study by HPLC for the comparison. Both ELISA methods were performed according to manufacturers' instructions; the HPLC method was conduced according to the method presented previously. One hundred common extracts (methanol-water, 80:20; 2 ml/g) in the critical range of 0-50 ppb were analyzed in duplicate by all methods.

Each ELISA assay properly identified 95% of samples containing no detectable aflatoxin as negative and >97% of samples containing >10 ppb aflatoxin as positive. The EZ-Screen Quick Card, which had a 20 ppb detection threshold, identified as positive 32 of 34 samples in the 11-20 ppb range. This indicated that the card test might actually have a detection threshold closer to 10 ppb. Most of the errors associated with both assays occurred on samples containing <10 ppb aflatoxin. The cup and card tests identified 76 and 67% of samples, respectively, as negative in the range of 4-10 ppb.

The objective of an aflatoxin testing program is to identify positive loads of farmers stock peanuts and to divert these to oil stock or possible cleanup. Results from this study indicated that either the card test or the cup test would reliably (>95%) identify samples of peanuts containing >10 ppb of aflatoxins. Likewise, both methods were reliable (95%) in properly identifying samples that were negative for aflatoxin. Because of this degree of accuracy, both methods were deemed well-suited for use as rapid screening tools at peanut grading points.

The USDA/Federal Grain Inspection Service recently completed a thorough study which compared six commercially available aflatoxin test kits to the currently used Holaday-Velasco minicolumn method for determining aflatoxin in corn (5). The study included five ELISA test kits. These were the Afla-20 Cup, Aflatest (VICAM, Somerville, MA), Agriscreen (Neogen, East Lansing, MI), EZ-Screen Quick Card and Oxoid (not an ELISA test). Criteria used for evaluation included accuracy, safety, user performance, speed, and variable and fixed costs. The objective of the FGIS study was to determine if a single alternate screening test could be used to replace the two currently used screening tests, the blacklight test and Holaday-Velasco minicolumn method (HV). The study was designed to compare the effectiveness of the alternate tests with the HV method. Three sample sets of corn containing spiked, naturally contaminated and negative samples were used in this study. The results showed that, with one exception (Agriscreen test), all tests evaluated were capable of providing the same reliability as the HV method. The performance of the Afla-20 Cup Test was the only test rated as better than the HV minicolumn. The Afla-20 Cup also scored highest on user preference where each analyst ranked the tests according to ease of use (Table 1). The average length of time required per assay, excluding sample preparation and extraction, was 5.71 min compared to 7.42 min for EZ-Screen, 9.09 min for AgriScreen and 10.33 min for Aflatest (Table 2). The Afla-Cup Test had the lowest variable costs at $4.68, while the EZ-Screen had the lowest fixed costs/site at $121.36 (Table 3).

Table I.   Ranking based on user preference scores

Ranking	Test
First	Afla-Cup
Second	EZ-Screen
Third	Aflatest
Fourth	Agriscreen
Fifth	Oxoid

Taken from FGIS study

Table 2. Average length of time required per assay

Test	Average time per test (min)
Afla-Cup	5.71
EZ-Screen	7.42
Agriscreen	9.09
Aflatest	10.33
Oxoid	15.27

Taken from FGIS study

Table 3. Variable and fixed costs for each test.

Test	Variable costs/test	Fixed costs/site
Afla-Cup	$4.68	$184.50
EZ-Screen	6.02	121.36
Aflatest	6.05	2,681.56
Agriscreen	7.00	495.00
Oxoid	18.04	901.36

Taken from FGIS study

The results of this study further substantiated previous studies that the new ELISA screening tests provide an excellent opportunity to implement a direct analytical method at peanut grading points.

Literature Cited

1. Cole, R. J. In *Mycotoxins and Phycotoxins '88*; Natori, S.; Hashimoto, K.; Ueno, Y., Eds.; Elsevier Sci. Pub., Amsterdam, The Netherlands, 1989; pp. 177-184.
2. Cole, R. J.; Dorner, J. W.; Kirksey, J. W.; Dowell, F. E. *Peanut Sci.* 1989, 15, 61-63.
3. Dorner, J. W.; Cole, R. J. *J. Assoc. Off. Anal. Chem.* 1988, 71, 43-47.
4. Dorner, J. W.; Cole, R. J. *J. Assoc. Off. Anal. Chem.* 1989, 72, 962-964.
5. Koeltzow, D. E.; Tanner, S. N. *Comparative Evaluation of Commercially Available Aflatoxin Test Methods*, U. S. Department of Agriculture, Federal Grain Inspection Service, Quality Assurance and Research Division, Research and Development Branch, 1989.

RECEIVED August 30, 1990

Chapter 14

Field Evaluation of Immunoassays for Aflatoxin Contamination in Agricultural Commodities

Douglas L. Park, Henry Njapau, Sam M. Rua, Jr., and Karen V. Jorgensen

Department of Nutrition and Food Science, University of Arizona, Tucson, AZ 85721

Four commercially available aflatoxin test kits based on immunochemical technology were used to monitor aflatoxin contamination in whole cottonseed under non-laboratory conditions. The kits, utilizing monoclonal antibodies to measure aflatoxin levels, can be divided into two categories: enzyme-linked immunosorbent assay (ELISA) and affinity columns. Assays were performed at cotton gins, dairy farms, feed mills, ammoniation plants and cotton oil processing plants under varying environmental conditions. Results from these locations were compared to those obtained in the laboratory using the same kits and thin layer chromatography (TLC). Preliminary results show good agreement between on-site and laboratory kit performance. Agreement between kits was 92% on-site and 83% in the laboratory. These results indicate that the kits are suitable for use under non-laboratory conditions. Aflatoxin stability in aqueous methanol extracts was determined using the immunochemical kits and TLC. Aflatoxin concentration decreased by almost 40% after 4 days.

Natural toxicants occurring in human foods and animal feeds present a potential health hazard to man. The significance of the risk due to mycotoxin contamination is dependent on the toxicological properties of the compound (acute, sub-acute or chronic toxicity, reproductive, mutagenic or teratogenic) as well as the extent of human exposure (occurrence, incidence and level of contamination). Aflatoxins, characterized as B_1, B_2, G_1, and G_2, are potent hepatocarcinogens and toxins. Their metabolites and reaction products, including aflatoxin M_1, a metabolite of aflatoxin B_1 found in milk of dairy cows exposed to aflatoxin-contaminated feed, have also demonstrated toxic potentials (1). Prevention of exposure to such toxic agents, which is dependent on rapid screening of food and feedstuffs, is more desirable than undertaking curative measures after toxicity has occurred. Such prophylactic measures are often in the form of surveys or monitoring programs which involve assaying large quantities of samples.

On a global scale, surveys and monitoring programs have had limited effectiveness due to various constraints, among them: the need for a laboratory set-up (often only one facility exists in a developing economy), transportation and methodology requiring expensive equipment. Thin layer (TLC) and liquid chromatographic (e.g. high performance liquid chromatography (HPLC)) methods have been developed, validated and are still used in a variety of situations (2,3). These methods are unsuited for rapid screening because they involve expensive and sophisticated equipment, extensive sample cleanup limiting the number

0097–6156/91/0451–0162$06.00/0
© 1991 American Chemical Society

of assays that can be performed and the need for a laboratory set-up. A number of immunochemical assays using aflatoxin-specific polyclonal antisera have been developed (4,5). The advantages offered by these assays include high specificity and sensitivity, relatively inexpensive equipment, and the capability to routinely screen large numbers of samples due to a reduction in assay time. The potential of immunochemical methods, particularly the enzyme-linked immunosorbent assay (ELISA) and affinity columns to rapidly assay for low levels of aflatoxin in milk, corn, cottonseed, peanuts and peanut products has been evaluated. In these studies their performance has been comparable to, and often more sensitive than, TLC (6), HPLC (7), and radioimmunoassay (8). Commercial production of stable, solid-phase-coupled, aflatoxin antibody kits (Aflatest-10, Vicam Corp.; Afla-20 Cup, International Diagnostics Systems Corp.; Agri-Chek, IDEXX Corp.; and Agri-Screen, Neogen Corp.) has greatly increased the potential of immunoassays as a rapid screening tool. The performance of some of these kits (Afla-20 Cup, Aflatest-10 and Agri-Screen) under laboratory conditions has been demonstrated in inter-laboratory validation studies for corn, peanuts, and cottonseed (9-11). The United States Department of Agriculture (USDA) has adopted the use of these techniques for screening aflatoxin in corn and peanuts (12). Studies evaluating the performance of these immunological based methods under non-laboratory conditions, which would enhance the efficiency of surveillance programs, have not been conducted.

This study focused on the applicability of the commercially available immunochemical test kits to adequately screen for aflatoxin in cottonseed under non-laboratory conditions by comparing their performance to laboratory results. On a global scale, TLC and immunochemical methods would have highest potential use. TLC was used to confirm the presence and identity of aflatoxins in the test samples.

Materials and Methods

Sample Collection and Preparation. Cottonseed samples of unknown aflatoxin contamination levels were collected from cotton gins, ammoniation plants (ammonia is used to decontaminate aflatoxin-contaminated seed), a dairy farm, cottonseed oil processing plants, and feed mills in the areas of Stanfield, Casa Grande and Tucson, Arizona. At each location six to ten three-pound samples of whole cottonseed were collected using a 3-inch vacuum probe (Probe-A-Vac) (Park, D.L. J. Assoc. Off. Anal. Chem., in press). Approximately 1 Kg of each sample was dehulled in a ultracentrifugal mill (Retsch, Brinkmann, ZM 1, Westbury, NY) with the grinding screen removed, followed by fine grinding in the same mill with a 2.0 mm grinding screen installed such that the ground material passed a No. 18 sieve. The entire ground sample was mixed in a twin shell blender for 20 minutes prior to taking a test portion.

The test portion (50g) was extracted with 250 mL methanol:water (80+20) for 1.0 minute at high speed in a Waring blender. The mixture was gravity settled for 15 minutes and filtered through Whatman No. 4 filter paper. This common filtrate was used for all immunochemical and TLC analyses.

Aflatoxin Analysis. Immunoassays were performed at the various locations where samples were collected and repeated approximately 24 hours later in the laboratory (Food Toxicology Research Laboratory, Department of Nutrition and Food Science, University of Arizona). The four commercially available (US market) test kits, listed below, were used:

    1. Aflatest-10 - Vicam Corp., Somerville, MA 02145.
    2. Afla-20 Cup - International Diagnostic Systems Corp.
          St. Joseph, MI 49085
    3. Agri-Chek - Idexx Corp. Portland, ME 04101.
    4. Agri-Screen - Neogen Corp. Lansing MI 48912.

Each test kit is available with reagents and accessories sufficient for analysis. Upon receipt, the Afla-20 Cup, Agri-Chek and Agri-Screen test materials were kept refrigerated (4° C) until use. The Aflatest-10 kit does not require refrigeration. The assays were conducted according to manufacturers' specifications (as summarized in Table I) with the relative concentrations of methanol in the extract adjusted accordingly. Assay readings were classified as negative or positive depending on whether they were below or above 20 ppb, respectively. The Agri-Screen test kit used in this study was Version I. Version II, currently available commercially, has improved sensitivity and specificity and is currently under study.

Table I. Comparison of Procedural Steps of Immunochemical Test Kits for Aflatoxin

	Immunochemical Kit			
	Aflatest-10 (Affinity Column)	Afla-20 Cup (ELISA)	Agri-Screen (ELISA)	Agri-Chek (ELISA)
MeOH:H$_2$O Ratio	70+30	80+20	55+45	70+29[a]
Extract Amount Used in Analysis	20 mL	100 μL	100 μL	100 μL
Incubation Time (minutes)	NA[b]	1.0	5.0	30.0
Washing	2 x 10 mL H$_2$O	1.5 mL reagent[c]	10 x 0.5 mL H$_2$O	5 x 0.5 mL H$_2$O
Color Development Time	15-30 secs.[d]	1.0 min.	5.0 min.	10 min.
Total # of Analytical steps	6	5	10	10
Assay Time (minimum)	12 min.	5 min.[e]	15 min.[f]	45 min.[f]
Nature of Test	semi-quantitative	positive/negative	semi-quantitative	semi-quantitative

[a]Methanol:water:dimethyl formamide (70+29+1)
[b]NA = not applicable
[c]Washing reagent supplied with kit
[d]Reaction time with dilute bromine developing solution
[e]Simultaneous analysis of up to three samples possible
[f]Simultaneous analysis of up to 10 samples possible (with multichannel pipette)

TLC analysis was performed in the laboratory approximately 24 hours after extraction in the field due to long distances from on-site locations (25-100 miles). The assays were based on adaptation of the method of Thean et al., (13) and AOAC sections 26.031 and 26.083. Fifty (50) mL from the common extract was mixed with 50 mL of 0.6M (NH$_4$)$_2$SO$_4$ and 5g of Celite 545, then stirred for 2 minutes and filtered through fluted filter paper. The filtrate (100mL) was placed in a 250 mL separatory funnel and partitioned twice into 5 mL portions of chloroform. The solution was evaporated to dryness under nitrogen and reconstituted in 0.5 mL chloroform. After dilution with 0.5 mL hexane the material was applied to a 1.0 cm (i.d.) column (4 cm bed, 37-75 μm Porasil A silica gel). The column was washed sequentially with 2 mL hexane and 2 mL ether. Aflatoxin was eluted from the column with 5 mL chloroform:ethanol (95+5) and evaporated to dryness. The dried extract was dissolved in 200 μL of benzene:acetonitrile (98+2) and varying amounts of sample and aflatoxin standard were spotted on a pre-coated TLC plate (Kieselgel 60, E. Merck) and developed in chloroform:acetone (9+1). Quantities were estimated visually under long wave UV light (365 nm). Aflatoxin identity was confirmed by trifluoroacetic acid derivatization (AOAC 26.083).

In order to determine whether the aflatoxin content of the 80% methanol extracts remained stable over the 24 hour period between extraction and laboratory analysis, 3 extracts from naturally contaminated cottonseed were stored at 4° C for a period of 10 days. The extracts were analyzed on days 0, 1, 2, 3, 4, 5, 6, 7, and 10 by both TLC and immunochemical methods. Reductions in aflatoxin concentration were recorded as a percentage of that obtained from day zero (Figure 1).

Results and Discussion

Results in Table II demonstrate that the performance of all kits was comparable under both conditions. There was an apparent difference in the performance of the individual kits, with Agri-Screen showing the highest variation. This discrepancy may be attributed to the underlying principle of each test kit and the ability of the analyst to perceive changes in color development. The detection phase of the Aflatest-10 affinity column is based on the measurement of the fluorescence of a reaction product between aflatoxin and bromine. For this specific test, extract age and site conditions may account for the variation in readings since a fluorometer was used on-site and in the laboratory. The other three immunochemical kits are based on color development. For the Afla-20 Cup, blue coloration implied an aflatoxin concentration below 20 ppb. This distinct end point greatly simplified the reading of the results hence minimizing analyst error. The Agri-Chek and Agri-Screen kits also employ color development but in a graded fashion where a small variation in the amount of color produced was dependent on the amount of aflatoxin in the extract. This compromised visual estimation as it was sometimes difficult to determine whether a sample was equal to, less than, or slightly above the concentration of the standard it was being compared to. There were no apparent difference between the individual kits because comparable results were observed on-

TABLE II. Comparison of On-Site and Laboratory Results of Immunochemical Methods

Immunochemical Kit	Number of Analyses	Percent (%) Agreement[1]
Afla-20 Cup	72	85
Aflatest-10 (affinity column)	62	94
Agri-Screen (visual[2])	60	67
(reader[3])	46	70
Agri-Chek (visual[2])	51	96
(reader[3])	47	85

[1] Agreement between on-site and laboratory results
[2] Analytical result determined visually
[3] Analytical result determined with Titertek Multiscan III

site (92%) and in the laboratory (83%). This observation may be attributed to the fact that although antibody specificity of the kits varies, they all have high reactivity with aflatoxin B$_1$, the major contaminant in cottonseed. Antibodies used in these kits cross-react with other

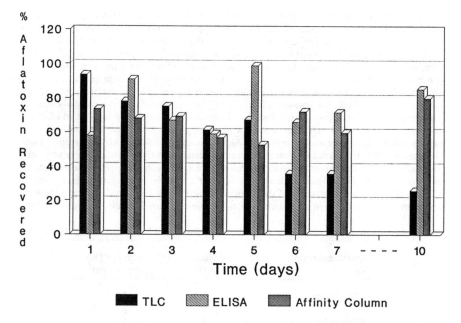

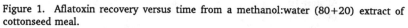

Figure 1. Aflatoxin recovery versus time from a methanol:water (80+20) extract of cottonseed meal.

aflatoxins to varying degrees; therefore, inter-kit variation may be different in commodities with higher levels of aflatoxins B_2, G_1, or G_2.

Ammoniated cottonseed samples were also analyzed in this study. Ammoniation is a process that has successfully reduced (>99%) aflatoxin levels in corn, peanut and cottonseed products (14). Arizona uses this process to treat aflatoxin-contaminated, whole cottonseed and cottonseed meal. This process had no apparent effect on the function of any of the immunochemical test kits.

Manufacturer's specifications for the kits included an operating temperature range of 23-29° C (room temperature). Temperatures at the testing sites varied from 26° C to 38° C; only the laboratory conformed to the recommended range. As the results in Table II indicate, these differences do not seem to affect kit performance to a great extent, as long as kits requiring refrigeration are kept at low temperature, i.e. in an ice chest, until an hour before assay time. Reports by other workers indicate that portions of immunochemical assay (enzyme-conjugate incubation) could be carried out at 37° C (15,16). An investigation into the effect of prolonged exposure to such high temperatures was outside the scope of this study. However, Koeltzow and Tanner (12) reported that the Afla-20 Cup, Aflatest-10 and Agri-Screen performed equally well at 18, 24 and 30° C.

Compared to TLC values, the kits seem sparingly accurate, i.e., 67% on-site and 68% in the laboratory (Table III). The laboratory performance of the kits used in this study has been evaluated by several investigators. Dorner and Cole (17) and Trucksess et al., (9) reported the satisfactory performance of the Afla-20 Cup test kit in assaying for aflatoxin in corn, cottonseed and peanuts. This kit has been adopted as official first action by the AOAC for corn (≥30 ppb) and cottonseed and peanut butter (≥20 ppb). Similarly the effectiveness of the affinity column test kit for monitoring total aflatoxin in corn, peanuts and peanut butter, has been validated through a collaborative study, and was found suitable for levels of 10 ppb or more (Trucksess, M.W. J. Assoc. Off. Anal. Chem., in press). This kit has also been designated interim first action by the AOAC. In other validation studies, Park et al., (10,11) evaluated the performance of the Agri-Screen test kit, which is also AOAC approved, and reported a 35.2 ± 15.9 mean when compared to a TLC value of 36 ppb in corn and peanuts. The difference between our results and those reported by other investigators (preceding discussion) could be attributed to the design of each study. We tested cottonseed of unknown, naturally-incurred contamination levels whereas spiked (known concentration) samples were used in the other studies. In this report inherent errors and variations in readings in both TLC and immunoassays (Table IV) were ignored because TLC was not intended to provide the correct aflatoxin concentrations to which the immunoassays were to be compared. TLC was solely used to confirm the presence and identity of aflatoxin in the samples. The kits are therefore more accurate than is reflected in this report.

Because of long distances between the laboratory and on-site locations, laboratory analyses had to be performed a day later. We investigated the effect of extract storage on the values obtained in the laboratory. Our results show that aflatoxin content decreased with time, to about 60% after four days, in methanol:water (80+20) when stored at 4° C (Figure 1). Contrasting findings have been reported by other workers. Dorner and Cole (17) reported that aflatoxin extracted into methanol from peanuts remained stable for at least three months. On the other hand, Kiemeier and Meshaley (18) reported aflatoxin M_1 decreased by 11-25% at 5° C after three days and by as much as 80% after six days at 0° C. Although we did not specifically investigate the changes occurring to the aflatoxin molecule, it was interesting to note that the decrease observed using immunochemical methods was less than TLC (Figure 1). We postulate that there may be co-extractants that bind the aflatoxin molecule making it less available for TLC but still recognizable by antibodies.

Summary and Conclusions

Current emphasis in aflatoxin monitoring programs is on large-scale screening. Using these immunochemical methods on-site offers the advantage of speed and prompt "go/no go" decisions. The kits are versatile and easily transported to testing locations with accessories

Table III. Percent (%) Agreement of Immunochemical Results When Compared to TLC[1]

Testing Location	Aflatoxin Concentration Range (μg/kg)	Afla-20 Cup	Aflatest-10	Agri-Screen	Agri-Chek	Mean % Agreement (per site)
On-site	<20	67 (54)[2]	77 (44)	56 (48)	67 (39)	
	>20	78 (18)	56 (18)	78 (18)	59 (12)	
	Mean	70	71	62	65	67
Laboratory	<20	65 (54)	80 (64)	60 (48)[3]	72 (39)[3]	
	>20	78 (18)	50 (18)	67 (12)[3]	59 (12)[3]	
	Mean	68	73	61	69	64
Overall Kit Agreement		69	72	62	67	

[1] Modified Thean et al. (1981) and AOAC CB method (26.031).
[2] Number of analyses in parentheses.
[3] Based on visual estimates only.

Table IV. Percent (%) False Readings[1] by Immunochemical or TLC Methods on Analyses Performed in the Laboratory[2]

	False Positives	False Negatives
Afla-20 Cup	8.7	4.3
Aflatest-10	0	0
Agri-Screen Visual[3]	0	8.8
Reader[4]	34.8	0
Agri-Chek Visual[3]	0	0
Reader[4]	17.4	0
TLC	4.3	21.7

[1] Based on one method (kit or TLC) in disagreement with the rest.
[2] Comparisons based on 23 samples.
[3] Analytical result determined visually.
[4] Analytical result determined using Titertek Multiscan III.

and solvents necessary for analyses easily fitting the trunk of any vehicle. With these immunoassays, the longest technique takes about one hour to run 10 samples (excluding sample preparation and extraction). The same number of samples would require 6-8 hours by TLC. The cost in terms of analyst wages is therefore greatly reduced. The preliminary results from this study demonstrate that the immunoassay kits can be satisfactorily used in varying non-laboratory environments for semi-quantitative screening of cottonseed for aflatoxins, provided extraction and analysis are done on the same day. For strict enforcement purposes, official quantitative methods should be used to confirm positive samples.

Acknowledgments

This project was partially funded by the National Cottonseed Products Association, Office of the State Chemist (Arizona), Arizona Cotton Research and Protection Council, Food and Drug Administration (1-RO1-FD01461-01) and the following commercial organizations in Arizona: Walker's Cottonseed (Stanfield), Arizona Grain Inc. (Casa Grande), Casa Grande Oil Mill (Casa Grande), Producers Cotton Oil Co. (Phoenix), Valley Industries (Peoria), Western Cotton Services (Phoenix) and Wilber-Ellis Co. (Chandler).

Literature Cited

1. Lee, L.S., Dunn, J.J., Delucca, A.J. and Ciegler, A. Experientia 1981, 37, 16-17.
2. Park, D.L. and Pohland, A.E. In Foodborne Microorganisms and their Toxins, Developing Methodology, Pearson, M.D. and Stern, N.J. Eds.; Marcel Dekker New York, 1986; 425-438.
3. Sheppard, M.J. In Modern Methods in the Analysis and Structural Elucidation of Mycotoxins. Cole, R.J. Ed.; Academic, New York, 1986; 294-334.
4. Chu, F.S. In Modern Methods in the Analysis and Structural Elucidation of Mycotoxins. Cole, R.J. Ed.; Academic, New York, 1986; 207-239.
5. Ram, B.P., Hart, L.P., Shortwell, O.L. and Pestka, J.J. J. Assoc. Off.Anal. Chem. 1986 69, 904-907.
6. Kawamura, O., Kajii, H., Nagayama, S., Ohtani, K. Chiba, J. and Ueno, Y. Toxicon 1989, 27 887-897.
7. Warner, R., Ram, B.P., Hart, P.L. and Pestka, J.J. J. Agric. Food Chem. 1986, 34 714-717.
8. Lee, R.C., Wei, R-D. and Chu, F.S. J. Assoc.Off. Anal. Chem. 1989, 72, 345-348.
9. Trucksess, M.W., Stack, M.E., Neshiem, S., Park, D.L. and Pohland, A.E. J. Assoc.Off. Anal. Chem. 1989, 72, 957-964.
10. Park, D.L., Miller, B.M., Hart, P.L., Yang, G., McVey, J., Page, S.W., Pestka, J.J. and Brown, L.S. J. Assoc. Off. Anal. Chem. 1989, 72, 326-332.
11. Park, D.L., Miller, B.M., Nesheim, S., Trucksess, M.W., Vekich, A., Bidigare,   B., McVey, J. and Brown, L.H. J. Assoc. Off. Anal. Chem. 1989, 72, 638-643.
12. Koeltzow, D.E. and Tanner, S.N. Comparative Evaluation of Commercially Available Test Methods. USDA Grain Inspection Service. 1989.
13. Thean, J.J., Lorenz, D.R., Wilson, D.M., Rodgers, K. and Gueldner, R. J. Assoc. Off. Anal. Chem. 1980, 63, 631-634.
14. Park, D.L., Lee, L.S., Price, R.L. and Pohland, A.E. J. Assoc. Off. Anal. Chem. 1988, 71, 685-703.
15. El-Nakib, O., Pestka, J.J. and Chu, F.S. J. Assoc. Off. Anal. Chem. 1981, 64, 1077-1082.
16. Sing, P. and Jang, L. Intl. J. Food Microbiol. 1987, 5, 73-80.
17. Dorner J.W. and Cole, R.J. J. Assoc. Off. Anal. Chem. 1989, 72, 962-964.
18. Kiermeier, F. and Meshaley, R. Z. lebensm. unters Forsch. 1977, 164, 183-187.

RECEIVED August 30, 1990

Chapter 15

Immunoassay for Detection of Zearalenone in Agricultural Commodities

Glenn A. Bennett

Agricultural Research Service, U.S. Department of Agriculture, Northern Regional Research Center, 1815 North University Street, Peoria, IL 61604

A direct competitive enzyme-linked immunosorbent assay for the detection of zearalenone in corn, wheat, and feed has been evaluated by 23 collaborators in an international collaborative study. Both visual and spectrophotometric determinations of zearalenone were done on blind duplicates of spiked and naturally contaminated samples of each commodity. Frequency of false negatives was 25% at the target level of 500 ng zearalenone/g commodity, but was only 3.4% at 800 ng zearalenone. No noticeable matrix effect was observed for the samples tested. Coefficients of variation for repeatability were 11.6% and 11.7% for spiked and naturally contaminated samples, respectively. Coefficients of reproducibility were 25.1% and 33.1% for spiked and naturally contaminated samples, respectively. This study demonstrates the reliability of the immunoassay procedure as a screening method for zearalenone at $\geq$800 ng/g in corn, wheat, and feed.

Trans-zearalenone, also know as F-2 toxin, is a non-steroidal estrogenic compound produced by several Fusarium species which colonize a number of agricultural commodities. Zearalenone has been implicated in numerous cases of mycotoxicoses in farm animals, especially pigs. The toxin causes "estrogenic syndrome" in pigs and is an economically important mycotoxin. Although corn is the most often contaminated grain in the United States, recent surveys demonstrate that zearalenone is found in other commodities world-wide (1). Analytical methodology for detection of zearalenone include TLC, GC, HPLC, and GC-MS (2-5). These procedures are sensitive but require extensive sample cleanup or require expensive instrumentation for detection and quantitation. The development of

enzyme-linked immunosorbent assays (ELISA) as an alternative
procedure, promises to reduce both assay expense and time. This
paper describes the evaluation of an ELISA procedure tested in a
collaborative study to screen for the presence of zearalenone in
corn, wheat, and animal feed.

Description of Study

Twenty-three collaborators received 18 coded samples, which
included six sets of blind duplicates of spiked and naturally
contaminated corn, wheat, and feed (Table I). All collaborators

Table I. Commodity and Zearalenone Levels Used for
Enzyme-linked Immunosorbent Assay

	Zearalenone (ng/g)					
Commodity	Spiked			Naturally Contaminated[1]		
Corn	0	800	800	247	2570	2570
Wheat	300	300	1000	215	1027	1027
Feed	0	500	500	352	352	1295

[1]Levels of zearalenone determined by liquid chromatography (AOAC Method 26:A09-26:A16) (8).

were also supplied with ELISA Test Kit components (available
commercially from Neogen Corp., Lansing, MI 48912. Mention of
companies or products by name does not imply their endorsement by
the U.S. Department of Agriculture over others not cited.).
Zearalenone is extracted with methanol-water (70 + 30) by
shaking ground, blended samples (20 g) on a wrist-action shaker for
3 min. A portion of the filtered extract is mixed with an equal
volume of zearalenone-enzyme conjugate and this mixture allowed to
react competitively for receptor sites of zearalenone-specific,
monoclonal antibodies coated on the surface of microtiter wells
(6). After a 15-min incubation time, the wells are washed (10X) to
remove unbound zearalenone and zearalenone-enzyme conjugate. A
pre-set volume of activated enzyme substrate is added to each well
and allowed to incubate for an additional 15 min. Sample extracts
containing less than 500 ng/g zearalenone will show development of
a blue color, darker than the control, which results from the
reaction between activated enzyme substrate and zearalenone-enzyme
conjugate bound to antibody receptor sites. Sample extracts
containing 500 ng/g or more zearalenone result in little or no
color development. Reduced color development is due to the binding
of free zearalenone in the sample extract to antibody receptor
sites, thus reducing the number of sites available for zearalenone-
enzyme conjugate binding. After the second 15-min incubation time,
a colored enzyme-stopping reagent is added. Color development in
standard and sample wells is determined visually or spectrophotometrically. In the visual method, the level of zearalenone in
samples is judged to be less than (-), equal to (=), or greater
than (+) 500 ng/g, by comparing color development in samples to
color development in controls which contain 500 ng/g zearalenone.
In the spectrophotometric method, approximate levels of zearalenone

are interpolated from a standard curve constructed from spiked
extracts of each commodity.

Results and Discussion

Visual Method. Results from the visual determination of
zearalenone are presented in Table II. At 300 ng/g, only 63%

Table II. Results of Visual Method for Immunochemical
Detection of Zearalenone in Blind Duplicate Samples

Commodity[1]		Zearalenone (ng/g)	No. Assays	% Correct	False Pos.	False Neg.
Corn	A	800	51	96	--	1
	B	2570	52	100	--	--
Wheat	A	300	54	63	10	--
	B	1027	54	100	--	0
Feed	A	500	52	75	--	8
	B	352	52	83	3	--

[1]A = Spiked samples; B = Naturally contaminated samples.

(34/54) of assays were correct and 10 false positives were
reported. At 500 ng/g, 75% (39/52) of assays were correct,
however, 8 false negatives were reported. When the level of
zearalenone was 800 ng/g, only one sample (3.4%) was
misidentified. All samples containing >800 ng/g were correctly
identified. Outliers were identified as samples which gave
conflicting results on duplicate assays and were not included in
determining number of false negative or false positive results.
The number of incorrect answers, especially false negatives,
affects the usefulness of this screening procedure. Figure 1 shows
the rate of positive answers at the different levels of zearalenone
in spiked samples. As the level increases from 500 to 800 ng/g,
the percent correct results increase from 75% to 96%. These data
show that the level of zearalenone must be at least 800 ng/g in
order to achieve a 95% confidence level for the assay.

Spectrophotometric Method. The accuracy of the screening procedure
depends on the construction of standard curves and interpolation of
levels in samples from these curves. Table III lists the average

Table III. Summary of Percent Absorbance Values of Spiked
Extracts from Extracts from Corn, Wheat, and Feed Used to
Construct Standard Curves Used to Estimate Levels
in Test Samples

Commodity	Zearalenone (ng/g)					
	0	200	500	1000	1500	3000
Corn	100	82.9+6.1	66.4+7.1	44.3+7.6	38.9+7.3	28.4+6.4
Wheat	100	82.4+7.9	63.9+5.9	48.0+5.4	39.5+5.2	26.4+4.8
Feed	100	82.9+6.6	66.7+6.7	52.0+6.4	42.1+5.6	29.6+4.2

absorbance values reported for 48 standard curves constructed by eight collaborators. Curves were constructed for each set of samples assayed and each set of samples was assayed in duplicate. The average value of duplicate assays was reported as the value for that sample. This technique, although more time-consuming than the visual method, provides quantitative data and can also be used to estimate the rate of microtiter well or test component failure. A typical standard curve is shown in Figure 2 and represents a "best fit" line through the data points for 20-300 ng/ml zearalenone (which corresponds to 200 to 3000 ng/g). Assay results from the spectrophotometric method are shown in Table IV. Very similar

Table IV. Results of Immunochemical Determination Spectrophotometric Method of Zearalenone in Blind Duplicate Samples of Corn, Wheat, and Feed

Commodity[1]		Zearalenone (ng/g)	Amount Found (ng/g)[2]	False Pos.	False Pos.
Corn	A	800	940	--	0
	B	2570	3191	--	0
Wheat	A	300	452	5	--
	B	1027	1482	--	0
Feed	A	500	613	--	4
	B	352	255	1	--

[1]A = Spiked samples; B = Naturally contaminated samples.

[2]Value is average value of duplicate assays reported by each collaborator.

rates of false positives and false negatives were reported as was reported for the visual method, however, no false negatives were reported at levels >500 ng/g. At the target level (500 ng/g), four false negatives were reported even though the method appears biased to give high values since >100% recovery was reported for all spiked samples. Analysis of variance for results from spiked and naturally contaminated samples were calculated and the relative standard deviations (repeatability) were almost identical for both type samples (11.6% and 11.7%). Relative standard deviations (reproducibility between labs) were 25.1% and 33.1% for spiked and naturally contaminated samples, respectively.

Summary and Conclusions

These data indicate that the enzyme-linked immunosorbent assay procedure tested can be used with confidence to screen for the presence of zearalenone in agricultural commodities at >800 ng/g. Both visual and spectrophotometric methods can be used to significantly reduce the number of samples that need to be assayed by more precise, time-consuming procedures by identifying contaminated (>800 ng/g) samples. Samples that are identified as positive by this screening procedure should be analyzed by an official method to determine exact levels of contamination. This

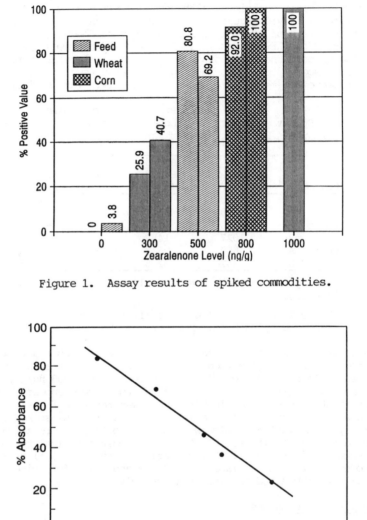

Figure 1.   Assay results of spiked commodities.

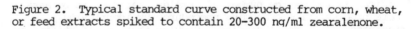

Figure 2.   Typical standard curve constructed from corn, wheat, or feed extracts spiked to contain 20-300 ng/ml zearalenone.

technique can be used to enhance surveillance of both animal and human foods for the mycotoxin zearalenone (7).

Literature Cited

1. Tanaka, T.; Hasegawa, A.; Yamamoto, S.; Lee-U-S.; Sugiura, Y.; Ueno, Y. J. Agric. Food Chem. 1988, 36, 979.
2. Bennett, G. A.; Shotwell, O. L. J. Am. Oil Chem. Soc. 1979, 56, 812.
3. Bata, A.; Vanyi, A.; Lasztity, R. J. Assoc. Off. Anal. Chem. 1983, 66, 577.
4. Bennett, G. A.; Shotwell, O. L.; Kwolek, W. F. J. Assoc. Off. Anal. Chem. 1985, 68, 958.
5. Plattner, R. D.; Bennett, G.A. J. Assoc. Off. Anal. Chem. 1983, 66, 1470.
6. Dixon, D. E.; Warner, R. L.; Ram, B. P.; Hart, L. P.; Pestka, J. J. J. Agr. Food Chem. 1989, 35, 122.
7. Pestka, J. J. J. Assoc. Off. Anal. Chem. 1988, 71, 1075.
8. Official Methods of Analysis (1984) 14th Ed., AOAC, Arlington, VA.

RECEIVED August 30, 1990

Chapter 16

A Pyrrolizidine Alkaloid Enzyme-Linked Immunosorbent Assay Detection Strategy

Mary A. Bober[1], Mark J. Kurth[1], Larry A. Milco[1], David M. Roseman[1], R. Bryan Miller[1], and Henry J. Segal[2]

[1]Department of Chemistry and [2]Department of Veterinary Medicine, Pharmacology, and Toxicology, University of California, Davis, CA 95616

A two-pronged *class-specific* and *compound-specific* immunochemical approach targeting naturally occurring macrocyclic pyrrolizidine alkaloids was developed. Antibodies have been developed against a retronamine-BSA conjugate and then used to detect the necine base retronecine, a common substructural feature of many naturally occurring macrocylcic pyrrolizidine alkaloids. In addition, three macrocyclic alkaloids isolated from the plant *Senecio vulgaris* (retrorsine, senecionine, and seneciphylliine) were detected using antibodies to retronamine. A comparison of antibodies produced against the retronamine conjugate in both rabbits and mice showed that the mouse antibodies possessed a slightly higher affinity (I_{50} = 1.1 ppm) for the retronecine analyte than antibodies raised in rabbits (I_{50} = 1.5 ppm). Antibodies produced to this retronamine conjugate may prove useful in assaying the toxic macrocyclic pyrrolizidine alkaloids in plant extracts. Antibodies were also developed to monocrotaline and can be used to detect quaternarized monocrotaline (I_{50} = 0.25 ppm at pH 7.6), N-methylated monocrotaline (I_{50} = 5.3 ppm at pH 7.6), and protonated monocrotaline (I_{50} = 6.0 ppm at pH 6.0). Antibodies to monocrotaline do not cross-react with retrorsine, retrorsine N-oxide, ridelliine or retronecine.

The pyrrolizidine alkaloids (PAs) constitute a class of secondary plant metabolites of notably wide geographical and botanical distribution, occurring in numerous plant families, (including Boraginaceae, Compositae, Gramineae, Leguminosae, Orchidaceae, Rhizophoraceae, Santalaceae, and Saptoaceae) which are indigenous to various environments throughout the world [1]. Indeed, the number of PA-producing plant species may approach 6,000, accounting for 3% of all known flowering plants [2] -- many of which produce more than one PA. Frequently these compounds co-exist in the plant with varying amounts of their corresponding N-oxide derivatives.

Most of the more than 200 known PAs are toxic to mammals [3,4], exhibiting a broad range of cytotoxic and pathological actions including hepatotoxic, pneumotoxic, embryotoxic, mutagenic, carcinogenic, and teratogenic effects [5,6,7]. Chronic gastrointestinal [8], cardiopulmonary [9], and central nervous system [10] disorders are further manifestations of PA poisoning. The toxicity plus the ubiquitous nature of these alkaloids makes them a world-wide health concern.

While non-competitive enzyme-linked immunosorbent assays (ELISAs) have been used to detect snake venom protein in whole blood in experimental [11] and clinical settings [12], *competitive* ELISAs have been recently developed against toxic alkaloids. The ability to detect ergotamine in spiked grain samples in concentrations of 10 ng/g [13] and tropane alkaloids as a class (with atropine in concentrations as small as 10 ng/mL) have been reported [14]. Thus the proven clinical, field, and research applications of ELISAs in both the detection and quantification of toxins prompted us to develop immunoassay techniques for the rapid, reliable screening of biological samples for PA contamination. As presented below, our strategy is to develop a two-pronged *class-specific* and *compound-specific* immunochemical approach to PA detection.

The suitability of the ELISA in PA detection encounters certain limitations which must be addressed and which have posed significant problems in the development of a broad spectrum immunoassay. Two important issues which must be considered in developing an immunoassay for the detection of PAs are: (i) selection of a hapten and linker arm that constitute a suitable immunogen, and (ii) use of an amplified immunoassay versus an indirect non-amplified system. While these issues are important in any immunoassay, we have found them to be critical in establishing a sensitive PA immunoassay.

Class-specific Immunoassay: In reviewing the structural features of the known macrocyclic PAs, one common similarity becomes obvious: retronecine, the necine base, is found in alkaloids from six botanical families and 26 genera [15]. Thus, one aspect of our approach to hapten design was based on the premise that antibodies produced in response to PA analogues possessing the retronecine moiety would deliver a class-specific immunoassay capable of detecting most macrocyclic PAs: for example, retrorsine, senecionine, and seneciphylliine are all PAs from the plant *Senecio vulgaris* which contain the retronecine substructural unit (**Figure 1**). In previous work from our laboratory [16], we coupled retronecine to bovine serum albumin (BSA) and produced retronecine antibodies that competed for retronecine and monocrotaline. More recently, we have developed anti-retronamine to a retronamine hapten, an antigen with improved *in vitro* (and presumably *in vivo*) stability, which also target the retronecine moiety. **Figure 2** presents a comparison in the ability of anti-retronecine and anti-retronamine to detect the analyte retronecine in a competitive ELISA.

Compound-specific Immunoassay: A second aspect of our PA detection strategy targeted developing a compound-specific immunoassay -- a strategy which required presenting the necic acid moiety as well as the necine base moiety to the immune system. Monocrotaline, which possesses the necine base retronecine and, by bis-esterification (i.e., macrolactonization), the necic acid monocrotalic acid, is present in numerous plants making it an ideal substrate for *compound-specific* ELISA development. Further, monocrotaline occurs in a sufficiently high concentration in some plants (over 9% by weight in the seeds of *Crotalaria retusa,* for example [17]) to make ELISA detection useful for some samples without the need for concentrating extracts before obtaining a sufficient mass of alkaloid to provide a working I_{50}, the alkaloid concentration at which 50% antibody inhibition is seen.

Materials and Methods

Antibodies were developed against three different hapten-BSA conjugates in rabbits: retronecine-BSA, retronamine-BSA, and monocrotaline-BSA. Antibodies to each of the hapten conjugates were assayed using the antibody detection ELISA. While each hapten was capable of stimulating an immune response, the ability of

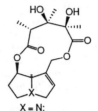

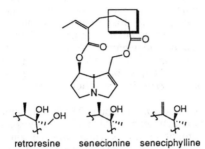

retroresine   senecionine   seneciphylline

X = N:
monocrotaline
X = N⁺CH₂CH=CHCO₂H
monocrotaline immunogen

R = H
retronecine
R = C(O)(CH₂)₂CO₂H
retronecine immunogen

R = H
retronamine
R = C(O)(CH₂)₄CO₂H
retronamine immunogen

Figure 1. Immunogen strategy based on the premise that antibodies produced in response to the retronecine moiety would deliver a class-specific immunoassay capable of detecting retrorsine, senecionine, seneciphylliine, and monocrotaline.

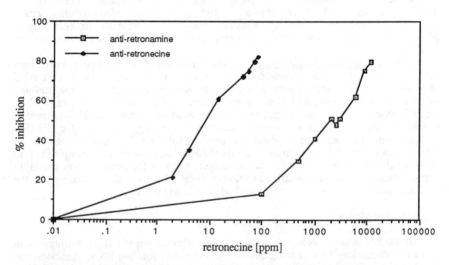

Figure 2. Rabbit antibodies to retronecine–BSA (1 μg/well retronecine–OVA coating antigen) and retronamine–BSA (10 μg/well retronamine–OVA coating antigen).

antibodies to detect retronecine and/or monocrotaline [Trans World Chemicals] in a competitive ELISA varied considerably.

The amplified assay system for retronecine was based upon the work of Laurent *et al.* [18] who were able to detect desmosine, a cross-linked amino acid present in minute quantities in rat connective tissue and urine using the avidin-biotin amplified ELISA. While it is possible to use a non-amplified indirect ELISA for antibody detection, such a strategy has the disadvantage of requiring increased substrate incubation time and generally results in lower overall absorbance values.   As discussed below, the amplified ELISA has also proven superior in assays using antisera developed against retronamine and monocrotaline haptens.

Retronamine and monocrotaline antibodies were used to detect retronecine and monocrotaline. In an attempt to improve on the relatively high I_{50} value of our anti-retronecine (1.68 ppt), we conjugated retronamine, a synthetic derivative of retronecine, to BSA by acylation of the retronamine primary amine with hexanedioic anhydride giving a hapten intermediate which was then coupled to BSA by water-soluble 1-ethyl-3-(3-dimethylaminopropyl)carbodiimide (EDAC). The tertiary nitrogen of monocrotaline was alkylated with the N-hydroxysuccinimide ester of 4-bromocrotonic acid giving a hapten intermediate which was then coupled to BSA via N-acylation of the lysine primary amines.

Microtiter plates were coated with 1 or 10 µg/well of hapten-chicken egg ovalbumin conjugates (hapten-OVA). The retronamine conjugate provided a more sensitive assay with a more concentrated coating solution (10 µg/well) than the retronecine conjugate, which was used at a 1 µg/well concentration. Both concentrations were tested with each assay. The concentration which provided the more sensitive assay was selected for later experiments. Standard concentrations of the hapten were assayed on each microtiter plate. The standards (retronecine and monocrotaline) were incubated with various dilutions of antisera for one hour at 37°C in culture tubes. Retrorsine [Aldrich], senecionine, and seneciphylliine were tested at a single concentration for cross-reactivity against the retronamine antiserum. A 100 µL aliquot of the standard solution was added to each well on the microtiter plate. Following a one hour incubation the plates were washed and biotinylated goat-antirabbit antibody [Sigma] was added. This step was followed by the addition of avidin labelled horseradish peroxidase. After the final washing the enzyme substrate *o*-phenylenediamine was added. The development of colored product was measured using a Molecular Devices UVmax microtiter plate reader.

Results and Discussion

Results from the assay using anti-retronecine are provided in our prior publication [16]. Under neutral conditions (pH 7.6) the I_{50} for retronecine detection using anti-retronamine was 1.5 ppm; a significant improvement when compared to our previous results using anti-retronecine (I_{50} = 1.68 ppt). For comparative purposes, it is useful to mention here that anti-retronecine was not as sensitive for detecting retronecine as the homologous system which uses anti-retronamine. In both cases, the antisera did not cross-react with monocrotaline. However, anti-monocrotaline was found to detect unconjugated monocrotaline to a far greater extent than retronecine. This would be expected since the necic acid portion of the molecule is most likely important to antibody-antigen interaction with anti-monocrotaline.

Anti-retronamine was used in a competitive ELISA for the detection of purified retronecine and for the detection of retronecine-containing PAs (i.e., retrorsine, senecionine, and seneciphylline) from *S. vulgaris*. Indeed, these studies have established that anti-retronamine is superior to anti-retronecine in detecting not only purified PAs, but also the mixture of PAs found in chloroform extracts of *S. vulgaris*.

A complete description of the hapten conjugation, antibody production, and assay development has been published [16]. It is worthwhile to reiterate that although antiserum to retronecine can detect the homologous hapten (retronecine) as well as a heterologous hapten (i.e., monocrotaline I_{50}'s of 1.68 vs. 1.21 µg/µL, respectively), we consistently find that the homologous system gives a significantly higher maximum inhibitory response than the heterologous system. While this retronecine assay indicates that the ELISA technique might prove useful in screening samples for the presence of the necine base, our initial results also indicated that assay sensitivity needed to be enhanced. Since the presentation of the retronecine moiety to the immune system is paramount to obtaining a sensitive ELISA, we investigated different sites for hapten-protein conjugation.

An important question considered in our immunochemical study was whether significant benefit would be derived from an amplified immunoassay. Indeed, previous work from our laboratory has shown that, when using retronecine as an immunizing hapten, an avidin-biotin amplified system is far superior in the competitive immunoassay [16]. This amplified ELISA, which takes advantage of the high binding affinity of avidin labelled horseradish-peroxidase for biotinylated second antibody (avidin/biotin $K_{D=10^{-15}}$), provides a useful means for detecting antigen-antibody interactions of both the homologous and the heterologous competitive assays.

As presented here (see **Figure 2**), rabbit anti-retronamine can detect retronecine with over a thousand-fold greater sensitivity than rabbit anti-retronecine. This is an intriguing result, especially in light of the structural similarity between the two haptens. While conclusive evidence is not available, the decreased sensitivity with anti-retronecine is consistent with partial hydrolysis of the retronecine-BSA ester bond and concomitant loss of the retronecine moiety -- an unwanted difficulty which is circumvented by the amide linkage of retronamine-BSA.

Given the sensitivity of rabbit anti-retronamine, we decided to immunize mice with a retronamine-BSA conjugate and were pleased to find the mice did produce polyclonal

Table I. Mouse Competitive ELISA Using Anti-retronamine

Retronecine	Anti-retronamine*
(µg/mL)	(O.D. at 490 nm)
0.0	1.84
0.5	0.66
1.0	0.27
10.0	0.19
15.0	0.12
20.0	0.08
Pyrrolizidine Alkaloids from	Anti-retronamine*
Senecio vulgaris (5-7 µg/mL)	(O.D. at 490 nm)
retrorsine	0.23
senecionine	0.01
seneciphylline	0.01

*Antibodies to retronamine were raise in mice against retronamine-BSA and then used in the competitive ELISA at a 1:1,335 dilution. Microtiter plates were coated with 1µg/well of retronamine-OVA and a standard curve of retronecine was assayed with each microtiter plate.

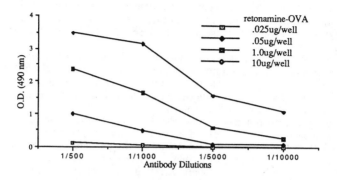

Figure 3. Dilutions of mouse antibodies to retronamine–BSA.

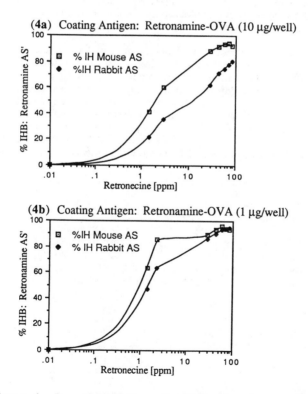

Figure 4. Percent inhibition of polyclonal antisera against retronamine versus retronecine (analyte). A 50% inhibition was observed at retronecine concentrations of 1 to 2 ppm.

antibodies (**Figure 3**) which could be detected at serum dilutions up to 1:10,000. In competition studies, we found that mouse anti-retronamine provides a slightly better I_{50} (1.1 µg/mL) than the corresponding rabbit anti-retronamine (1.5 µg/mL). In addition, mouse polyclonal anti-retronamine detects the macrocyclic PAs (senecionine, seneciphylline and retrorsine) isolated from *Senecio vulgaris* in a range of 5 - 7 µg/µL (**Table I**).

Figure 4 shows a comparison between rabbit and mouse antisera developed against the retronamine-OVA immunogen. In addition, **Figure 4** illustrates that assay sensitivity increases slightly when the coating antigen is decreased from 10 µg/well to 1 µg/well. These promising results suggested the possibility of raising monoclonal antibodies to the retronamine-BSA conjugate.

The antibody detection assay indicated that the monocrotaline-BSA immunogen had elicited a specific immune response in rabbits and that, at serum dilutions as high as 1:150,000, antibodies to monocrotaline could be detected. Moreover, these antibodies to monocrotaline are capable of detecting monocrotaline at a dilution of 1:50,000. These initial results were exciting since they demonstrated the first immunogenic response to a naturally occurring macrocyclic PA. The next steps undertaken were (i) screening available macrocyclic PAs (i.e., monocrotaline, ridelliine, retrorsine, and retrorsine N-oxide) in competitive assays, and (ii) assaying for the detection of retronecine in a competitive assay. Our results establish that anti-monocrotaline is specific for monocrotaline and does not compete with the other macrocyclic alkaloids tested. Antibodies to monocrotaline did not compete for retronecine.

Hapten design and selection of a hapten that provides the desired sensitivity and immunogenicity is critical to establishing a useful PA immunoassay. Since protonation of the tertiary amine of monocrotaline in neutral buffer appears important in antibody-antigen recognition, our next strategy was to investigate the use of quaternarized monocrotaline (i.e., N-alkylated monocrotaline) as a competitor. N-(Methyl)- and N-(crotonic acid) monocrotaline were analyzed under neutral conditions (pH 7.6) providing I_{50}'s of 0.53 and 0.25 ppm, respectively, while monocrotaline (pH 6.0) gave an I_{50} of 6.0 ppm. The increased sensitivity obtained with N-(crotonic acid)monocrotaline over N-(methyl)monocrotaline reflects linker-arm recognition of the crotonic acid moiety.

Conclusions

The competitive, amplified ELISA can be used to detect monocrotaline and retronecine and within the range of 5-7 µg/µL of retrorsine, seneciphylline and senecionine. Assay sensitivity is dependent on the immunizing hapten and the type of ELISA selected. Retronamine and monocrotaline are currently our haptens of choice for detecting a broad spectrum of naturally occurring PAs, although, many more alkaloids need to be screened for cross-reactivity against these antisera.

Acknowledgments

We thank Professor H. J. Segall of the U. C. Davis Department of Veterinary Pharmacology and Toxicology for the retrorsine, senecionine, and seneciphylliine standards, and are pleased to acknowledge support of this work by the U. S. Department of Agriculture (86-CRSR-2-2852) and the National Institutes of Environmental Health Sciences by ES04274, ES04699, and ES00182. MJK is a Sloan Foundation Fellow, 1987-1991.

Literature Cited

1. Smith, L. W.; Culvenor, C. C. J. *J. Nat. Prod* 1981, *44*, 129-152.
2. Cullvenor, C. C. J. In: *Toxicology in the Tropics*; Smith, R. L., Bababunmi, E. A., Eds. Taylor and Francis: London 1980.
3. Kumana, C. R.; Ng, M.; Lin, H. J.; Ko, W.; Wu, P. C.; Todd, D. *Gut* 1985, *26*, 101-104.
4. Tandon, B. N.; Tandon, H. D.; Tandon, R. K.; Narndranathan, M.; Joshi, Y. K. *The Lancet* 1976, August, 271-272.
5. Bull, L. B.; Culvenor, C. C. J.; Dick, A. T. In: *The Pyrrolizidine Alkaloids*; pp. 133-225. North-Holland: Amsterdam 1968.
6. McLean, E. K. *Pharmocol. Rev.* 1970, *22*, 429-483.
7. Peterson, J. E.; Culvenor, C.C.J. In: *Handbook of Natural Toxins*; R.F. Keeler and A.T. Tu, Eds.; Vol. 1, pp 637-671. Dekker: New York 1983.
8. Hooper, P. T. *J. Comp. Path.* 1975, *85*, 341-349.
9. Chesney, C. F.; Allen, J. R. *Am. J. Pathol.* 1973, *70*, 489-492.
10. Hooper, P.T. *Vet. Rec.* 1972, *90*, 37-38.
11. Labrousse, H.; Nishikawa, A. K.; Bon, C.; Avrameas, S. *Toxicon* 1988, *26*, 1157-1167.
12. Chandler, H. M.; Hurrell, J. G. *Clinica Chimica Acta* 1982, *121*, 225-230.
13. Shelby, Richard Al.; Kelley, Virginia C. *J. Agric Food Chem.* 1990, *38*, 1130-4.
14. Fliniaux, M. A.; Jacquin-Dubreuil, A. *Planta Med.* 1987. *53*, 87-90.
15. Mattocks, A. R. Toxicity of pyrrolizidine alkaloids. In: *Chemistry and Toxicology of Pyrrolizidine Alkaloids*; pp 1-14. Academic Press: San Diego 1986.
16. Bober, M. A.; Milco, L. A.; Miller, B. R.; Mount, M.; Wicks, B.; Kurth, M. J. *Toxicon* 1989, *27*, 1059-1064.
18. Laurent, P.; Magne, L.; DePamas, J.; Bignon, J.; Jaurance, M. C. *J. Immun. Meth.* 1988, *107*, 1-11.
17. Kumari, S.; Kapur, K. K.; Atal, C. C. *Curr. Sci.* 1967, *35*, 546-47.

RECEIVED August 30, 1990

IMMUNOASSAYS
FOR MONITORING HUMAN
EXPOSURE TO TOXIC CHEMICALS

Chapter 17

Molecular Epidemiology
Dosimetry, Susceptibility, and Cancer Risk

P. G. Shields, A. Weston, H. Sugimura, E. D. Bowman, N. E. Caporaso[1],
D. K. Manchester[2], G. E. Trivers, S. Tamai, J. H. Resau[3], B. F. Trump[3],
and C. C. Harris[4]

Laboratory of Human Carcinogenesis, National Cancer Institute, National
Institutes of Health, Bethesda, MD 20892

Molecular epidemiology is a multi-disciplinary field
including internal dosimetry assessments of carcinogen
exposure, involving metabolic phenotyping and examination
of genetic alterations. Carcinogen-adduct studies are
becoming increasingly chemically specific and multiple
corroborative assays are more frequently performed. These
should facilitate interlaboratory comparisons and
ultimately improve measurements of biologically effective
doses. Metabolic phenotyping can be predictive of cancer
risk and available evidence indicates that, when combined
with an exposure assessment (PAH, asbestos), predictabil-
ity is enhanced. Investigations of pediatric tumors have
documented that mutation and chromosome loss either inherited or acquired increases cancer risk. Other acquired
defects as identified through restriction fragment length
polymorphism analysis for HRAS-1 and MYCL relates to tumor
incidence and metastasis. Analyses of lung cancer cases
demonstrate that loss of heterozygosity of allelic DNA
sequences is non-random in both small-cell and non-small-
cell lung cancer. This paper will review our experience
in molecular epidemiology as well as summarize significant
accomplishments of other investigators.

[1]Current address: Environmental Epidemiology Branch, National Cancer Institute, National
Institutes of Health, Bethesda, MD 20892

[2]Current address: Division of Genetics, The Children's Hospital, Denver, CO 80218

[3]Current address: Department of Pathology, University of Maryland School of Medicine, and
Baltimore Veterans Administration Hospital, Baltimore, MD 21201

[4]Address correspondence to this author: Laboratory of Human Carcinogenesis, National
Cancer Institute, National Institutes of Health, Building 37, Room 2C05, Bethesda, MD
20892

A major goal of molecular epidemiology is to identify and quantify
carcinogenic risk in an individual. Risk is a function of exposure
to carcinogens as well as acquired inherited predispositions.
Clinical cancer results from multiple steps involving both genetic
and epigenetic alterations (Figure 1) (1-3). The development of
methods to identify and quantitate these alterations will enhance
the assessment of cancer risk.
     Molecular dosimetry quantitates carcinogen exposures at the
level of DNA (or its surrogate) and has been the focus of research
in numerous laboratories (4). Adduct levels depend upon the
interaction between environmental exposure, absorption, metabolism,
DNA repair and can be associated with promutagenic lesions.
Interspecies, interindividual and intertissue variations have been
well documented (5,6). Studies using laboratory animals generally
demonstrate a relationship between dose, DNA adduction and
carcinogenicity; some of these studies will be described in this
review. Human pilot studies are now beginning to demonstrate the
application of molecular dosimetry for individual exposure
assessments, but this has proven difficult due to complexities of
multiple carcinogen exposures and inter-individual variation in
metabolism and repair. Methods need to be developed which can
identify these adducts with chemical specificity, adequate
sensitivity and quantitative reliability.
     Genetic predispositions to cancer can be inherited and
acquired. Models of pediatric tumorigenesis postulate the
accumulation of at least two mutations within a single cell (7,8).
These models are exemplified by studies of retinoblastoma,
osteosarcoma and Wilms' tumors (9-12). Family members of patients
with either retinoblastoma or osteosarcoma may be at risk for one or
both of these tumors (9). These tumors are the direct result of an
autosomally dominant trait with 90% penetrance. A homozygous
deletion, detected by a restriction fragment length polymorphism
(RFLP) specific for chromosome 13q is often found in these tumors.
This deletion also includes a loss of the retinoblastoma gene (RB1),
a putative tumor suppressor gene. In the familial form, loss of one
allele is inherited while the loss of the remaining functional copy
occurs during life via several possible mechanisms (Figure 2).
Other tumors such as Wilms' or bladder are associated with reduction
to homozygosity of chromosome 11 (13-17), while acoustic neuroma is
associated with loss in chromosome 22 (18). Loss of chromosome 17p
and mutations in p53, a putative tumor suppressor gene on that arm,
have been detected during progression of human colon carcinoma (19),
breast cancer, lung cancer, brain cancer, neurofibrosarcomas and
other human malignancies (20). A recent report associating allelic
deletions of chromosome 18 in colorectal cancers has shown that the
DCC gene from this region codes for a protein similar to neural cell
adhesion molecules and other related cell surface glycoproteins
involved in growth regulation (21). The expression at this gene is
greatly reduced in colorectal carcinomas and several somatic
mutations were observed (deletions, point mutations and insertions).
Genetic alterations have been shown to be important in both early
and late stages of carcinogenesis (Figure 1). Vogelstein et al.,
have shown that activation of the KRAS oncogene by base substitution

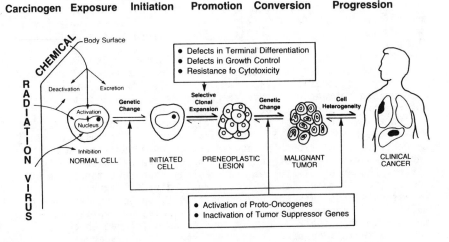

Figure 1. Multistage carcinogenesis.

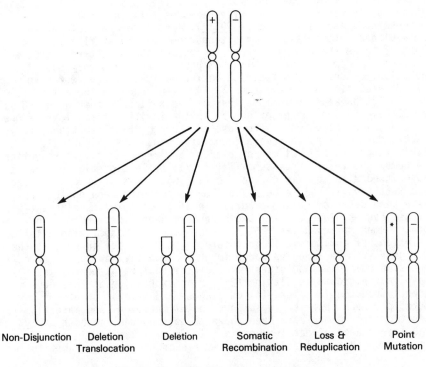

Figure 2. Mutational mechanisms.

is an early event in human colon carcinogenesis (22) while
amplification of N-myc is associated with tumor progression of human
neuroblastoma (23). Detection of genetic alterations in
preneoplastic lesions and carcinoma in situ may aid our assessment
of individual cancer risk.
Inheritance of certain metabolic phenotypes is also associated
with cancer risk. Rapid hydroxylation of debrisoquine, a medication
originally used to treat hypertension, is associated with increased
risk of lung cancer (24). Similarly, individuals that have a slow
acetylation rate for caffeine are at increased risk of bladder
cancer (25-27), while those with a rapid acetylation phenotype are
more susceptible to colon cancer (25). The combined effect of these
phenotypes with environmental exposures further increases cancer
risk.
Lung cancer serves as a model for the study of human
carcinogenesis, molecular epidemiology and cancer risk. Cigarette
smoking is the leading risk factor for lung cancer (28,29). Tobacco
smoke is a complex mixture comprised of numerous carcinogens
resulting in multiple effects (30). Consequently, our laboratory,
as well as others, have studied the interaction of tobacco smoke
carcinogens and genetic predispositions in relationship to molecular
epidemiology. This report summarizes these efforts.

Methods in Molecular Dosimetry

A variety of assays are available to identify carcinogen adducts in
DNA and proteins. Enzyme immunoassays (31-40), ^{32}P-
postlabeling/nucleotide chromatography (41-43), fluorescence
spectroscopy (44), synchronous fluorescence spectroscopy (SFS)
(45-48), gas chromatography/mass spectroscopy (GC/MS) (48,49) and
electrochemical detection (50) have been applied to the analysis of
human tissues. In some cases, these have been combined with
preparative techniques such as high performance liquid
chromatography (HPLC) or immunoaffinity chromatography (IAC). While
these techniques have generally been validated by laboratory animal
and cell culture experiments, their use with human samples has been
limited in sensitivity, specificity and quantitative reliability.
Previous research has demonstrated the need to develop
chemically specific assays dependent upon the use of synthetic
standards and leading to accurate detection and quantitation of
adducts. Moreover, the techniques should be sufficiently specific
to allow for separation and identification of individual adducts.
Preparing samples with HPLC followed by the ^{32}P-postlabeling assay,
which relies on three different separations (HPLC and 2 dimensional
thin layer chromatography) can detect O^6methyl-2'-deoxyguanosine
(51), O^6ethyldeoxyguanosine (O^6ethyldG) (51) and
N7methyldeoxyguanosine (N7methyldG) (Shields, P. G., Cancer Res., in
press) (Figure 3) at levels as low as 1 adduct in 10^7
2'deoxyguanosine residues in human tissues. Other laboratories have
utilized HPLC after ^{32}P-postlabeling (52). Another approach is to
combine HPLC and/or IAC with SFS for the detection of polycyclic
aromatic hydrocarbon (PAH) adducts (Figure 4) (45,48). The use of
IAC is not limited by a lack of chemical specificity within a

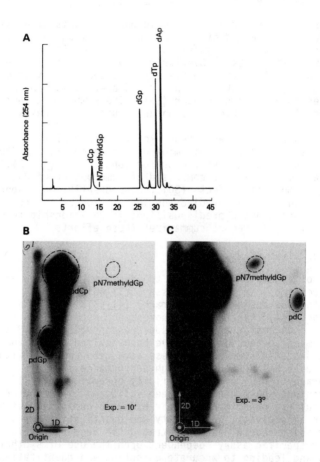

Figure 3.
Combined HPLC and the ^{32}P-postlabeling assay for N7methyldGp in human lung: **A)** HPLC elution profile of 3'-monophosphate nucleotide digest with the position of N7methyldGp shown; **B)** Autoradiography of ^{32}P-postlabeled fractions exposed for 10 minutes; and **C)** 3 hours. Fractions corresponding to dGp were pooled, diluted and added back to the fractions for N7methyldGp. The mixture was ^{32}P-postlabeled and separated by 2-dimensional thin layer chromatography as bisphosphate-mononucleotides. This sample was from a 57 year old painter with a 40 pack year smoking history. The N7methyldGp level was determined to be 70×10^{-7} per normal 2'-deoxyguanosine.

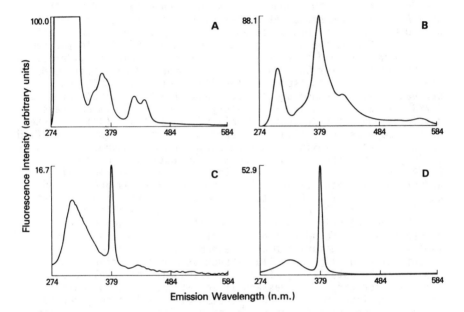

Figure 4.
Synchronous fluorescence spectra (wavelength difference = 34nm). A) Acid-hydrolyzed isoamyl alcohol extracts of human placental DNA pooled from 8 donors. Emission wavelength at 344 corresponds with signature signal of BP tetrols. B) Eluant of immunoaffinity column chromatography of acid-hydrolyzed pooled placental DNA. C) Eluant of immunoaffinity column chromatography after additional clean-up with HPLC. D) Authentic sample of BP-7,10/8,9-tetrol.

chemical class, as are immunoassays that employ the same antibodies
(4), because the use of HPLC with IAC followed by detection with SFS
and/or the ^{32}P-postlabeling assay can provide corroborative and
specific chemical analyses (47,48,53). HPLC has also been combined
with immunoassays (38,39).
    Methods in molecular dosimetry need to be validated by
complementary assays. For example, we have used ultrasensitive
enzyme radioimmunoassay (USERIA) (54,55), SFS (46,56), GC/MS (45,48)
and ^{32}P-postlabeling assay for the detection of PAH-DNA adducts; and
we have also investigated levels of human serum antibodies that
recognize PAH-adducts (31). In a group of 41 coke oven workers, 75%
had detectable PAH-DNA adduct levels in lymphocytes by SFS, 64% by
USERIA and 27% had serum antibodies (32). Six of nine persons who
were USERIA negative had detectable levels by SFS and five of eight
who were SFS negative had detectable levels by USERIA. In the same
group of workers and another small set of cancer patients, detection
by HPLC with SFS for DNA adducts was corroborated by use of GC/MS
(45). The use of IAC (specific for PAHs) followed by SFS and
derivative fluorescence spectroscopy has also been corroborated by
GC/MS (47,48). In an interlaboratory comparison of immunoassays
developed against BPDE-modified DNA, Santella et al. (34), reported
that as much as an 11-fold difference in adduct level was identified
among laboratories using either an enzyme linked immunosorbent assay
(ELISA) or the USERIA. Comparing the ^{32}P-postlabeling assay to
ELISA for PAHs failed to agree on the presence of smoking-related
adducts in placental tissue (57). These differences are due, in
part, to a lack of appropriate and common standards (i.e., degree of
modification of BPDE-DNA) and lack of chemical specificity.
    Tissue selection, identification of alternate sources of
exposures and confounders impact on study design and results.
Tables I and II summarize available human tissue studies. It would
be optimal to utilize body fluids such as blood or urine. It
remains to be established if blood testing will be a reliable
surrogate for target tissues and if so, if one blood component is
more appropriate than another. We also will need to consider half-
lives of blood cells (lymphocytes and red blood cells survive longer
than granulocytes), differences in repair and ability to separate
blood cells. It has been reported that the initial adduction of
lymphocytes and granulocytes is similar but that adduct levels are
more persistent in the latter (58). Our laboratory has used
lymphocytes (31-33,45) while others have studied total white blood
cells (predominantly granulocytes) (36,59).
    The placenta is useful to study because it is generally
available in large quantities and it has inducible cytochrome P450
enzymes (48,57,60). Twenty-eight placentas were studied using IAC
separation followed by SFS. Detectable levels of BPDE-DNA were
identified in 10 samples, which included smokers and non-smokers
(48). The ^{32}P-postlabeling assay (57) and SFS also have been
utilized for the detection of PAH-DNA adducts (Manchester D.K.
Carcinogenesis, in press) while an immunoassay has been used for
detection of O[6]methyldG (37). Whether or not these types of assays
will be useful for maternal and/or fetal exposure assessments and

Table I: DNA Adducts in Normal Human Tissue[a,b]

Adduct	Placenta	Esophagus	Lymphocytes	Buffy Coat	Liver	Lung	Oral Mucosa	Method[c]	Cohort[d]	Ref
PAH	33 (7/31)							E	G	(57)
	16 (7/30)							E	G	(57)
	10 (10/28)							S	G	(48)
			23 (13/13)					U	Co	(33)
			276 (18/27)					U	Co	(32)
			+[e] (31/41)					S	Co	(32)
			7 (7/20)					U	Fo	(123)
			45 (7/28)					U	R	(123)
				18 (30/35)				E	Fo	(36)
				65 (15/43)				E	Fi	(59)
						28 (4/27)		E	G	(124)
						+ (9/17)		U	G	(125)
						0.32 (5/12)		E	G	(87)
O[6]methyldG	0.01 (5/20)							E	G	(126)
		2 (16/22)						RIA	G	(90)
					590 (1/1)[f]			F	DMN	(44)
						110 (16/17)		P	G	(125)
							0.05 (4/20)	E	G	(39)
O[4]ethyldT					12 (11/12)			RIA	G	(91)
O[6]ethyldG						14 (17/17)		P	G	(125)
N7methyldG					1368 (1/1)[f]			F	DMN	(44)
^{32}P-PL[g]	+ (19/31)							P	G	(57)
			5 (7/17)					P	F	(127)
			2 (9/24)					P	G	(41)
						9 (29/29)		P	G	(128)
							647 (7/19)	P	G	(42)
							888 (7/22)	P	B	(42)
							936 (7/22)	P	TC	(42)
							4.7 (35/35)	P	G	(61)
							0.28 (25/25)	P	G	(61)
4-Aminobiphenyl						9 (10/17)		E	G	(125)
Aflatoxin B[1]						53 (7/8)		E	HC	(87)

[a] Parentheses indicate number of individuals positive/total examined.
[b] Adduct levels are 10^8 per nucleotide except where noted
[c] E=ELISA; F=Fluorescence; P=^{32}P-postlabeling; RIA=Radioimmunoassay; U=USERIA; S=Synchronous Fluorescence.
[d] B=Betel nut chewers; Co=Coke-oven workers; DMN=N-nitrosodimethylamine poisoning; Fi=Firefighters; Fo=Foundry workers; G=General population (including smokers and non-smokers); HC=Hepatocellular carcinoma patients; R=Roofers; TC=Tobacco chewers
[e] +denotes adduct detected but quantitative date not available from publication.
[f] per mol guanine
[g] ^{32}P-PL=^{32}P-postlabeling assay without micropreparative technique or chemical specificity.

Table II: Protein Adducts in Humans

Adduct	Hemoglobin	Albumin[b]	Method[c]	Cohort[d]	Ref
Aflatoxin B$_1$		102 (42/42)	RIA	C	(88)
4-Aminobiphenyl	51 (25/25)[a]		GC/MS	NSm	(49)
	176 (43/43)[a]		GC/MS	Sm-Blo	(49)
	288 (18/18)[a]		GC/MS	Sm-Bla	(49)
	28 (25/25)[a]		GC/MS	NSm	(129)
	154 (19/19)[a]		GC/MS	Sm-Blo	(129)
PAH	+ (2/2)		HPLC/SFS	Co	(45)
Eto	3.7 (4/4)[e]		GC/MS	ETO-1	(130)
	0.3 (6/6)[e]		GC/MS	ETO-2	(130)
	0.39 (11/11)[e]		GC/MS	Sm-Blo	(131)
	0.058 (14/14)[e]		GC/MS	NSm	(131)

[a]pg/gm

[b]ng/gm

[c]GC/MS = Gas chromatography/mass spectroscopy; HPLC/SFS = High performance liquid chromatography/ Synchronous fluorescence spectroscopy; RIA = Radioimmunoassay.

[d]C = Chinese residents; Nsm = non-smokers; Sm-Blo = Blond tobacco smokers; Sm-Bla = Black tobacco smokers; Co = Coke-oven workers; ETO-1 = Ethylene-oxide exposed workers (high and intermediate exposed); ETO-2 = Controls and low exposed ethylene-oxide workers.

[e]nmol/gm

cancer risk estimates will need to be investigated in prospective studies.

Oral mucosal cells are another relatively non-invasive source of DNA. Immunoassays detect O^6methyldG and N7methyldG (39) while the ^{32}P-postlabeling assay detects putative carcinogen-DNA adducts (42). Investigations of other tissues such as liver or lung are highly dependent upon availability (e.g. autopsy or surgical samples). While providing important information relevant to elucidating mechanisms of carcinogenesis, these tissues cannot be useful for sampling cohorts or used prospectively to assess an individuals exposure and cancer risk.

Molecular epidemiologic studies need to assess multiple sources of exposure and modifying behaviors. For example, Foiles and coworkers reported lower adduct levels as detected by ^{32}P-postlabeling assay in oral mucosal cells of persons consuming alcohol (61) while Liou et al., (59) found no statistically significant difference in occurrence of PAH-DNA adducts using similar methods. Tobacco smoke can influence PAH-DNA adduct levels as well as other adducts identified by immunoassays (32,59) and ^{32}P-postlabeling assay (57,61). Persons consuming blond or black tobacco have demonstrably different levels of 4-aminobiphenyl or 3-aminobiphenyl adducts when compared to each other and with non-smokers (49). Several studies did not identify smoking related differences for PAH, alkyl and ^{32}P-postlabeled putative adducts in placenta (37,57,62). These phenomena might be tissue specific as adduct differences are observed more consistently in blood than placenta.

Laboratory Animal and Human Pilot Studies

Laboratory animal experiments demonstrate that dose-response relationships exist between carcinogen exposure and adduct levels, suggesting the utility of these studies in human exposure assessments. In animals, DNA and protein adduct levels correlate with dose of PAHs (63-65), tobacco-smoke condensate (43,60,66), N-nitrosamines (67-70), aromatic amines (71) (Beland, F. A., unpublished data; Poirier, M. C., unpublished data) and heterocyclic amines (72,73).

Adduct levels correlate with *in vitro* mutagenicity assays (68,71,74-76) and laboratory animal cancer incidence (65,71,75,77-84) (Beland, F. A., unpublished data; Poirier, M. C., unpublished data). Carcinogen binding to DNA in rat liver is directly related to carcinogenic potency although the liver is not always the target organ (77). In other tissues, tumor incidence correlates with benzo[a]pyrene (BP) adducts in mouse skin (78) as does methylation caused by 1,2-dimethylhydrazine in intestine (79) and persistence of O^6ethylguanine caused by ethylnitrosourea in rat brain (80). Benzidine treatment following partial hepatectomy in mice led to a high correlation of adduct levels measured by the ^{32}P-postlabeling assay in association with clastogenic effects in its target organ (71). Animals sensitive to the carcinogenic effects of various PAHs can have higher DNA adduct levels than those caused by non-carcinogenic PAH exposure (81). Chronic dosing leads to higher levels of DNA adducts than single exposures although this is not necessarily the case with hemoglobin adducts (65,82). Target organs also tend to have higher levels (75,80,83,84). However, adduct levels do not always correlate with tumorigenicity (85,86), but this might reflect the complexity of cancer mechanisms.

Pilot human dosimetry studies have demonstrated the usefulness of adduct measurements in exposure assessments. Higher adduct levels in heavily exposed coke oven workers compared to those with lower exposures or controls have been reported using USERIA (32), ELISA (36) or the ^{32}P-postlabeling assay (41). Tobacco consumption is also positively correlated with adduct formation (49). Chemotherapeutic trials are among the best evidence that adduct levels are related to exposure because in these cases, negative controls are truly negative. The development of immunoassays for cisplatin-DNA adducts by Poirier et al., have led to internal dosimetry estimates and predictions for prognosis (35). Levels of aflatoxin B_1-adducts have also been identified in human liver and breast with ELISA (40,87), in serum albumin with RIA (88) and ELISA (38). Multiple adduct assessments are also feasible (Table III) (89).

Cancer risk studies are now being formulated as methodologies are becoming sophisticated enough to reliably detect and quantitate adducts. Evaluations of cancer patients have shown that Chinese people with esophageal cancer and presumed higher exposures to dietary N-nitrosamines have higher levels of O^6methyl-2'-deoxyguanosine adducts than European cancer patients (90). Radioimmunoassay for O^4methylthymine in human liver DNA showed higher levels in autopsy samples of cancer patients compared to controls and also a further subdivision for those with and without liver cancer (91). In addition, urinary excretion of aflatoxin B_1-guanine adducts was correlated with cancer risk of certain ethnic groups in Kenya using HPLC/UV and confirmation with spectral fluorescence (92).

Genetic Predispositions

Metabolic phenotyping has suggested an inherited predisposition for cancer. Ayesh et al., (24) reported that among a cohort of smokers in London, England, poor metabolizers of debrisoquine, an antihypertensive medication, had a reduced risk for lung cancer. A concomitant four-fold increased risk for extensive metabolizers was identified. The poor metabolic phenotype is an autosomally recessive trait occurring in 5-10% of whites. The findings of Ayesh and coworkers have been confirmed in a second study of lung cancer patients participating in a case-control study evaluating cancer risk and prognosis (93). These data have been extended to suggest an interactive effect with pulmonary carcinogens (asbestos or PAHs) (93-95). In addition to lung cancer, weak associations have also been found with liver and advanced bladder carcinoma risk and extensive debrisoquine metabolizer phenotype (96,97). The hepatic mono-oxygenase (P450$_{II}$D$_6$) responsible for the debrisoquine phenotype has been cloned and expressed in mammalian cell culture (98,99). Studies are now under way to correlate restriction fragment length

Table III: Analysis of Human Peripheral Lung DNA

Donor No.	Age/Sex/Race	Occupation	Smoking 'Pack' Year	PAH[a]	BP-tetrol[b]	4-ABP[c]	O6mdG[d]	O6etdG[d]	N7mdG[e]
1	23/M/W	Auto mechanic	1	−	Broad	180	1±1	36±32	ND[f]
2	34/M/W	Salesperson	1.5	−	Broad	40	15±11	29±29	ND
3	39/F/W	Accountant	8	+	Specific	50	7±7	23±13	41
4	61/M/W	Shipping clerk	10	+	Broad	40	11±6	16±9	ND
5	28/M/W	Mason	30	+	Broad	<10	6±1	11±5	ND
6	71/M/W	Cook	30	−	−	70	1±1	5±1	33
7	35/F/W	Salesperson	30	−	Specific	<10	5±3	13±12	37
8	61/M/B	Stockhandler	30	+	Broad	40	4±4	11±5	14
9	57/M/W	Painter	40	−	−	20	11±2	26±26	70
10	52/M/W	(not known)	75	−	Specific	<10	12	6	ND
11	67/M/B	Taxi driver	−	+	−	<10	3±2	4±4	ND
12	54/M/W	Auto mechanic	−	−	ND	850	10±1	21±11	ND
13	39/F/W	Therapy assist.	−	+	−	ND	1	4	ND
14	20/M/W	Student	−	+	ND	ND	22	2	ND
15	15/F/W	Student	−	+	Broad	<10	<1	11±10	ND
16	22/M/W	Student	−	−	−	20	24	7	ND
17	68/M/W	Inspector	*[g]	+	−	110	52±3	8±8	ND

[a]USERIA. Positives denote the detection of >1 PAH-DNA adduct per 10^7 nucleotides.

[b]HPLC-linked SFS. Specific and broad denote the character of the SFS spectra observed in the samples when the results suggested the presence of BPDE-DNA adducts at a level >1 adduct per 5×10^6 nucleotides.

[c]ELISA values represent the number of 4-ABP-DNA adducts per 10^7 nucleotides.

[d]HPLC/^{32}p-Postlabeling values represent the number of O6alkyldG adducts per 10^7 deoxyguanosine residues.

[e]HPLC/^{32}p-Postlabeling values represent the number of N7methyldG $\times 10^{-7}$ adducts per deoxyguanosine residue.

[f]ND, not determined.

[g]Cigar and pipe smoker.

polymorphisms with metabolic phenotypes (99). Although the
phenotype is associated with the hydroxylation of a number of
medications, no candidate carcinogen has been identified as a
substrate for this enzyme. Whether the cytochrome $P450_{IID_6}$ enzyme
is responsible for activating procarcinogens or its respective gene
is in linkage disequilibrium with another gene related to cancer
risk is unknown.

An investigation involving the metabolism of a medication used
in the treatment of tuberculosis, isoniazid, has led to the
identification of an acetylator phenotype for N-acetyltransferase.
This enzyme, in addition to metabolizing medications and caffeine,
also acetylates aromatic amines linked to bladder cancer. Slow
acetylation has been associated with bladder cancer in aromatic
amine exposed workers (25), as well as the general population (26).
It is also positively correlated with bladder tumor invasion (27)
and laryngeal cancer (27). Slow acetylators are homozygous for an
autosomal recessive gene. In contrast to slow acetylators, rapid
acetylation is associated with colon cancer (27,100). The phenotype
of an individual can be readily determined by a urine assay for
caffeine metabolites or other drug testing (101).

Acquired genetic abnormalities include activation of
protooncogenes, loss of suppressor genes and other chromosomal
alterations. Our laboratory has investigated the incidence of rare
alleles of the HRAS-1 protooncogene locus in lung cancer patients
and controls. Rare nucleotide tandem repeat alleles detected by
DNA-RFLP analysis have been previously detected in bladder cancer
(102) and breast cancer patients (103). The HRAS-1 protooncogene
contains a hypervariable insertion/deletion polymorphism with 30 to
100 tandem nucleotide repeats of a 28 base pair consensus sequence
aligned head-to-tail downstream from the 3' end of the structural
gene of HRAS-1. The function of this variable tandem repeat is
unknown but it is possible that this region contains enhancer
activity, and may be a "hot spot" for the effects of chemical
carcinogens or reflects genomic instability. In an interlaboratory
study, we have been able to unambiguously distinguish more than 20
different restriction enzyme patterns at the HRAS-1 gene locus in
our subjects and found that "rare" alleles occurred more frequently
than in controls (Sugimura, H. Cancer Res., in press).

We have also investigated the association of lung cancer and
MYCL genetic phenotype using RFLP. The S allele has been detected
more frequently in males with soft tissue sarcoma (104), and the
presence of metastasis in gastric (105) and lung cancer patients
(106). While the presence of the S allele was not informative in
renal cell cancer patients, the L-L allelotype was more closely
associated with the lack of metastases and lower grade tumors (107).
A recent analysis of our lung cancer cohort failed to reveal
predictive information of MYCL alleles (Tamai, S., unpublished
data). Interestingly, racial differences were found whereby the S-S
allele was detected more frequently in American blacks than whites.
In other tumors, such as colon (108) or breast (105), no association
with MYCL genotypes was found.

It is commonly accepted that lung cancer results from exposure
to multiple carcinogenic agents found in tobacco smoke and various

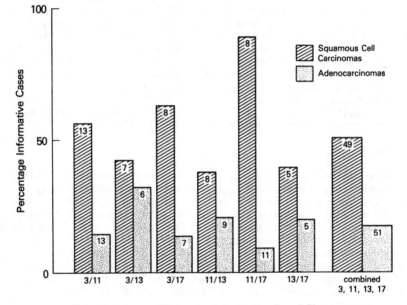

Coincident Loss of Heterozygosity for Numbered Chromosomes

Figure 5. Complex genetic alterations in lung cancer.

environmental sources (109). Karyotype data indicates that genetic abnormalities are predictably complex (110). In small-cell and non-small-cell carcinoma of the lung, a loss of chromosome 3p21 has been noted (111-114) albeit less consistently in non-small cell carcinoma (111,112). The long arm of chromosome 3 is also lost in some non-small-cell tumor types (115). We have observed abnormalities in other genes and have noted associations with specific histological types (Figure 5). In this case, fifty-four non-small lung cancer cases were examined for loss of heterozygosity by DNA-RFLP analysis (23 squamous cell carcinomas, 23 adenocarcinomas and 8 large cell carcinomas). Of these, only 20 did not show loss of heterozygosity at chromosome 11 and 10 of those 20 had abnormalities at other chromosomal loci (116). Loss of heterozygosity from chromosome 17 was identified in 8/9 of squamous cell carcinomas while loss from both chromosomes 17 and 11 were observed in 7/8 cases. Whereas loss of heterozygosity at chromosome 17p (D17S1) was detected in only 2 of 11 cases of adenocarcinoma of the lung. Loss of heterozygosity in 17p is particularly interesting because the p53 gene, a putative tumor suppressor, is located on this chromosome (119-121) and p53 mutations are found in lung cancer (20,122). These studies indicate that the spectrum of mutations and deletions in oncogenes and putative tumor suppressor genes is complex in lung cancer carcinogenesis. These genetic changes are likely to be involved in a "multi-hit" framework similar to the specific changes that have been described for other malignancies (9,11,117,118).

Conclusion

Molecular epidemiology continues to evolve by developing multiple corroborative and chemically specific carcinogen adduct assays. These are more apt to provide reliable exposure assessments. The ease of metabolic phenotyping, combined with exposure assessments, will enhance cancer risk estimates. The application of a variety of molecular biological techniques is providing new insights into both carcinogenic mechanisms and elucidating at-risk inherited or acquired genetic predispositions. The integration of adduct assays, phenotyping and other molecular studies is now being realized so that individual cancer risk assessments might be feasible.

Acknowledgments

We gratefully acknowledge the editorial and typing assistance of Mr. Bob Julia.

Literature Cited

1. Barbacid, M. Carcinogenesis 1986, 7, 1037.
2. Bishop, J. M. Leukemia 1988, 2, 199.
3. Weinstein, I. B. Cancer Res. 1988, 48, 4135.
4. Weston, A.; Manchester, D. K.; Povey, A. C.; Harris, C. C. J. Am. Coll. Toxicol. 1989, 8, 913.
5. Harris, C. C.; Trump, B. F.; Grafstrom, R. C.; Autrup, H. J. Cell Biochem. 1982, 18, 285.

6. Harris, C. C. Carcinogenesis 1989, 10, 1563.
7. Knudson, A. G., Jr. Proc. Natl. Acad. Sci. USA 1971, 68, 820.
8. Knudson, A. G., Jr. Cancer Res. 1985, 45, 1437.
9. Hansen, M. F.; Koufos, A.; Gallie, B. L.; Phillips, R. A.; Fodstad, O.; Brogger, A.; Gedde-Dahl, T.; Cavenee, W. K. Proc. Natl. Acad. Sci. USA 1985, 82, 6216.
10. Fung, Y. K.; Murphree, A. L.; T'Ang, A.; Qian, J.; Hinrichs, S. H.; Benedict, W. F. Science 1987, 236, 1657.
11. Cavenee, W. K.; Murphree, A. L.; Shull, M. M.; Benedict, W. F.; Sparkes, R. S.; Kock, E.; Nordenskjold, M. N. Engl. J. Med. 1986, 314, 1201.
12. Friend, S. H.; Bernards, R.; Rogelj, S.; Weinberg, R. A.; Rapaport, J. M.; Albert, D. M.; Dryja, T. P. Nature 1986, 323, 643.
13. Koufos, A.; Hansen, M. F.; Lampkin, B. C.; Workman, M. L.; Copeland, N. G.; Jenkins, N. A.; Cavenee, W. K. Nature 1984, 309, 170.
14. Orkin, S. H.; Goldman, D. S.; Sallan, S. E. Nature 1984, 309, 172.
15. Reeve, A. E.; Housiaux, P. J.; Gardner, R. J.; Chewings, W. E.; Grindley, R. M.; Millow, L. J. Nature 1984, 309, 174.
16. Fearon, E. R.; Vogelstein, B.; Feinberg, A. P. Nature 1984, 309, 176.
17. Fearon, E. R.; Feinberg, A. P.; Hamilton, S. H.; Vogelstein, B. Nature 1985, 318, 377.
18. Seizinger, B. R.; Martuza, R. L.; Gusella, J. F. Nature 1986, 322, 644.
19. Baker, S. J.; Fearon, E. R.; Nigro, J. M.; Hamilton, S. R.; Preisinger, A. C.; Jessup, J. M.; vanTuinen, P.; Ledbetter, D. H.; Barker, D. F.; Nakamura, Y. Science 1989, 244, 217.
20. Nigro, J. M.; Baker, S. J.; Preisinger, A. C.; Jessup, J. M.; Hostetter, R.; Cleary, K.; Bigner, S. H.; Davidson, N.; Baylin, S.; Devilee, P.; Glover, T.; Collins, F. S.; Weston, A.; Modali, R.; Harris, C. C.; Vogelstein, B. Nature 1989, 342, 705.
21. Fearon, E. R.; Cho, K. R.; Nigro, J. M.; Kern, S. E.; Simons, J. W.; Ruppert, J. M.; Hamilton, S. R.; Preisinger, A. C.; Thomas, G.; Kinzler, K. W.; Vogelstein, B. Science 1990, 247, 49.
22. Vogelstein, B.; Fearon, E. R.; Hamilton, S. R.; Kern, S. E.; Preisinger, A. C.; Leppert, M.; Nakamura, Y.; White, R.; Smits, A. M.; Bos, J. L. N. Engl. J. Med. 1988, 319, 525.
23. Brodeur, G. M.; Hayes, F. A.; Green, A. A.; Casper, J. T.; Wasson, J.; Wallach, S.; Seeger, R. C. Cancer Res. 1987, 47, 4248.
24. Ayesh, R.; Idle, J. R.; Ritchie, J. C.; Crothers, M. J.; Hetzel, M. R. Nature 1984, 312, 169.
25. Cartwright, R. A.; Glashan, R. W.; Rogers, H. J.; Ahmad, R. A.; Barham-Hall, D.; Higgins, E.; Kahn, M. A. Lancet 1982, 2, 842.
26. Mommsen, S.; Barfod, N. M.; Aagard, J. Carcinogenesis 1985, 6, 199.
27. Hein, D. W. Biochim. Biophys. Acta 1988, 948, 37.

28. International Agency on the Research of Cancer IARC Monographs on the Evaluation of the Carcinogenic Risk of Chemicals to Humans: Tobacco Smoking. Vol. 38; IARC: Lyon, France, 1986;
29. Wynder, E. L.; Graham, E. A. J. A. M. A. 1950, 143, 329.
30. Grafstrom, R. C.; Willey, J. C.; Sundqvist, K.; Harris, C. C. In Banbury Report #23: Mechanisms of Tobacco Carcinogenesis; Hoffmann, D.; Harris, C. C., Eds.; Cold Spring Harbor Press: Cold Spring Harbor, New York, 1986; p 273.
31. Newman, M. J.; Light, B. A.; Weston, A.; Tollurud, D.; Clark, J. L.; Mann, D. L.; Blackmon, J. P.; Harris, C. C. J. Clin. Invest. 1988, 82, 145.
32. Harris, C. C.; Vahakangas, K.; Newman, M. J.; Trivers, G. E.; Shamsuddin, A. K. M.; Sinopoli, N. T.; Mann, D. L.; Wright, W. E. Proc. Natl. Acad. Sci. USA 1985, 82, 6672.
33. Haugen, A.; Becher, G.; Benestad, C.; Vahakangas, K.; Trivers, G. E.; Newman, M. J.; Harris, C. C. Cancer Res. 1986, 46, 4178.
34. Santella, R. M.; Weston, A.; Perera, F. P.; Trivers, G. E.; Harris, C. C.; Young, T. L.; Nguyen, D.; Lee, B. M.; Poirier, M. C. Carcinogenesis 1988, 9, 1265.
35. Poirier, M. C.; Reed, E.; Ozols, R. F.; Fasy, T.; Yuspa, S. H. Prog. Exp. Tumor Res. 1987, 31, 104.
36. Perera, F. P.; Hemminki, K.; Young, T. L.; Brenner, D.; Kelly, G.; Santella, R. M. Cancer Res. 1988, 48, 2288.
37. Foiles, P. G.; Miglietta, L. M.; Akerkar, S. A.; Everson, R. B.; Hecht, S. S. Cancer Res. 1988, 48, 4184.
38. Wild, C. P.; Jiang, Y. Z.; Montesano, R.; Parkin, M.; Khlat, M.; Srivatanakul, P. Proc. Am. Assoc. Cancer Res. 1989, 30, 317.
39. Wild, C. P.; Stich, H. F.; Montesano, R. Proc. Am. Assoc. Cancer Res. 1989, 30, 318.
40. Wild, C. P.; Lu, S. H.; Montesano, R. IARC Sci. Publ. 1987, 534.
41. Phillips, D. H.; Hemminki, K.; Alhonen, A.; Hewer, A.; Grover, P. L. Mutat. Res. 1988, 204, 531.
42. Dunn, B. P.; Stich, H. F. Carcinogenesis 1986, 7, 1115.
43. Chacko, M.; Gupta, R. C. Carcinogenesis 1988, 9, 2309.
44. Herron, D. C.; Shank, R. C. Cancer Res. 1980, 40, 3116.
45. Weston, A.; Rowe, M. L.; Manchester, D. K.; Farmer, P. B.; Mann, D. L.; Harris, C. C. Carcinogenesis 1989, 10, 251.
46. Vahakangas, K.; Haugen, A.; Harris, C. C. Carcinogenesis 1985, 6, 1109.
47. Weston, A.; Manchester, D. K.; Poirier, M. C.; Choi, J. S.; Trivers, G. E.; Mann, D. L.; Harris, C. C. Chem. Res. Toxicol. 1989, 2, 104.
48. Manchester, D. K.; Weston, A.; Choi, J. S.; Trivers, G. E.; Fennessey, P.; Quintana, E.; Farmer, P. B.; Mann, D. L.; Harris, C. C. Proc. Natl. Acad. Sci. USA 1988, 85, 9243.
49. Bryant, M. S.; Vineis, P.; Skipper, P. L.; Tannenbaum, S. R. Proc. Natl. Acad. Sci. USA 1988, 85, 9788.
50. Shigenaga, M. K.; Gimeno, C. J.; Ames, B. N. Proc. Natl. Acad. Sci. USA 1989, 86, 9697.
51. Wilson, V. L.; Basu, A. K.; Essigmann, J. M.; Smith, R. A.; Harris, C. C. Cancer Res. 1988, 48, 2156.

52. Gorelick, N. J.; Wogan, G. N. Carcinogenesis 1989, 10, 1567.
53. Weston, A.; Beland, F. A.; Manchester, D. K.; Parker, N. B.;
    Harris, C. C.; Poirier, M. C. Proc. Am. Assoc. Cancer Res.
    1989, 30, 134.
54. Poirier, M. C.; Santella, R. M.; Weinstein, I. B.; Grunberger,
    D.; Yuspa, S. H. Cancer Res. 1980, 40, 412.
55. Hsu, I. C.; Poirier, M. C.; Yuspa, S. H.; Grunberger, D.;
    Weinstein, I. B.; Yolken, R. H.; Harris, C. C. Cancer Res.
    1981, 41, 1091.
56. Vahakangas, K.; Trivers, G. E.; Rowe, M. L.; Harris, C. C.
    Environ. Health Perspect. 1985, 62, 101.
57. Everson, R. B.; Randerath, E.; Santella, R. M.; Cefalo, R. C.;
    Avitts, T. A.; Randerath, K. Science 1986, 231, 54.
58. Schutte, H. H.; Van der Schans, G. P.; Lohman, P. H. Mutat.
    Res. 1988, 194, 23.
59. Liou, S. H.; Jacobson-Kram, D.; Poirier, M. C.; Nguyen, D.;
    Strickland, P. T.; Tockman, M. S. Cancer Res. 1989, 49, 4929.
60. Randerath, E.; Avitts, T. A.; Reddy, M. V.; Miller, R. H.;
    Everson, R. B.; Randerath, K. Cancer Res. 1986, 46, 5869.
61. Foiles, P. G.; Miglietta, L. M.; Quart, A. M.; Quart, E.;
    Kabat, G. C.; Hecht, S. S. Carcinogenesis 1989, 10, 1429.
62. Everson, R. B.; Randerath, E.; Santella, R. M.; Avitts, T. A.;
    Weinstein, I. B.; Randerath, K. JNCI 1988, 80, 567.
63. Lee, B. M.; Santella, R. M. Carcinogenesis 1988, 9, 1773.
64. Schoepe, K. B.; Friesel, H.; Schurdak, M. E.; Randerath, K.;
    Hecker, E. Carcinogenesis 1986, 7, 535.
65. Schoket, B.; Hewer, A.; Grover, P. L.; Phillips, D. H.
    Carcinogenesis 1988, 9, 1253.
66. Gupta, R. C.; Sopori, M. L.; Gairola, C. G. Cancer Res. 1989,
    49, 1916.
67. Becker, R. A.; Barrows, L. R.; Shank, R. C. Carcinogenesis
    1981, 2, 1181.
68. Van Zeeland, A. A. Mutagenesis 1988, 3, 179.
69. Kleihues, P.; Magee, P. N. J. Neurochem. 1973, 20, 595.
70. Koepke, S. R.; Kroeger-Koepke, M. B.; Bosan, W.; Thomas, B. J.;
    Alvord, W. G.; Michejda, C. J. Cancer Res. 1988, 48, 1537.
71. Talaska, G.; Au, W. W.; Ward, J. B., Jr.; Randerath, K.;
    Legator, M. S. Carcinogenesis 1987, 8, 1899.
72. Schut, H. A.; Putman, K. L.; Randerath, K. In Carcinogenic and
    Mutagenic Responses to Aromatic Amines and Nitroarenes; King,
    C. M.; Romano, L. J.; Schuetzle, D., Eds.; Elsevier: Amsterdam,
    1988; p 265.
73. Turesky, R. J.; Skipper, P. L.; Tannenbaum, S. R.
    Carcinogenesis 1987, 8, 1537.
74. Arce, G. T.; Allen, J. W.; Doerr, C. L.; Elmore, E.; Hatch, G.
    G.; Moore, M. M.; Sharief, Y.; Grunberger, D.; Nesnow, S.
    Cancer Res. 1987, 47, 3388.
75. Pegg, A. E. Rev. Biochem. Toxicol. 1983, 5, 83.
76. Bignami, M.; Vitelli, A.; Di Muccio, A.; Terlizzese, M.;
    Calcagnile, A.; Zapponi, G. A.; Lohman, P. H.; Den Engelse, L.;
    Dogliotti, E. Mutat. Res. 1988, 193, 43.
77. Lutz, W. K. J. Cancer Res. Clin. Oncol. 1986, 112, 85.

78. Ashurst, S. W.; Cohen, G. M.; Nesnow, S.; DiGiovanni, J.; Slaga, T. J. Cancer Res. 1983, 43, 1024.
79. James, J. T.; Autrup, H. J. Natl. Cancer Inst. 1983, 70, 541.
80. Goth, R.; Rajewsky, M. F. Proc. Natl. Acad. Sci. USA 1974, 71, 639.
81. Daniel, F. B.; Joyce, N. J. J. Natl. Cancer Inst. 1983, 70, 111.
82. Gorelick, N. J.; Hutchins, D. A.; Tannenbaum, S. R.; Wogan, G. N. Carcinogenesis 1989, 10, 1579.
83. Schurdak, M. E.; Stong, D. B.; Warshawsky, D.; Randerath, K. Carcinogenesis 1987, 8, 1405.
84. Schurdak, M. E.; Randerath, K. Carcinogenesis 1985, 6, 1271.
85. Dipple, A.; Pigott, M. A.; Bigger, A. H.; Blake, D. M. Carcinogenesis 1984, 5, 1087.
86. Daniel, F. B.; Joyce, N. J. Carcinogenesis 1984, 5, 1021.
87. Garner, R. C.; Dvorackova, I.; Tursi, F. Int. Arch. Occup. Environ. Health 1988, 60, 145.
88. Gan, L. S.; Skipper, P. L.; Peng, X.; Groopman, J. D.; Chen, J. S.; Wogan, G. N.; Tannenbaum, S. R. Carcinogenesis 1988, 9, 1323.
89. Wyndham, C.; Devenish, J.; Safe, S. Res. Commun. Chem. Pathol. Pharmacol. 1976, 15, 563.
90. Umbenhauer, D.; Wild, C. P.; Montesano, R.; Saffhill, R.; Boyle, J. M.; Huh, N.; Kirstein, U.; Thomale, J.; Rajewsky, M. F.; Lu, S. H. Int. J. Cancer 1985, 36, 661.
91. Huh, N. H.; Satoh, M. S.; Shiga, J.; Rajewsky, M. F.; Kuroki, T. Cancer Res. 1989, 49, 93.
92. Autrup, H.; Seremet, T.; Wakhisi, J.; Wasunna, A. Cancer Res. 1987, 47, 3430.
93. Caporaso, N.; Hoover, R.; Eisner, S.; Resau, J.; Trump, B. F.; Issaq, H.; Morshik, G.; Harris, C. C. ASCO Proceedings 1988, 7, 336.
94. Caporaso, N.; Hayes, R. B.; Dosemeci, M.; Hoover, R.; Ayesh, R.; Hetzel, M.; Idle, J. R. Cancer Res. 1989, 49, 3675.
95. Caporaso, N.; Pickle, L. W.; Bale, S.; Ayesh, R.; Hetzel, M.; Idle, J. Genet. Epidemiol. 1989, 6, 517.
96. Kaisary, A.; Smith, P.; Jaczq, E.; McAllister, C. B.; Wilkinson, G. R.; Ray, W. A.; Branch, R. A. Cancer Res. 1987, 47, 5488.
97. Ritchie, J. C.; Idle, J. R. IARC Sci. Publ. 1982, 381.
98. Gonzalez, F. J. Pharmacol. Rev. 1988, 40, 243.
99. Skoda, R. C.; Gonzalez, F. J.; Demierre, A.; Meyer, U. A. Proc. Natl. Acad. Sci. USA 1988, 85, 5240.
100. Ilett, K. F.; David, B. M.; Detchon, P.; Castleden, W. M.; Kwa, R. Cancer Res. 1987, 47, 1466.
101. Weber, W. W. The Acetylator Genes and Drug Response; Oxford University Press: New York, 1987;
102. Krontiris, T. G.; DiMartino, N. A.; Colb, M.; Parkinson, D. R. Nature 1985, 313, 369.
103. Lidereau, R.; Escot, C.; Theillet, C.; Champeme, M. H.; Brunet, M.; Gest, J.; Callahan, R. JNCI 1986, 77, 697.

104. Kato, M.; Toguchida, J.; Honda, K.; Sasaki, M. S.; Ikenaga, M.; Sugimoto, M.; Yamaguchi, T.; Kotoura, Y.; Yamamuro, T.; Ishizaki, K. Int. J. Cancer 1990, 45, 47.
105. Ishizaki, K.; Kato, M.; Ikenaga, M.; Honda, K.; Ozawa, K.; Toguchida, J. J. Natl. Cancer Inst. 1990, 82, 238.
106. Kawashima, K.; Shikama, H.; Imoto, K.; Izawa, M.; Naruke, T.; Okabayashi, K.; Nishimura, S. Proc. Natl. Acad. Sci. USA 1988, 85, 2353.
107. Kakehi, Y.; Yoshida, O. Int. J. Cancer 1989, 43, 391.
108. Ikeda, I.; Ishizaka, Y.; Ochiai, M.; Sakai, R.; Itabashi, M.; Onda, M.; Sugimura, T.; Nagao, M. Gann 1988, 79, 674.
109. Doll, R. In Biochemical and Molecular Epidemiology of Cancer; Harris, C. C., Ed.; Alan R. Liss, Inc.: New York, 1985; p 111.
110. Zech, L.; Bergh, J.; Nilsson, K. Cancer Genet. Cytogenet. 1985, 15, 335.
111. Brauch, H.; Johnson, B.; Hovis, J.; Yano, T.; Gazdar, A. F.; Pettengill, O. S.; Graziano, S. L.; Sorenson, G. D.; Poiesz, B. J.; Minna, J.; Linehan, M.; Zbar, B. N. Engl. J. Med. 1987, 317, 1109.
112. Kok, K.; Osinga, J.; Carritt, B.; Davis, M. B.; van der Hout, A. H.; van der Veen, A. Y.; Landsvater, R. M.; de Leij, L. F.; Berendsen, H. H.; Postmus, P. E.; Poppema, S.; Buys, C. H. Nature 1987, 330, 578.
113. Naylor, S. L.; Johnson, B. E.; Minna, J. D.; Sakaguchi, A. Y. Nature 1987, 329, 451.
114. Johnson, B. E.; Sakaguchi, A. Y.; Gazdar, A. F.; Minna, J. D.; Burch, D.; Marshall, A.; Naylor, S. L. J. Clin. Invest. 1988, 82, 502.
115. Rabbitts, P.; Douglas, J.; Daly, M.; Sundaresan, V.; Fox, B.; Haselton, P.; Wells, F.; Albertson, D.; Waters, J.; Bergh, J. Genes Chromosomes Cancer 1989, 1, 95.
116. Weston, A.; Willey, J. C.; Modali, R.; Sugimura, H.; McDowell, E. M.; Resau, J.; Light, B.; Haugen, A.; Mann, D. L.; Trump, B. F.; Harris, C. C. Proc. Natl. Acad. Sci. USA 1989, 86, 5099.
117. Gusella, J. F. Annu. Rev. Biochem. 1986, 55, 831.
118. Hansen, M. F.; Cavenee, W. K. Cancer Res. 1987, 47, 5518.
119. Finlay, C. A.; Hinds, P. W.; Levine, A. J. Cell 1989, 57, 1083.
120. Eliyahu, D.; Michalovitz, D.; Eliyahu, S.; Pinhasi-Kimhi, O.; Oren, M. Proc. Natl. Acad. Sci. USA 1989, 86, 8763.
121. Isobe, M.; Emanuel, B. S.; Givol, D.; Oren, M.; Croce, C. M. Nature 1986, 320, 84.
122. Takahashi, T.; Nau, M. M.; Chiba, I.; Birrer, M. J.; Rosenberg, R. K.; Vinocour, M.; Levitt, M.; Pass, H.; Gazdar, A. F.; Minna, J. D. Science 1989, 246, 491.
123. Shamsuddin, A. K. M.; Sinopoli, N. T.; Hemminki, K.; Boesch, R. R.; Harris, C. C. Cancer Res. 1985, 45, 66.
124. Perera, F. P.; Poirier, M. C.; Yuspa, S. H.; Nakayama, J.; Jaretzki, A.; Curnen, M. M.; Knowles, D. M.; Weinstein, I. B. Carcinogenesis 1982, 3, 1405.
125. Wilson, V. L.; Weston, A.; Manchester, D. K.; Trivers, G. E.; Roberts, D. W.; Kadlubar, F. F.; Wild, C. P.; Montesano, R.; Willey, J. C.; Mann, D. L.; Harris, C. C. Carcinogenesis 1989, 10, 2149.

126. Nebert, D. W.; Negishi, M. Biochem. Pharmacol. 1982, 31, 2311.
127. Savela, K.; Leppala, S.; Hemminki, K. Arch. Toxicol. Suppl.
     1989, 13, 101.
128. Phillips, D. H.; Hewer, A.; Martin, C. N.; Garner, R. C.; King,
     M. M. Nature 1988, 336, 790.
129. Bryant, M. S.; Skipper, P. L.; Tannenbaum, S. R.; Maclure, M.
     Cancer Res. 1987, 47, 602.
130. Farmer, P. B.; Bailey, E.; Gorf, S. M.; Tornqvist, M.;
     Osterman-Golkar, S.; Kautiainen, A.; Lewis-Enright, D. P.
     Carcinogenesis 1986, 7, 637.
131. Tornqvist, M.; Osterman-Golkar, S.; Kautiainen, A.; Jensen, S.;
     Farmer, P. B.; Ehrenberg, L. Carcinogenesis 1986, 7, 1519.

RECEIVED August 30, 1990

Chapter 18

Immunoaffinity-Based Monitoring of Human Exposure to Aflatoxins in China and Gambia

John D. Groopman and Audrey Zarba

Department of Environmental Health Sciences, School of Hygiene and Public Health, The Johns Hopkins University, 615 North Wolfe Street, Baltimore, MD 21205

Monoclonal antibody immunoaffinity column chromatography (IAC) used in combination with high performance liquid chromatography (HPLC) is a useful technique that permits the concentration, resolution, quantification of aflatoxin B1 (AFB1), its metabolites, and the aflatoxin-DNA adducts from various body fluids and food samples of people chronically exposed to aflatoxin. Two population groups of men and women living in Guangxi Province, People's Republic of China and the village of Keneba, The Gambia were studied and samples measured by combined IAC/HPLC. In both studies samples of blood, urine, and human milk were collected. The average daily intakes of AFB1 of men and women studied in Guangxi Province were 48.4 ug and 92.4 ug, respectively. Partial results of analyses of blood and urine for aflatoxin-DNA adducts are reported here. We found two advantages to using IAC/HPLC methods: (1) IAC/HPLC permits efficient analysis of many samples of body fluids and food or grain samples in a short time; (2) IAC/HPLC is non-destructive to the aflatoxin molecule so the same sample aliquot can be used for confirmatory analysis.

Using epidemiological methods, the relationship between aflatoxin exposure and human liver cancer has been hindered by inadequate data on aflatoxin consumption, excretion, metabolism, and the general poor quality of world-wide cancer morbidity and mortality statistics. Molecular dosimetry methods are needed to help accurately assess an individual's exposure to aflatoxins. This is especially important because of the recent reclassification by the

0097–6156/91/0451–0207$06.00/0

International Agency for Research on Cancer (IARC) of aflatoxin B1 (AFB1) to a Category I, known human carcinogen (1).
The development of molecular dosimetry methods to permit the
monitoring of an individual's aflatoxin induced genotoxic
burden will lead to the identification of people at high
risk for developing disease long before clinical manifesta-
tion occurs. In this chapter, we will describe the use of
immunoaffinity/HPLC techniques for urine, serum albumin, and
food analysis and how this method has been applied to the
biological monitoring of people exposed to aflatoxins in
China and The Gambia.
     Extensive literature reviews of the aflatoxin field
have been published and the reader should refer to the
following recent reviews that describe the toxicology (2);
biological monitoring (3) and epidemiological aspects of
aflatoxins and liver cancer (4) for more complete descriptions of research in these areas.

Immunoaffinity Chromatography and Molecular Dosimetry of
Aflatoxin

Urine The initial pilot study was conducted in Guangxi
Province, People's Republic of China. This investigation
explored both the dietary consumption of aflatoxin B1 and
the urinary excretion of oxidative aflatoxin metabolites in
the same person. The methods employed immunoaffinity
chromatography with subsequent HPLC quantitation. These
pharmacokinetic data are essential in determining the relationship among aflatoxin exposure, biologically effective
doses, and liver cancer (5-8). Twenty individuals were
selected from a presumptive high aflatoxin exposure group
and urine samples were obtained in the morning. The daily
intake of AFB1 from the diet, which is primarily corn contaminated with aflatoxin B1, ranged from 13.4 to 87.5 ug
AFB1. Urine samples from four individuals who had been
exposed to the highest level (87.5 ug) the previous day were
prepared with the monoclonal antibody immunoaffinity column
and then individual aflatoxin metabolites were separated by
analytical HPLC. These prove the presence of the major
AFB1-DNA adduct, AFB1-N7-Gua, and indicated that the monoclonal antibody columns, in combination with HPLC, can
quantify aflatoxin-DNA adducts in human urine samples obtained from individuals environmentally exposed to AFB1.
     The above experience stimulated a more extensive fol-
low-up study in Guangxi Province. To assess the dose and
excretion of AFB1 and its adducts in chronically exposed
people, the following protocol was developed. The diets of
30 males and 12 females, ages ranging from 25 to 64 years,
were studied for one week and the total aflatoxin intake was
determined for each day. Urine was obtained in two twelve
hour fractions for three consecutive days during the one

week collection period. These urine samples were obtained only after dietary aflatoxin levels had been measured for at least three consecutive days and the urine collections were initiated on the fourth day of the protocol. The average male intake of AFB1 was 48.4 ug per day and the mean total intake for the seven day period was 276.8 ug AFB1. The average female daily intake was 92.4 ug AFB1 per day. Preparative immunoaffinity chromatography was performed on aliquots of the twelve hour urines. In total, 256 urine samples were analyzed.

Complete oxidative aflatoxin metabolite excretion into urine for each twelve hour sample period was calculated by multiplying the urine volume by the concentration of aflatoxins determined by radioimmunoassay in the aliquot of urine. **Figure 1** depicts a scatterplot comparison of aflatoxin intake with total oxidative aflatoxin metabolite excretion. All the male and female data were combined for this presentation. The aflatoxin intake data represents the total integrated ingestion by an individual for the day before urine collection and during the three day urine collection. The excretion data are the composite of all aflatoxin metabolites excreted into the urine during the three days of urine sampling. These data reveal that despite a twenty fold range of aflatoxin B1 dietary intake the amount of aflatoxin excreted generally varied only over a three fold range.

All urine samples were then analyzed by HPLC and AFM1, AFP1 and the major AFB-DNA adducts quantified **(Figure 2)**. When the aflatoxin DNA adduct excretion levels in urine were correlated to aflatoxin B1 dietary intake amount in ug per day, a dose dependent excretion is seen. Because the aflatoxin-DNA adduct has a relatively short half life in DNA (2) this dose dependent excretion pattern reflects exposures during the previous few days. The correlation coefficient for this association of adduct excretion and AFB1 intake is 0.6 with a P value of less than 0.00001. Taken together, it appears that the quantifying of aflatoxin DNA adducts in urine is a valid compartment to sample for aflatoxin exposure, but more data must be collected for developing a risk model for people.

Serum Albumin Another application of the immunoaffinity chromatography technique was in the analysis of AFB1 bound to serum albumin (9). Blood samples were collected on Day 5 from the same individuals who participated in the study described above. Serum albumin was selectively isolated from blood and subjected to enzymatic proteolysis using Pronase. Aflatoxin specific adducts were purified by immunoaffinity chromatography and quantified by competitive radioimmunoassay. A highly significant correlation of adduct level with AFB1 intake (r = 0.69, P less than 0.000001) was observed. From the slope of the regression

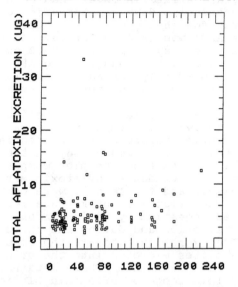

1. Scatterplot of total male and female intake compared to
excretion of aflatoxin metabolites of people living in
Fusuai County, Guangxi Province, P.R.C.

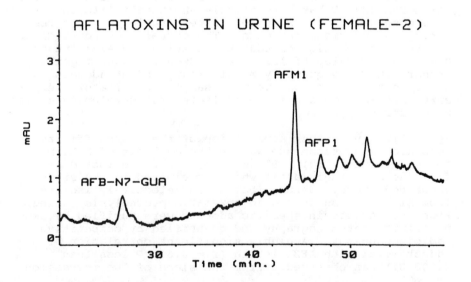

2. IAC/HPLC profile of the aflatoxin-DNA adduct and other
aflatoxin metabolites of a person living in Fusuai County,
Guangxi Province, P.R.C.

line for adduct level as a function of intake, we calculated that 1.4-2.3% of ingested AFB1 becomes covalently bound to serum albumin. These combined studies indicate the amount of serum albumin adduct formed and the amount of DNA adduct excreted can be used as a way to identify populations at risk and that quantitation of these adducts are useful dosimeters in exposed people.

Food Analysis Preparative immunoaffinity chromatography techniques are also applicable to the analysis of food samples for the presence of parent aflatoxins. To validate this method, naturally contaminated corn and peanut product samples were obtained and extracts made using methanol-water (60%:40%, vol/vol). An aliquot of extract was applied to the monoclonal antibody affinity column, aflatoxins eluted, and the product measured by reversed phase HPLC. In **Figure 3** is depicted the HPLC profile from a naturally contaminated peanut sample. This sample when analyzed contained 943 ppb (ug/kg) total aflatoxins. This chromatogram demonstrates that aflatoxins can be identified without any further derivatization by using a diode array detector after the affinity clean-up procedure and reversed phase HPLC. This detector obtains the spectrum of chromatographic peaks. Since as little as 2 ppb in the diet of rats (2) has been found to produce liver tumors following life time exposure, the levels of aflatoxins in the diets of people eating this level of contaminated foods would be of serious concern.

The Gambia Aflatoxin-Hepatitis B Study The preliminary data gathered in China encouraged the establishment of a longer term study in The Gambia, where liver cancer is a major disease. This is a large collaborative study involving the IARC, the Massachusetts Institute of Technology (MIT) and The Johns Hopkins University and is ancillary to The Gambia Hepatitis Intervention Program. In the first year, dietary intake of aflatoxin of twenty individuals was determined during a consecutive eight day period and the excretion of aflatoxin metabolites in 24 hour urine collections from days 4 to 8 was measured. The individuals selected for this study included both chronic hepatitis B virus (HBV) carriers and non-carriers. This is the first time HBV status, also thought to be an important risk factor in the etiology of liver cancer, has been integrated into a molecular epidemiology study with aflatoxin B1. We hope to obtain information about the interaction of AFB1 and HBV and its relation to AFB1 metabolism. In addition, insights into the role of specific cytochrome P450 metabolism of aflatoxin may be gained.
     This four year molecular epidemiological study in The Gambia will help elucidate the role of hepatitis B virus and aflatoxins in the etiology of liver cancer by the analysis and quantitation of specific biomarkers. Both seasonal and

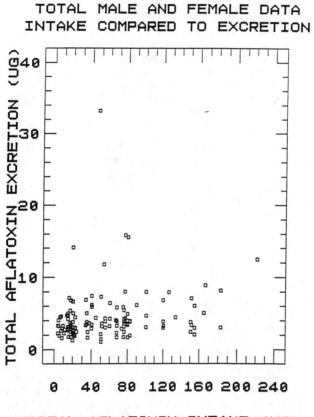

TOTAL MALE AND FEMALE DATA
INTAKE COMPARED TO EXCRETION

TOTAL AFLATOXIN INTAKE (UG)

3. Reversed phase HPLC profile of naturally contaminated
peanut sample prepared by monoclonal antibody immunoaffinity
chromatograph.

annual variabilities of aflatoxin consumption, metabolism, urinary excretion and albumin adducts in people, both hepatitis B surface antigen (HBsAg) carriers and non-carriers are being examined. Therefore, these studies will systematically evaluate the relationship between dietary intake of aflatoxins and biological markers, such as, excreted AFB-DNA adducts, and oxidative metabolites in urine, aflatoxin M1 in human milk, and covalently bound aflatoxin to serum albumin, to determine which of these markers has utility for noninvasively assessing the exposure status of people at high risk for liver cancer.

All collections are carried out in the village of Keneba. Keneba is a typical rural African village with a population of approximately 600. In Keneba, the Dunn Nutrition Unit has established housing facilities for visiting scientists as well as a sophisticated laboratory with the requisite facilities for aliquoting the food, urine, milk and blood samples that will be collected during this study.

Presently the first year collection of samples has occurred and preliminary data on aflatoxin consumption, albumin adduct binding and urine excretion of adducts analyzed. The data are very preliminary in nature, but we hope that over the next three years a more complete picture of aflatoxin metabolism in people will be elucidated.

Summary Primary hepatocellular carcinoma is one of the most common cancers in the world and is prevalent on the continents of Africa and Asia. A number of classical epidemiological studies have determined that the exposure status of people to aflatoxin B1 is an important risk factor in the etiology of liver cancer. However, these studies have only relied upon the criteria of presumptive intake data, rather than information obtained from quantitative analyses of food samples, biological fluids and from people exposed to aflatoxin. Information obtained by monitoring exposed individuals for specific DNA adducts and metabolites will define the pharmacokinetics of aflatoxin B1 in people, thereby facilitating risk assessments. Preliminary data, reported here, support the concept that measurement of the major, rapidly excised AFB-N7-Gua adduct in urine and quantification of the more persistent aflatoxin albumin adduct are appropriate dosimeters for estimating exposure status and possibly risk in individuals consuming this mycotoxin.

Acknowledgments: This work was supported in part by grant from the USPHS 1 U01 CA48409 and ES 07141 from NIEHS.

Literature Cited

1. IARC Monographs of the Evaluation of the Carcinogenic risk of Chmeicals to Man, Supplement 7, Lyon, France,1988.

2. Busby, W.F. and Wogan, G.N. Aflatoxins. (1985) In: Chemical Carcinogens, 2nd Ed. C.E. Searle,(Ed.), pp. 945-1136, American Chemical Society, Washington,D. C.

3. Groopman, J.D., Cain, L.G. and Kensler, T.W. (1988) CRC Critical Reviews in Toxicology 19:113-145.

4. Bosch,F.X. and Munoz, N.   (1988) IARC Sci. Publ. 89:427-38.

5. Groopman, J.D., Donahue, P.R., Zhu, J., Chen, J., and Wogan, G.N. (1985) Proc. Nat'l. Acad. Sci., USA. 82: 6492-6497.

6. Groopman, J.D., Donahue, P.R., Zhu, J., Chen, J. and Wogan, G.N. (1987) Proc. Amer. Assoc. Cancer. Res. 28:36.

7. Groopman, J., Gan, L., Skipper, P., Chen, J., Tannenbaum, S. and Wogan, G. (1989) Proc. Amer. Assoc. Cancer Res. 30:317.

8. Groopman, J.D., Donahue, P.R., Zhu, J., Chen, J. and Wogan, G.N. (1990) Cancer. Res., submitted.

9. Gan,L.S. Skipper,P.L., Peng,X., Groopman,J.D., Chen,J-S, Wogan,G.N., Tannenbaum,S.R. (1988) Carcinogenesis. 9:1323-5.

RECEIVED August 30, 1990

Chapter 19

Immunological Quantitation of Human Exposure to Aflatoxins and *N*-Nitrosamines

Christopher P. Wild and Ruggero Montesano

Unit of Mechanisms of Carcinogenesis, International Agency for Research on Cancer, 150 cours Albert Thomas, 69372 Lyon Cedex 08, France

Methods for measuring exposure to aflatoxins (AF) and N-nitrosamines (NNO) have been developed and applied to the analysis of human samples. A combination of immunoassay and hplc-fluorescence techniques permits the quantitation of AF-albumin adducts in serum samples from fingerprick blood samples and this approach is applicable to field studies. AF-albumin adducts have been measured in various African and Asian populations, with no detectable levels found in European populations. Antibodies to the methylated DNA bases, 06-methyldeoxyguanosine and 7-methyldeoxyguanosine, have allowed the detection of these adducts in peripheral blood cells in experimental animals following exposures to NNO. The potential application of these methodologies to epidemiological studies of the etiology of various human cancers is discussed.

The understanding of the etiology of a large proportion of human cancer requires an integrated effort of laboratory scientists and epidemiologists because of the multifactorial origin of most tumours and the multistage process of neoplastic development. Man is exposed to a variety of carcinogens (1) and among these aflatoxins (AF) and N-nitrosamines (NNO) could be particularly relevant as human exposure to these carcinogens is extensive and their carcinogenicity has been observed in a variety of experimental systems (2, 3).

Epidemiological studies implicate AF as of importance in the induction of liver cancer in man, particularly in certain parts of the world (e.g. South-east Asia and sub-Saharan Africa), however, it is not evident whether this carcinogen induces cancer alone or through an interaction with hepatitis B virus (HBV) infection (4). Although human exposure to NNO is well documented (see 3, 5), it is not evident which human cancer(s) could be directly attributed to this group of compounds.

One significant limitation in clearly demonstrating the causal association between exposure to the above compounds and the etiology of disease has been the lack of methods to measure exposure to carcinogens, or their interaction with other risk factors, at an

0097–6156/91/0451–0215$06.00/0
© 1991 American Chemical Society

individual level (6). In response, significant methodological
advances have been achieved in the quantitation of markers of
exposure in human tissues and body fluids for both AF (see 7, 8) and
NNO (see Wild C.P. and Montesano R, in: Molecular Dosimetry of Human
Cancer: Epidemiological, Analytical and Social Considerations,
Telford Press, 1990, in press). This paper reviews and discusses the
properties of some of the most promising methods now available and
assesses their suitability for integration into epidemiological
studies.

Aflatoxin

Methodological Approaches. The major approaches to measuring
individual human exposure have included the measurement of AF
metabolites and DNA or protein adducts in urine, serum, tissues and
milk (see 7, 8). The most widely applied techniques have involved
the use of antibodies both to purify the AF molecules by
immunoaffinity chromatography and to quantitate by radioimmunoassay
or ELISA (see 7, 8, Groopman J.D., Sabbioni G. and Wild C.P., in:
Molecular Dosimetry of Human Cancer: Epidemiological, Analytical and
Social Considerations, Telford Press, 1990, in press, and Groopman et
al., this volume). Recently the measurement of AF-albumin adducts in
serum has emerged as a most promising marker for determining
individual exposure to this carcinogen in field studies for a number
of reasons.
    First, it has been demonstrated that AFB_1 binds covalently to
albumin after single and multiple doses (9, 10) in rats in a
dose-dependent manner and in the latter case an accumulation of
binding to a steady state level is achieved. In man, with a
half-life of albumin of approximately 20 days, an accumulation of
AF-albumin adducts to a level 30 times that of a single exposure can
be calculated (11). Second, Sabbioni et al. (11) specifically
identified AFB_1-lysine as a major adduct in rat albumin and this
adduct has recently been observed in human sera (12). The adduct is
a result of the same metabolic activation which leads to nucleic acid
adduct formation (11) and therefore (i) may reflect the level of DNA
damage in the hepatocyte, the target cell for experimental AFB_1
carcinogenesis (10); (ii) gives a reliable indirect measurement of
biologically effective dose at an individual level (13). Moreover,
data are also available which demonstrate that AFB_1 intake in food is
reflected by binding to serum albumin (14, Wild et al. unpublished
data); and (iii) may provide a more reliable basis for extrapolation
of dosimetry/carcinogenicity data in animal models to the human
situation. Third, antibodies have been developed and used to detect
AF-albumin adducts including AFB_1-lysine in peripheral blood by an
enzyme-linked immunosorbent assay following albumin hydrolysis (12).
The assay has been validated in experimental systems in comparison
with other methods of detection (hplc-fluorescence, radioactive
carcinogen) and has been shown to be highly sensitive and specific.
The immunoassay can be performed on as little as 1 mg hydrolysed
albumin (50 µl serum) which also makes it feasible to apply in field
studies.

Applications to Human Sera. The immunoassay for AF-albumin adducts
has been validated in terms of specificity and reproducibility using

sera from subjects in various parts of the world.   In total, around
80 serum samples from France and Poland have been found to be
negative for the presence of AF-albumin adducts by ELISA (detection
limit: 5 pg AFB_1-lysine per mg albumin).   This provides one
indication that the assay has a very low false positive rate.   The
specificity has been further examined by measuring AF-albumin by both
the immunoassay and an hplc-fluorescence technique (12).   Whilst the
immunoassay approach involves measuring total AF moieties present in
the hydrolysed albumin sample, the hplc-fluorescence detects one
specific adduct, AFB_1-lysine, and therefore the two techniques used
together ensure a high specificity to further eliminate the risk of
false positive results.   Figure 1 shows the correlation between the
two methods using human sera from an African population.   From
experiments in rats using [^{14}C]AFB_1 it was shown that the immunoassay
was more sensitive than hplc-fluorescence with low quantities of
albumin (12) and this should be the predominant explanation for the
fact that samples below about 50 pg AFB_1-lysine per mg albumin in
ELISA are undetectable by hplc-fluorescence.   In addition,
quantitatively the ELISA gives higher levels probably because as
mentioned above, it is measuring any AF moieties in hydrolysed
albumin which are recognised by the antibody whilst hplc-fluorescence
is measuring only AFB_1-lysine.   Another contributing factor may be
that in addition to AFB_1 in food samples one also finds AFG_1 (see 2).
This aflatoxin can cause DNA and protein modification in a similar
way to AFB_1 (15) and the antibody used in ELISA will react with
AFG_1-DNA (16) and -albumin (Sabbioni, G., and Wild, C.P., unpublished
data) adducts so that any AFG_1 residues present in human albumin
hydrolysates could also be measured.   In contrast, the hplc-
fluorescence technique separates AFB_1 and AFG_1 adducts.   We examined
this question by treating rats with either AFB_1 or AFG_1 and comparing
the inhibition of the albumin hydrolysates in ELISA (Table I).   We

Table I.   Levels of AF-Albumin Adducts in Rat Plasma
following Treatment with Aflatoxins B_1 and G_1

Dose (mg/kg)	Aflatoxin	Mean AF-albumin level (pg AF-lysine / mg albumin)[a]
1.20	B_1	9855
0.97	G_1	229
0.32	B_1	2885
0.27	G_1	82

[a]   BDIV rats were treated by gavage with AFB_1 or AFG_1 in olive oil.
Plasma was collected 48h after treatment and assayed for
AF-albumin adducts in ELISA using an AFB_1-lysine standard as
described (12)

observed about a 35-fold lower level by immunoassay for AFG_1-albumin adducts compared to AFB_1-albumin adducts after the same dose of AF when quantitating in both cases against an AFB_1-lysine standard in ELISA. Consequently in human sera, the contribution of AFG_1 albumin-adducts to the total quantitated AF-albumin adduct level would be expected to be relatively small, particularly considering that AFG_1 levels are generally lower than AFB_1 in the same food sample (see 2). It should not be concluded, however, that AFG_1 gives a 35-fold lower absolute level of adduct because the specificity of the antibody varies for the AFG_1- and AFB_1-albumin adducts and quantitation was made against an AFB_1-lysine standard curve in ELISA. In fact studies by hplc-fluorescence, or ELISA quantitating against an AFG_1-lysine standard, indicate that the albumin binding is only between two- and five-fold lower with AFG_1 compared to AFB_1 (Sabbioni, G. and Wild, C.P., unpublished observations).

Data are accumulating on the prevalence and levels of AF-albumin adducts in several populations using the above techniques and a summary of some of these data is given in Table II. It is clear from the samples analysed to date that AF exposure appears to be higher in Africa compared to Thailand whilst in Europe we have not observed any positive sera. Although these data do not provide a complete geographical or temporal picture of AF exposure in these countries they do provide (i) valuable data concerning the level of adduct and exposure which is occurring in humans and (ii) demonstrate that the approach has the required sensitivity and specificity to be applied in field studies.

Table II. Aflatoxin Albumin Adducts in Human Sera

Country	No. analysed	Percent positive	AF-albumin adducts (pg AF-lysine/mg albumin) mean ± SD (range)
Kenya	61	49	45.2 ± 70.0 (4.6 – 338)
Thailand	84	19	10.7 ± 11.5 (3.6 – 50)
The Gambia	20	100	44.1 ± 51.6 (6.8 – 176)
France	14	0	–

N-nitrosamines

Methodological Approaches. In analogy with AF, several approaches to measuring exposure to NNO have been taken including excised DNA adducts excreted in urine, DNA adducts in tissues and haemoglobin adducts (see Wild C.P. and Montesano R., in: Molecular Dosimetry of Human Cancer: Epidemiological, Analytical and Social Considerations, Telford Press, 1990, in press, 13, 17, 18). In the case of NNO our work has been focused on methylating agents and specifically to detecting two DNA adducts, 06-methyldeoxyguanosine (06-medG) and 7-methyldeoxyguanosine (7-medG), using a combination of hplc and immunoassay (19, 21). Methylation adducts were studied because frequent exposure to methylating agents occurs from tobacco, diet and through drug exposures (3, 22, 23). Using monoclonal antibodies to 06-medG we were able to measure this adduct in surgical samples of oesophagus and stomach from patients in high and low risk areas for cancer at these two sites (20). These studies have been extended by ourselves and others to show the presence of 06-medG in other human tissues, including colon, liver, placenta, lung, oral mucosa (Wild, C.P. and Montesano R., in: Molecular Dosimetry of Human Cancer: Epidemiological, Analytical and Social Considerations, Telford Press, 1990, in press, 24, 25). Whilst these studies are important in demonstrating the formation and persistence of this promutagenic adduct in human tissues, the application to epidemiology has been limited because of the difficulty in obtaining tissue samples and the limitation to retrospective studies.

Developments in two directions have been made in order to circumvent these problems, one concerning improved sensitivity of assay techniques for 06-medG and other adducts and the second involving the use of more readily available human DNA.

Some promising approaches to improving sensitivity with small DNA samples (eg 100 µg or less) are the ^{32}P-post-labelling technique, immunoslot blot assays, immunohistochemistry and a competitive 0^6-meG repair assay.

Wilson et al. (26) used an hplc purification step, prior to ^{32}P-post-labelling, to detect 06-medG in human lung DNA samples. In the same study the small quantities of DNA (~100 µg) were insufficient to allow quantitation by hplc-radioimmunoassay. In addition by a similar approach the former method has been applied to 7-medG (27). These methods with sensitivities of around one adduct in 10^7 parent deoxynucleosides in a 100 µg DNA sample may permit the analysis of DNA from peripheral blood cells or biopsies. The use of antibodies to purify methylation adducts, as has been reported for aflatoxin (8, 28), benzo(a)pyrene (29) and 3-methyladenine (30), prior to analysis by ^{32}P post-labelling, is also a potentially useful approach which may improve both the specificity and the speed of analysis.

Another approach which requires further validation in human tissues was reported by Souliotis and Kyrtopoulos (31). The principle involves a two-step competition assay for 06-medG repair using AGT, DNA and radioactive 06-medG containing oligonucleotides. This method has been reported to detect 0.5 fmol 06-medG in less than 10 µg DNA.

In order to circumvent the limited sensitivity imposed by the smaller quantities of DNA which can be obtained from peripheral blood or biopsies compared to tissue samples, we have developed antibodies to imidazole ring-opened 7-methyldeoxyguanosine (iro-7-medG) (21) to permit quantitation of 7-medG. This latter adduct is formed in greater amounts and is generally repaired at a slower rate than 06-medG and consequently may be present at higher levels in DNA. In addition, as a result of these properties, 7-medG might accumulate in cellular DNA and thus give a more sensitive and informative measure of past exposure (Montesano, R., Hall, J., Hollstein, M., Mironov, N. and Wild, C.P. Proceedings of Symposium "DNA Damage and Repair in Human Tissues", Plenum Press, 1990, in press). To date, these antibodies have been used in an hplc-ELISA procedure (21, see below and Figure 2) but more recently they have proved to be applicable to

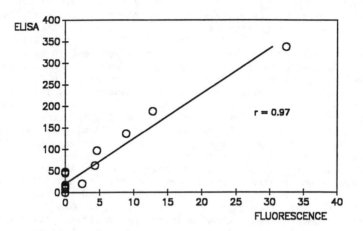

Figure 1. Levels of AF-albumin adducts (pg AFB_1-lysine/mg albumin) as measured by immunoassay and hplc-fluorescence using methods described previously (12).

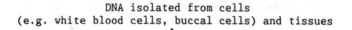

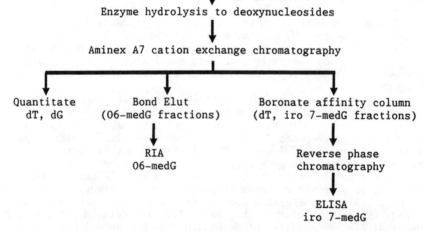

Figure 2. Analysis of Methylated Bases in DNA

in situ visualisation of adducts by immunohistochemistry (32) and in
an immunoslot blot procedure (unpublished data in collaboration with
Drs B. Ludeke and P. Kleihues). These latter two techniques require
significantly smaller samples than the hplc-ELISA approach.
The immunohistochemistry approach would be particularly
informative in examining the distribution of methylation adducts
amongst different cell types of a given tissue. In addition, if the
methods can be adapted to fixed paraffin embedded tissues, then
retrospective studies could be performed on samples already
available. At present the detection limits for methylation- and
aflatoxin-DNA adducts are of the order of 1 in 10^5 to 10^7 unmodified
nucleotides (16, 33) and further studies are required to see if this
level of sensitivity is sufficient for human studies.

DNA Alkylation in Peripheral Blood Cells. One of the most readily
available sources of human DNA is from peripheral blood cells.
Initially it was necessary to establish whether the above methylation
adducts were formed in vivo in leucocytes upon systemic
administration of methylating agents. At the same time it is
important to understand the relationship between the adduct levels in
internal organs and leucocytes if these cells are to be used as
surrogates for target organ exposure (see 34). There are very
limited data concerning these questions for methylating agents and
the information available is summarized in Table III. It is clear at
least with two of the three compounds tested in rats that (a)
methylation does occur in peripheral blood cell DNA and (b) that the
level of adduct is not greatly dissimilar to that seen in the

Table III. Methylation of Leucocyte DNA in Rats

Methylating Agent	Dose (mg/kg)	Time (h)	Internal organ(s)	Ratio of 06-medG internal organ: leucocytes
Dimethyl- nitrosamine	0.4; 1.0	6	Liver	1.0; 1.2[a]
		24	Liver	1.4; 0.9[a]
Dimethyl- nitrosamine	0.03-1.0	6	Liver	≃1.1[a]
Procarbazine	5	2	Liver	2.4
	10	2	Liver	1.9
NNK	30; 100	24	Lung	>1.2[b]

[a] Lymphocytes
[b] In experiments using 4-(N-nitrosomethylamino)-1-(3-pyridyl)-1-
butanone (NNK) the sensitivity of analyses in leucocytes only
permitted ratios of 1.2 or lower to be detected. Consequently
methylation in these cells may occur at ratios higher than 1.2.
Data from refs. 21, 31 and Wild et al., unpublished data

internal organs examined at least at the doses tested. Further
studies will show if such a relationship also occurs at lower dose
levels.
    The advantage of 7-medG as a marker of exposure is clearly
demonstrated in Table IV. In this experiment the level of 7-medG
accumulates throughout the treatment period but the O6-medG remains at
low levels probably due to the effective repair of this adduct.
Similar results of accumulation of 7-medG in rat liver, lung and
kidney after [^{14}C]dimethylnitrosamine treatment have been observed
over periods up to 24 weeks (35). These results are consistent with
observations of a low level of glycosylase repair activity for 7-medG
in rat tissues (36) and it is interesting that similarly low repair
has been observed in human liver samples (Hall, J. and Montesano, R.,
unpublished data).

Table IV. Methylated Bases in Rat Leucocyte DNA after Single
or Multiple Treatment with 1 mg/kg Dimethylnitrosamine (DMN)[a]

Treatment	(μmoles/mole dG)		Ratio: $\dfrac{O6\text{-medG}}{7\text{-medG}}$
	7-medG	O6-medG	
1 x 1 mg/kg	236	27.0	0.1027
5 x 1 mg/kg	1393	9.8	0.0070
15 x 1 mg/kg	4246	7.4	0.0017

[a]BDIV rats were administered DMN daily by gavage (5 days per week)
and were sacrificed 6 hours after the last treatment. Analyses
were performed as described previously (21).

    Based on these data we have begun to assay for the presence of
7-medG in human leucocytes, lymphocytes and some other tissues. Some
preliminary data for peripheral blood cells are shown in Table V. The
values range from non-detectable (< 1μmole 7-medG/mole dT) to 49μmole
7-medG/mole dT (a leucocyte sample). For comparison, data are
presented for adduct levels in leucocytes from cancer patients treated
with 300 or 600 mg N-methyl-N-nitrosourea which are one to two orders
of magnitude higher (37).
    A major concern with the analysis of 7-medG is the possibility of
contamination with 7-methylguanosine which is a normal constituent of
tRNA at levels of around 3.5 mmoles 7-methylguanosine per mole
ribonucleosides. As a consequence a 1% contamination of DNA with RNA
would result in an 'apparent' 7-medG level of around 35 μmoles adduct
per mole dT. In order to overcome this problem we have incorporated,
amongst other measures, a boronate affinity chromatography step
(aminophenylboronic acid-agarose, Sigma, St. Louis, MO, USA) which
specifically retains ribonucleosides whilst the DNA derived
deoxynucleosides elute in the void volume. Following this separation
on approximately 2 ml/gel columns the eluate containing dT and
iro-7-medG is separated by reverse-phase chromatography and ELISA is
performed on hplc fractions. The overall methodology is shown in

Table V. Levels of 7-Methyldeoxyguanosine
in Human Peripheral Blood Cells

Samples	Fraction of positive samples	μmole 7-medG / mole dT
White blood cells (Chemotherapy-NMU)	14/14	147 – 1162[a]
White blood cells (Smokers)	4/12	9; 18; 37; 49
Lymphocytes[b]	8/8	4 – 29

[a] μmole 7-medG/mole dG
[b] Each sample was pooled from four to six individuals

Figure 2. We have demonstrated the quantitative recovery of
thymidine by boronate affinity chromatography and the quantitative
retention of adenosine with up to 1 μmole quantities.

In four individuals from whom leucocytes were collected (see
Table V) sufficient DNA was available to repeat the analyses
following heating of the DNA at 100° for 30 min. at pH7. This latter
treatment should cause depurination of methylated bases (38) and thus
yield a sample free of 7-medG. Using this approach the two samples
having an apparent level of 7-medG of 37 and 18 μmoles/mole dT were
negative after heat depurination whilst two samples which were
negative prior to heat depurination were also negative after this
treatment.

The data from animal models and from human samples reported
above supports the rationale of measuring 7-medG in peripheral blood
cells and the use of hplc-ELISA together with other complementary
techniques appears feasible.

Possibilities of Integration into Epidemiological Studies

The above summary illustrates the advances which have been made in
developing ways of quantitating markers of human exposure to AF and
NNO and some considerations can be made regarding their application
to epidemiological studies. It should be noted, however, that the
integration of these methods into epidemiological studies has been so
far very limited.

For the assay of AF exposure the use of AF-albumin adducts is a
significant advance over previous methods of extrapolation from food
analysis and the use of questionnaires (see 4), giving a measure of
biologically effective dose which may also reflect DNA damage in the
liver (as discussed above). The availability of two complementary
methods constitutes a strong reliable approach although it is still
noteworthy that it is the immunoassay which permits the application
to field studies with the hplc-fluorescence analysis limited to tens

of samples rather than hundreds. The immunoassay procedure can be
performed on fingerprick blood samples and could be easily integrated
as one parameter of studies which also examine other end-points such
as serological markers of hepatitis. The application to correlation,
case-control and cohort (prospective) studies is therefore feasible.
Nevertheless the potential advantages for any one study should be
carefully assessed prior to application and in fact a good
understanding of the interpretation of the assay data is essential if
the epidemiologist is to take full advantage of the potential. For
example the use of AF-albumin adducts in cohort studies of hepatitis
B virus surface antigen (HBsAg) carriers using primary hepatocellular
carcinoma as the endpoint is potentially a powerful approach to
examining the role of AF in the progression of liver damage to liver
cancer. However, as the marker is probably only relevant to exposure
over the previous 2-3 months this implies repeated sampling of the
cohort for the marker to be of most value. Therefore it is
advantageous to actually design the study based on a knowledge of the
qualities of the available marker rather than adding the marker as a
component to an existing study. This implies an interaction of
epidemiologists with laboratory scientists at the point of study
design. For example, a single AF-albumin assay only gives a measure
of exposure in a small fraction of the lifespan of an adult and may
be limited on this basis, but if the assay was performed at birth
(cord blood) and, for example, at times of routine vaccination during
the first year of life a measure of 'lifetime' (in utero included)
exposure could be obtained for a particular individual. This type of
approach with a relatively short follow-up could certainly be used
for example to test whether AF exposure early in life increased the
chances of becoming an HBsAg carrier following HBV infection. In
contrast, the length of follow-up and the size of cohort required for
a study of liver cancer in, for example, non-HBsAg carriers, could be
argued to be prohibitive but other shorter term clinical end-points
could perhaps be used and a precedent has been set in the form of the
Gambia Hepatitis Intervention Study (GHIS) on HBV vaccination (39,
40) where a 40-year follow-up is envisaged.
     In the context of case-control studies of liver cancer the
limitation is more prominently the lack of information on past (i.e.
20-30 years) exposure. This again may be circumvented by examining
end-points other than cancer. Precancerous lesions or markers (e.g.,
alphafoetoprotein) could be used as end-points with which the
AF-albumin assay could be compared in a case-control approach. In
addition, a case-control approach could be employed to investigate
short-term effects of AF exposure such as aflatoxicosis.
     In correlation studies or for monitoring population exposures
the advantages of the assay are less evident because the major object
of developing the methodology i.e. to give a measure of individual
exposure, is not a requirement in such studies. As a consequence the
previous approach of assaying AF in food samples could be argued to
be appropriate. This decision may be dependent upon logistics of
sample collection (e.g., the use of blood samples collected for other
purposes), cost of analyses, completeness of knowledge of the
distribution of AF in different food types etc. However
notwithstanding these considerations, the knowledge gained from
measuring at an individual level can be valuable in (i) establishing
baseline or background levels of the marker in a population and (ii)

contributing to the planning of prospective studies, because the
number of individuals, number of repeat samplings, geographical area
in which to conduct the study etc. can be better decided once the
prevalence and level of exposure in a population is known.

The availability of individual markers of exposure could also
lead to the identification of a pathological condition previously
unsuspected of being associated with that exposure. For example, in
collaboration with Drs D. Forman, F. Sitas and G. Lachlan, ICRF,
London, UK, we have found suggestion of a strong association between
dyspepsia and AF-albumin adduct levels in Kenyan individuals (41).

In contrast to AF, the situation with NNO is much less advanced
in terms of integration of markers into field studies. Whilst the
knowledge concerning types of methylation adducts, their repair and
their mutagenic potential is considerable ($\underline{42, 43}$) the assay of low
levels (<1 adduct in 10^7 or 10^8 parent deoxynucleosides) in a way
which could be applied to field studies has not been realised. There
are some specific reasons which can be identified and for which
solutions can be envisaged:

Firstly, sensitivity and specificity have been difficult to
achieve. Athough low levels of 06-medG were found using
hplc-immunoassay (Wild C.P. and Montesano R., in: Molecular Dosimetry
of Human Cancer: Epidemiological, Analytical and Social
Considerations, Telford Press, 1990, in press, $\underline{20, 24, 25}$) this
method has insufficient sensitivity to apply to samples other than
surgical tissues. The immunological quantitation of more abundant
adducts such as 7-medG has to date been hampered by specificity (i.e.
RNA derived 7-methylguanosine). New methods now being validated
should solve these limitations with the goal of providing
complementary methods for the measurement of the same adduct (see
above) as has been done for AF-albumin analysis.

Secondly, environmental exposure to NNO is likely to be low ($\underline{3}$)
and therefore it is more difficult to identify populations known to
be exposed to elevated levels and which could be used to validate the
adduct analyses. One situation in which this problem can be solved
is with patients undergoing chemotherapy with methylating agents such
as N-methyl-N-nitrosourea ($\underline{37}$) or, more commonly, procarbazine
(Souliotis, V. and Kyrtopoulos, S., 1990, unpublished data). These
studies would be valuable to allow correlations between DNA adduct
levels and clinical response as well as validating the assays in
human samples as has been successfully done with cis-platin ($\underline{44}$).

In conclusion, considerable progress has been made towards the
eventual integration of laboratory methodologies into epidemiological
studies and as the interaction between these two disciplines
increases the understanding of the assay requirements becomes better
defined. In addition, the measurement of carcinogen macromolecular
adducts on which this review has focused should be recognized as one
tool of molecular epidemiology, to be used together with other
markers providing information on genetic susceptibility, mutation,
and genetic alterations (oncogene activation, allele loss) as has
been discussed previously ($\underline{45-47}$).

Acknowledgments

The authors express their thanks to Y.-Z. Jiang, B. Chapot, A. Munnia
at IARC for their contribution to the work discussed in this paper.

In addition the work is dependent upon continuing collaboration with
Dr A. Hall and the Gambia Hepatitis Intervention Study Group, Dr G.
Sabbioni, University of Würzburg, Würzburg, FRG, Dr J.D. Groopman,
Johns Hopkins University Baltimore, USA, Dr D. Forman, ICRF,
University of Oxford, UK, and Dr Petcharin Srivatanakul, NCI,
Bangkok, Thailand. The authors also appreciate the comments of Dr
F.X. Bosch, IARC, on the final draft of this manuscript. This work
was supported by USA NCI grant No. 1 U01-CA48409-02, USA NIEHS grant
No. 5-U01-ES04281-03 and the International Programme on Chemical
Safety, WHO.

Literature cited

1. Tomatis, L.; Aitio, A.; Wilbourn, J.; Shuker, L. Jpn. J. Cancer
   Res. 1989, 80, 795-807.
2. Busby, W.F.; Wogan, G.N. Aflatoxins; American Chemical Society
   Monograph 182, Washington DC, 1984; p 945.
3. Bartsch, H.; Montesano, R. Carcinogenesis, 1984, 5, 1381-1393.
4. Bosch, F.X.; Muñoz, N. In Liver Cell Carcinoma; Bannasch, P.;
   Keppler, D.; Weber, G., Eds.; Kluwer Academic: London, 1989; p 3.
5. International Agency for Research on Cancer The Relevance of
   N-Nitroso Compounds to Human Cancer: Exposures and Mechanisms
   Bartsch, H.; O'Neill, I.; Schulte-Hermann, R.; Eds.; IARC
   Scientific Publications No. 84, International Agency for Research
   on Cancer, Lyon, 1987.
6. IARC/IPCS Cancer Res. 1982, 42, 5236-5239.
7. Groopman, J.D.; Cain, L.G.; Kensler, T.W. CRC Critical Reviews in
   Toxicology 1988, 19, 113-145.
8. Wild, C.P.; Chapot, B.; Scherer, E.; Den Engelse, L.; Montesano,
   R. In Methods for detecting DNA damaging agents in humans:
   applications in cancer epidemiology and prevention Bartsch, H.;
   Hemminki, K.; O'Neill, I.K.; Eds.; IARC Scientific Publications
   No. International Agency for Research on Cancer, Lyon, 1988, pp
   67.
9. Skipper, P.L.; Hutchins, D.H.; Turesky, R.J.; Sabbioni, G.;
   Tannenbaum, S.R. Proc. Amer. Assoc. Cancer Res. 1985, 26, 90.
10. Wild, C.P.; Garner, R.C.; Montesano, R.; Tursi, F. Carcinogenesis
    1986, 7, 853-858.
11. Sabbioni, G.; Skipper, P.L.; Büchi, G.; Tannenbaum, S.R.
    Carcinogenesis 1987, 8, 819-824.
12. Wild, C.P.;Jiang, Y-Z.; Sabbioni, G.; Chapot, B.; Montesano, R.
    Cancer Res. 1990, 50, 245-251.
13. Skipper, P.L.; Tannenbaum, S.R. Carcinogenesis 1990, 11, 507-518.
14. Gan, L-S.; Skipper, P.L.; Peng, X.; Groopman, J.D.; Chen, J.;
    Wogan, G.N.; Tannenbaum, S.R. Carcinogenesis 1988, 9, 1323-1325.
15. Garner, R.C.; Martin, C.N.; Lindsay Smith, J.R.; Coles, B.F.;
    Tolson, M.R. Chem.-Biol. Interactions 1979, 26, 57-73.
16. Wild, C.P.; Montesano, R.; Van Benthem, J.; Scherer, E.; Den
    Engelse, L. J. Cancer Res. Clin. Oncol. 1990, 116, 134-140.
17. Shuker, D.E.G. Arch. Toxicol. 1989 Suppl. 13, 55-65.
18. Hecht, S.S.; Carmella, S.G.; Trushin, N.; Spratt, T.E.; Foiles,
    P.G.; Hoffmann, D. In Methods for Detecting Damaging Agents in
    Humans: Applications in Cancer Epidemiology and Prevention;
    Bartsch, H.; Hemminki, K.; O'Neill, I.K., Eds.: IARC Scientific
    Publications No. 89, Lyon, 1988, pp 121-128.

19. Wild, C.P.; Smart, G.; Saffhill, R.; Boyle, J.M. Carcinogenesis 1983, 4, 1605-1609.
20. Umbenhauer, D.; Wild, C.P.; Montesano, R.; Saffhill, R.; Boyle, J.M.; Huh, N.; Kirstein, U.; Thomale, J.; Rajewsky, M.F.; Lu, S.H. Int. J. Cancer 1985, 36, 661-665.
21. Degan, P.; Montesano, R.; Wild, C.P. Cancer Res. 1988 48, 5065-5071.
22. Hoffman, D.; Hecht, S.S. Cancer Res. 1985, 45, 935-942.
23. International Agency for Research on Cancer, Carcinogenicity of Alkylating Cytostatic Drugs Schmähl, D.; Kaldor, J.M., Eds.; IARC Scientific Publications No. 78, International Agency for Research on Cancer, Lyon, 1986.
24. Foiles, P.G.; Miglietta, L.M.; Akerkar, S.A.; Everson, R.B.; Hecht, S.S. Cancer Res. 1988 48 4184-4188.
25. Saffhill, R.; Badawi, A.F.; Hall, C.N. In Methods for Detecting Damaging Agents in Humans: Applications in Cancer Epidemiology and Prevention Bartsch, H.; Hemminki, K., O'Neill, I.K., Eds.: IARC Scientific Publications No. 89, International Agency for Research on Cance, Lyon, 1988 pp 301-305.
26. Wilson, V.L.; Weston, A.; Manchester, D.K.; Trivers, G.E.; Roberts, D.W.; Kadlubar, F.F.; Wild, C.P.; Montesano, R.; Willey, J.C.; Mann, D.L.; Harris, C.C. Carcinogenesis 1989, 10, 2149-2153.
27. Shields, P.G.; Povey, A.C.; Wilson, V.L.; Harris, C.C. Proc. Amer. Assoc. Cancer Res. 1989, 30, 316.
28. Groopman, J.D.; Donahue, P.R.; Zhu, J.; Chen, J.; Wogan, G.N. Proc. Natl. Acad. Sci. USA 1985, 82, 6492-6496.
29. Tierney, B.; Benson A.; Garner, R.C. J. Natl. Cancer Inst. 1986, 77, 261-267.
30. Shuker, D.E.G.; Friesen, M.D.; Garren, L.; Prevost, V. In Relevance to Human Cancer of N-Nitroso Compounds, Tobacco Smoke and Mycotoxins, IARC Scientific Publication No. 105, International Agency for Research on Cancer, Lyon, 1990.
31. Souliotis, V.L.; Kyrtopoulos, S.A. Cancer Res. 1990, 49, 6997-7001.
32. Van Benthem, J.; Wild, C.P.; Vermeulen, E.; Winterwerp, H.H.K.; Den Engelse, L.; Scherer, E. In Methods for Detection of DNA Damaging Agents in Humans: Applications in Cancer Epidemiology and Prevention, Bartsch, H.; Hemminki, K.; O'Neill, I.K., Eds.: IARC Scientific Publications No. 89, International Agency for Research on Cancer, Lyon, 1988, pp 102-150.
33. Den Engelse, L.; Bax, J.; Terheggen, P.M.A.B.; Van Schooten, F.-J.; Van Benthem, J.; Wild, C.P.; Scherer, E. Ann. Ist. Super. Sanità 1989, 25, 11-20.
34. Lucier, G.; Thompson, C.L. Env. Health Perspect. 1987, 76, 187-191.
35. Margison, G.P.; Margison, J.M.; Montesano, R. Biochem. J. 1977, 165, 463-478.
36. Margison, G.P.; Pegg, A.E. Proc. Natl. Acad. Sci. (USA) 1981, 78, 861-865.
37. Wild, C.P.; Degan, P.; Brésil, H.; Serres, M.; Montesano, R.; Gershanovitch, M.; Likhachev, A. Proc. Amer. Assoc. Cancer Res. 1988, 29, 260.
38. Berenek, D.T.; Weiss, C.C.; Swenson, D.H. Carcinogenesis 1980, 1, 595-606.

39. Hall, A.J.; Inskip, H.M.; Loik, F.; Day, N.E.; O'Conor, G.;
    Bosch, X.; Muir, C.S.; Parkin, M.; Muñoz, N.; Tomatis, L.;
    Greenwood, B.; Whittle, H.; Ryder, R.; Oldfield, F.S.,J.; N'jie,
    A.B.H.; Smith, P.G.; Coursaget, P. Cancer Res. 1987, 47,
    5782-5787.
40. Hall, A.J.; Inskip, H.M.; Loik, F.; Chotard, J.; Jawara, M.; Vall
    Mayans, M.; Greenwood, B.M.; Whittle, H.; Njie, A.B.H.; Cham, K.;
    Bosch, F.X.; Muir, C.S. Lancet 1989, 1057-1060.
41. Wild, C.P.; Sitas, F.; Lachlan, G.; Jiang, Y-Z.; Forman, D.
    Proc. Amer. Assoc. Cancer Res. 1990, 31, 231.
42. Saffhill, R.; Margison, G.P.; O'Connor, P.J. Biochim. Biophys.
    Acta 1985, 823, 111-145.
43. Lawley, P.D. Mutat. Res. 1989, 213, 3-25.
44. Poirier, M.C.; Egorin, M.J.; Fichtinger-Schepman, A.M.J.; Yuspa,
    S.H.; Reed, E. In Methods for Detecting Damaging Agents in
    Humans: Applications in Cancer Epidemiology and Prevention
    Bartsch, H.; Hemminki, K.; O'Neill, I.K. Eds: IARC Scientific
    Publications No. 89, International Agency for Research on Cancer,
    Lyon, 1988, pp 313-320.
45. Schulte, P.A. Amer. J. Epidem. 1987, 126, 1006-1016.
46. Bos, J.L. Cancer Res. 1989, 49, 4682-4689.
47. Montesano, R. In Complex Mixtures and Cancer Risk IARC Scientific
    Publications No. 104, International Agency for Research on
    Cancer, Lyon, 1989, pp 11-19.

RECEIVED August 30, 1990

Chapter 20

Immunological Methods for Monitoring Human Exposure to Benzo[a]pyrene and Aflatoxin B$_1$

Measurement of Carcinogen Adducts

Regina M. Santella, Yu Jing Zhang, Ling Ling Hsieh, Tie Lan Young, Xiao Qing Lu, Byung Mu Lee, Guang Yang Yang, and Frederica P. Perera

Cancer Center and Division of Environmental Science, School of Public Health, Columbia University, New York, NY 10032

Immunologic methods have been developed for the measurement of human exposure to environmental and occupational carcinogens by quantitation of carcinogen–DNA or protein adducts. Antibodies recognizing specific adducts have been used in highly sensitive competitive enzyme linked immunosorbent assays (ELISA) to detect femtomole levels of adducts in human samples. Antibodies recognizing benzo(a)pyrene diol epoxide–DNA have been used to quantitate adducts in white blood cells of foundry workers and coal tar treated psoriasis patients and in white blood cells and placenta of smokers and nonsmokers. These same antibodies have been used in indirect immuno-fluorescence staining of tissues to localize adduct formation to specific cell types. Albumin adducts have also been measured in several of these populations. Antibodies recognizing aflatoxin B$_1$–DNA adducts have been used to quantitate adducts in liver tissue from hepato-cellular cancer patients. Adduct measurement provides a relevant marker of carcinogen exposure but may also prove useful in identifying individuals at risk for cancer development.

Monitoring human exposure to environmental or occupational carcinogens can be carried out at several different levels ([1]). External exposure can be measured by methods which quantitate carcinogen levels in air, food and water. Internal dose can be estimated by measuring levels of the carcinogen in body fluids such as blood and urine. More recently, methods have been developed to measure the biologically effective dose of the chemical, defined as the amount of chemical reacting with critical cellular targets ([2]).

Covalent binding of the carcinogen to DNA is believed to be the
initial critical step in chemical carcinogenesis. Thus, measurement
of carcinogen–DNA adducts should be a more relevant marker of
exposure to carcinogens than measurement of the chemical itself in
the environment or body fluids. Such assays take into account
individual differences in absorption and metabolism of carcinogens
as well as repair of adducts once they are formed. In addition to
being a more relevant marker of exposure, it is hoped that such
measurements will be useful as markers of individual risk for
development of cancer.

   While DNA is believed to be the critical target for chemical
carcinogens, such agents also bind to RNA and proteins.
Quantitation of adducts on either hemoglobin or albumin has been
used as an alternate marker of exposure to environmental
carcinogens. In contrast to DNA adduct measurements, large amounts
of protein can be obtained from blood samples and no repair occurs
suggesting that chronic, low levels of exposure may be measurable.
Red blood cells have an average lifespan of 4 months while albumin
has a half life of 21 days. Thus, only recent exposure will be
quantifiable with protein adduct measurement.

   We have concentrated on the development of immunologic methods
for the measurement of carcinogen–DNA and protein adducts as well as
measurement of the carcinogen itself in body fluids such as urine
and sera. These antibodies can be used in highly sensitive
competitive enzyme–linked immunosorbent assays (ELISA) with color–
or fluorescence–endpoint detection to give femto (10^{-15}) mole
sensitivities. For DNA adduct measurement, with this level of
sensitivity, and the ability to assay 50ug of DNA per well adduct
levels in the range of $1/10^8$ nucleotides can be measured. Table I
lists the monoclonal antibodies recognizing carcinogen–DNA adducts
that we have developed to date. Several of these antibodies,
including those recognizing aflatoxin, benzo(a)pyrene diol epoxide
and 8–methoxypsoralen–DNA adducts, have been applied to adduct
detection in humans. A major advantage of immunologic methods is

Table I. Monoclonal Antibodies Recognizing Carcinogen–
DNA and Protein Adducts

Acetylaminofluorene–DNA	(3)
Aflatoxin–DNA	(4)
4–Aminobiphenyl–guanosine	unpublished
Aminopyrene–DNA	(5)
Benzo(a)pyrene diol epoxide–DNA	(6)
Benzo(a)pyrene diol epoxide–guanosine	(6)
Ethenoadenine	(7)
Ethenocytidine	(7)
7–(Hydroxyethyl)guanosine	unpublished
8–Methoxypsoralen–DNA	(8)
8–Oxoguanosine	unpublished
Trimethylangelicine–DNA	(9)
Aflatoxin B_1–albumin	unpublished
Benzo(a)pyrene diol epoxide–protein	(10)
Ethylene oxide–hemoglobin	unpublished

that once a sensitive and specific method has been developed it can easily be applied to the large number of samples that are collected in epidemiologic studies. However, before an immunoassay can be developed the structure of the adduct of interest must be known and it must be possible to synthesize an appropriate immunogen.

Also shown in Table I are the antibodies we have recently developed for measurement of carcinogen-protein adducts. Only the antibody recognizing benzo(a)pyrene diol epoxide-protein adducts has been applied to human samples (see below).

Measurement of Exposure to Polycyclic Aromatic Hydrocarbons: Antibodies to Benzo(a)pyrene diol epoxide-DNA

Benzo(a)pyrene (BP), a polycyclic aromatic hydrocarbon (PAH), is a ubiquitous environmental pollutant resulting from the incomplete combustion of organic material, including fossil fuels. It is found in urban air, sidestream and mainstream cigarette smoke and the food supply. BP is generally used as a representative indicator of total PAH concentration. Extensive studies from several laboratories have determined that it is metabolized in vivo to 7,8-dihydroxy-9,10-epoxy-7,8,9,10-tetrahydrobenzo(a)pyrene (BPDE-I), the major electrophilic, mutagenic and carcinogenic metabolite involved in DNA binding (Figure 1, reviewed in (11)). The complete structure and conformation of the adduct formed between BPDE-I and DNA in vitro is known and results from binding of the C-10 position of BPDE-I to the 2-amino group of guanosine (Figure 1). This adduct has been detected as the major adduct formed when a variety of human, bovine and rodent cells are exposed to BP in culture. We have developed polyclonal and monoclonal antibodies recognizing this adduct. Two types of antigens were used, the modified DNA electrostatically complexed to bovine serum albumin (BSA) or the ribose form of the nucleoside monoadduct covalently coupled through the adjacent hydroxides on the ribose ring to BSA (6,12). When first characterized, the antisera developed against BPDE-I-DNA were found to be highly specific for the modified DNA not recognizing BP itself or nonmodified DNA. In addition, there was no crossreactivity with several other carcinogen modified DNAs, including acetylaminofluorene and 8-methoxypsoralen-DNA. More recently, both the polyclonal and monoclonal antibodies were found to crossreact with structurally related diol epoxide adducts of several other PAHs including chrysene and benz(a)anthracene (13). For example, polyclonal antibody #29,(12) obtained from animals immunized with BPDE-I-DNA, recognizes DNA modified by chrysene-1,2-diol-3,4-epoxide more efficiently (50% inhibition at 18 fmol) than it recognizes BPDE-I-DNA (50% inhibition at 30 fmol). This antibody also binds to DNA modified by benz(a)anthracene-8,9-diol-10,11-epoxide (50% inhibition at 42 fmol) and 3,4-diol-1,2-epoxide (50% inhibition at 114 fmol). These results indicate that multiple adducts may be detected by the ELISA. Since humans are exposed to BP in complex mixtures containing a number of other PAHs, a number of different adducts may be present. The identity of the adducts cannot be determined and thus absolute quantitation of adducts is not possible. However, since a number of PAHs in addition to BP are

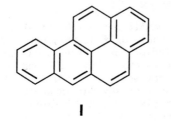 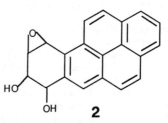

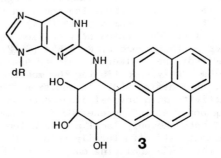

Figure 1. Structure of benzo(a)pyrene (1), BPDE-I (2) and the adduct of BPDE-I with guanine (3).

carcinogenic, the ELISA provides a biologically relevant general index of DNA binding by this class of compounds. Measured values are expressed as femtomole equivalents of BP adducts which would cause a similar inhibition in the assay.

We have also recently determined that antibody #29 detects adducts more efficiently in highly modified DNA (1.2 adducts/100 nucleotides) than in DNA modified to a lower level (1.5/10^5) (14). This efficiency also varied with the type of ELISA used. With the color endpoint ELISA there was a 2.5 fold difference between the high and low modified DNA samples but with the fluorescence endpoint ELISA the difference was 10 fold. These two assays differ in antibody dilution and concentration of antigen used for plate coating. The antigen used for antibody development was highly modified DNA. Clustering of adducts or some unique determinents present on highly modified DNA may be responsible for the higher sensitivity with these samples. Utilization of highly modified DNA in the standard curve in our original studies resulted in an underestimation of adduct levels. In contrast, antibodies recognizing 8-MOP-DNA have similar crossreactivity with adducts in high and low modified DNA (15). These results demonstrate the importance of thorough characterization of antisera before application to human samples and the utilization of appropriate standards for analyzing biological samples. Currently, we use polyclonal antibody #29 in a competitive ELISA with a low modified BPDE-I-DNA standard and fluorescence endpoint detection. This assay has a 50% inhibition of antibody binding with low modified DNA of about 44 fmol. If 20% inhibition is taken as the lower limit of detectability, adduct levels of about 1/10^8 nucleotides can be determined.

Detection of PAH-DNA Adducts in Humans

Occupational Exposure. Antibodies recognizing PAH-DNA have been used in several studies to quantitate adduct levels in humans. Lung, the target tissue for PAH carcinogenesis, is not available on a routine basis from healthy individuals. Therefore, white blood cells have been utilized in a number of biomonitoring studies as a surrogate source of DNA since it is readily and repeatedly available. A 30ml blood sample provides the 500ug of DNA required for duplicate assays in triplicate wells as normally carried out. In collaboration with K. Hemminki, Institute of Occupational Health, Finland, adducts were measured in white blood cell DNA from iron foundry workers and controls (16). Foundry workers are known to be at elevated risk of lung cancer (17) and constitute a model population for purposes of validating PAH-DNA adducts as a marker of biologically effective dose. Workers were classified into high, medium, or low exposure to BP based on air monitoring data and an industrial hygienist's evaluation of the job description (Table II). A dose response relationship was seen between estimated exposure to BP and adduct levels. Adducts in these samples were also analyzed by two other laboratories by [32P] postlabeling (18). In this method, the DNA is enzymatically digested to 3'monophosphates followed by labeling with T4 polynucleotide kinase

and gamma [^{32}P] ATP. Normal nucleotides are separated from adducted
ones by thin layer chromatography and adducts visualized by
autoradiography (19-20). While adduct levels were lower in the
postlabeling assay, there was a good correlation between the
immunoassay and postlabeling data. From a subset of 9 foundry
workers, we recently collected blood samples immediately after a one
month vacation and then again after the individuals had been working
for 3 months. Mean adduct levels increased 10 fold after return to
the work environment (Table II). Although the sample number is
small, these results suggest rapid repair of adducts in white
blood cells.

Table II. PAH-DNA adducts in white blood cells of foundry
oven workers determined by ELISA.

	Exposure level ug BP/m^3	N	mean adducts/ 10^7 nucleotides
High	>0.2	4	5.0
Medium	0.05-0.2	13	2.2
Low	<0.05	18	0.80
Control	<0.01	10	0.22
At work	0.05->0.2	9	3.1
After vacation	<0.01	9	0.39

Clinical Exposure to Crude Coal Tar. Topical application of 2-5%
crude coal tar followed by skin irradiation with UVB (Goeckerman
theraphy) is used as an effective therapy for psoriasis, a
hyperproliferative disease of the skin. Coal tars, including
pharmaceutical-grade preparations, are complex mixtures of PAHs and
well established mutagens and animal carcinogens (17). Isolated
case reports have also suggested that therapeutic coal tar treatment
can produce squamous and basal cell carcinomas (reviewed in
(21)). We have estimated that patients treated with total body
application of about 50gm of a 3% tar preparation are exposed to an
average of 1.5gm of tar per day.
    PAHs are known to be absorbed through the skin after topical
application as demonstrated by elevated levels of several PAHs in
blood (22). Urinary excretion of 1-hydroxypyrene, a fluorescent
metabolite of pyrene, has also been used as a marker of internal
dose of coal tar (23-24). Elevated levels of sister chromatid
exchange (SCE) and chromosomal aberrations in peripheral blood
lymphocytes and of urinary mutagens, measured by the Salmonella
mutagenesis assay, have also been found (24-26).
    We have carried out a small pilot study on blood samples from
coal tar treated psoriasis patients and controls, with a panel of
biomarkers for exposure to genotoxic agents as a model system for
skin exposure in the occupational setting. Blood samples were
obtained from 22 coal tar treated psoriasis patients and 5 controls.
Only one control had detectable adducts by ELISA while 13/22
patients were positive. Mean values for patients was 1.70 adducts

/10^7 nucleotides compared to 0.79/10^7 for controls. Mean values
for the controls are higher than seen in previous studies on
nonoccupationally exposed individuals (0.15-0.23/10^7). This was due
to the very high adduct level, comparable to that of foundry
workers, seen in the one positive control (3.3/10^7), a nurse on the
dermatology floor. However, her occupational exposure to coal tar
should have been minimal and it was not clear what exposures were
responsible for the high adduct levels. Sister chromatic exchange
(SCE) and micronuclei (MN) were also analyzed in a subset of 17 of
the patients and in the 5 controls. Mean SCE levels were higher for
patients (9.43) than controls (7.74) but no significant difference
was seen in MN between patients (13.59) and controls (13.75). We
are currently carrying out a larger scale study of patients and
controls with a panel of biomarkers.

Immunohistochemical Studies on Skin Biopsies. Antibodies to
particular carcinogen-DNA adducts can also be used to investigate
localization of adduct formation in specific cell or tissue types by
indirect immunofluorescence utilizing fluorescein isothiocyanate
(FITC) labeled secondary antibodies. We obtained 4mm punch biopsies
from several coal tar treated patients as well as control biopsies
from untreated volunteers. Sections were cut and stained with the
same antibody #29 used for ELISA quantitation of PAH-DNA. To
enhance accessibility of the antigen for antibody binding, slides
were first treated with RNase A, proteinase K and HCl. Staining of
skin biopsy sections from coal tar treated patients indicated
specific nuclear staining in the stratum spinosum and granulosum of
the epidermis (Figure 2). Staining was also scattered variably
throughout the dermis with some localization to fibroblasts and
vessels. No staining was visible in sections from an untreated
control.

With the conventional immunofluorescence methods utilized here,
quantitative adduct levels cannot be determined. However, they can
be estimated by comparison to previous studies carried out on
keratinocytes treated in culture with BP. In these studies, adduct
levels in the range of 1/10^6 gave detectable immunofluorescence
(27). This is in the same range as found in studies utilizing
antibodies to 8-methoxypsoralen-DNA adducts (15). Improvements in
sensitivity, coupled to methods giving quantitative data should
allow the application of the immunofluorescence method to detection
of adducts resulting from environmental exposures.

Environmental Exposure. In contrast to these results, several
studies on adducts in smokers and nonsmokers have not detected
significant differences between these populations. Mean adduct
levels in placental DNA of smokers were 1.9/10^7 nucleotides and in
nonsmokers 1.2/10^7 (28-29). White blood cell DNA adducts were
lower than in placenta but also were not significantly different in
smokers (0.15/10^7) compared to nonsmokers (0.12/10^7) (30). These
results probably reflect the ubiquitous exposure of the general
population to PAHs from a number of sources including air pollution
and, more importantly, diet. Recently, we have also found a
seasonal effect on adduct levels with blood samples collected during

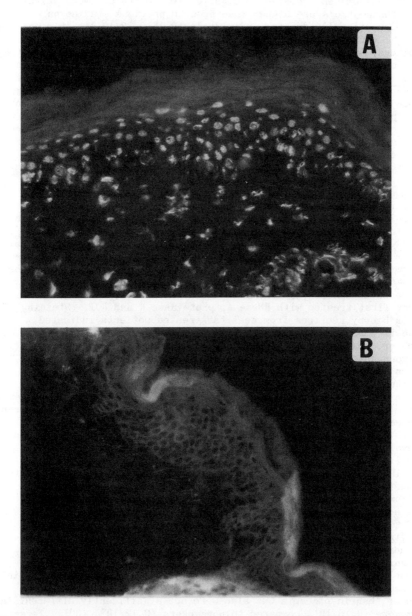

Figure 2. Indirect immunofluorescence staining of human skin biopsies from a patient treated topically with coal tar and a control. Slides were treated with RNase, proteinase K and 4N HCl before staining with antisera #29 (1:50 dilution) and goat anti-rabbit IgG antibody conjugated with fluorescein (1:30). Magnification was 220fold. A.) section from patient; B) section from control

the late summer and early fall showing higher levels than those collected during the other two seasons (31). This seasonal effect is consistent with observations of a peak in aryl hydrocarbon hydroxylase (AHH) inducibility during this period (32). The initial data are from samples collected throughout the year. Repeat measurements on the same individual during the year should provide further information about the relationship of adduct formation and season.

Detection of PAH-Protein Adducts. Quantitation of adducts on albumin has been used as an alternate marker of exposure to PAHs. To determine PAH-protein adduct levels, we have utilized a monoclonal antibody, 8E11, developed from animals immunized with BPDE-I-guanosine coupled to BSA (6). This antibody recognizes BPDE-I modified deoxyguanosine, DNA, and protein as well as BPDE-I-tetrols, the hydrolysis products of BPDE-I. More recently, we have further characterized this antibody in terms of crossreactivity with a number of other BP metabolites, other PAHs and other PAH diol epoxide modified DNAs. 8E11 crossreacts with a number of BP metabolites (Table III), with higher sensitivity for the diols and weak recognition of phenols. It also recognizes the diol epoxide DNA adducts of several other PAHs, and there is weak cross-reactivity with other PAHs including pyrene, aminopyrene and nitropyrene. Although not available for testing, this antibody will probably recognize a number of PAH protein adducts. As with the PAH-DNA immunoassay, multiple adducts will be measured and this assay will be a general index of exposure to this class of compounds.

Direct quantitation of adducts on protein cannot be carried out with precision due to inefficient detection of adduct in intact

Table III. Competitive Inhibition of Antibody 8E11
binding to BPDE-I-BSA

Competitors	50% Inhibitory Concentration femtomole
BP-9,10-diol	150
Chrysene diol epoxide-DNA	160
BP-7,8-diol	250
BPDE-I-tetrol	350
BPDE-I-DNA	350
BPDE-I-BSA digested	400
Benz(a)anthracene diol epoxide-DNA	1350
BPDE-I-BSA nondigested	1450
1-OH-pyrene	3400
BP	6000
1-nitropyrene	16000
Pyrene	16200
4-OH-BP	42700
1-aminopyrene	70000
5-OH-BP	$>1 \times 10^5$
Dimethylbenz(a)anthracene	$>1 \times 10^6$

protein. This low sensitivity is probably due to burying of the
adduct in hydrophobic regions of the protein. Others have suggested
that release of tetrols by acid hydrolysis, followed by quantitation,
is a sensitive method for determination of protein adducts (33-34).
While this method worked well on protein treated in vitro with BPDE-
I, our initial studies in mice treated with radiolabeled BP
indicated that only low levels of radioactivity could be released
from globin by acid treatment (35). For this reason, we used an
alternate approach for measurement of protein adducts. When protein
modified in vitro with BDPE-I was enzymatically digested to peptides
and amino acids before ELISA sensitivity was increased 3-4 fold
(Table III). This assay was validated using globin isolated from
animals treated with radiolabeled BP. The ELISA was able to detect
90-100% of the adducts measured by radioactivity (36). These animal
studies also indicated that adduct levels were about 10 fold higher
in albumin than in globin. For this reason, our initial work on
human samples has been with albumin isolated from workers
occupationally exposed to PAHs. Albumin was isolated by Reactive
blue 2-Sepharose CL-4B affinity chromatography and enzymatically
digested with insoluble protease coupled to carboxymethyl cellulose,
which was subsequently removed by centrifugation. Samples were then
analyzed by competitive ELISA with antibody 8E11. Initial studies
have been carried out on a small number of roofers occupationally
exposed during the removal of an old pitch roof and application of
new hot asphalt. These studies were carried out in collaboration
with R. Herbert, Mt. Sinai Medical Center, NY. Seventy percent of
the roofers samples were positive with a mean level of 5.4fmol/ug
while 62% of the controls had detectable adduct levels (mean of
4.0fmol/ug). In this small number of subjects there was a trend but
no significant difference between roofers and controls. However, we
are continuing studies of PAH-albumin adducts using a larger sample.

Monitoring Exposure to Aflatoxin B_1

Detection of Aflatoxin-DNA Adducts. Aflatoxin B_1 (AFB_1), is a
fungal metabolite produced by specific strains of Aspergillus
flavus. This mycotoxin contaminates the food supply in many regions
especially in Asia and Africa. Epidemiological studies have
demonstrated a strong association between exposure to this potent
liver carcinogen and human liver cancer incidence (37). AFB_1 is
metabolized in vivo by microsomal mixed-function oxidases to produce
a reactive electrophilic epoxide which binds to the N7 position of
guanine (Figure 3). This adduct is unstable having an apparent half
life of 7.5hr and either depurinates or imidazole ring-opens (Figure
3) (38). The imidazole ring opened (iro) AFB_1-FAPy adducts are
stable, have been shown to accumulate in treated animals and may
play an important role in hepatocarcinogenesis (39).
    We have developed monoclonal antibodies recognize AFB_1-FAPy
adducts from animals immunized with DNA modified in vitro with AFB_1
followed by base treatment to convert adducts to the FAPy form
(4). An immunogen was prepared by complexing this DNA with
methylated BSA. Competitive ELISAs using antibody 6A10 are shown in
Figure 4. This antibody has higher reactivity with denatured highly

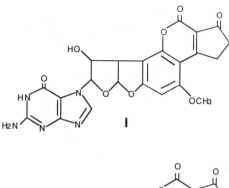

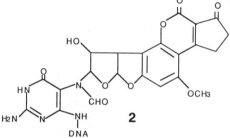

Figure 3. Structure of aflatoxin B₁-guanine (1) and imidazole ring opened aflatoxin B₁-guanine (AFB₁-FAPy), (2).

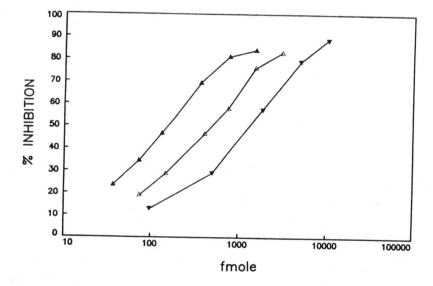

Figure 4. Competitive ELISA of antibody 6A10 binding to iro AFB₁-DNA. The competitors were highly modified denatured iro AFB₁-DNA (2.5 adducts/100 nucleotides) (▲), highly modified denatured AFB₁-DNA (7 adducts/100 nucleotides) (▼), and low modified denatured iro AFB₁-DNA (4 adducts/10⁵ nucleotides) (△). Reproduced with permission from Ref. 4, Copyright 1988, Cancer Research.

modified iro AFB_1-DNA (2.5 adducts/100 nucleotides, 50% inhibition at
160fmol) than with iro AFB_1-DNA modified to a lower level
(4 adducts/10^5 nucleotide, 50% inhibition 500fmol) or with denatured
highly modified AFB_1-DNA (50% inhibition at 1300fmol).
Since AFB_1-DNA contains both ring-opened and ring-closed adducts,
these results suggest that the antibody binds more strongly with the
ring opened adduct against which it was generated.   Adducts are
detected about 4-fold more efficiently in highly modified DNA than
in DNA with a lower modification level.   While there is no
crossreactivity with unmodified native calf thymus DNA, there is
minor crossreactivity with unmodified denatured DNA.   There is no
crossreactivity with free aflatoxins, aflatoxin-protein conjugates,
AFB_1-guanine or several other carcinogen modified DNAs including
DNAs modified by acetylaminofluorene, psoralen or BPDE-I.   The limit
of sensitivity, based on assaying 50ug of DNA/well and >20%
inhibition, is 5 adducts/10^7 nucleotides.
    To validate the immunoassay, antibody 6A10 was used to detect
adducts in liver and kidney of mice and rats treated with
radiolabeled AFB_1.   The ELISA detected 55-65% of total adducts
measured by radioactivity in both rat and mouse liver (4).
Relatively high adduct levels (>1/10^6) were also detectable in
2 of 8 tumor adjacent-normal tissues and 7 of 7 tumor tissues
obtained from liver cancer patients from Taiwan (Table IV).

Table IV. Levels of AFB_1-FAPY Adducts in Human Liver

Patient	Tumor	Adjacent-normal
	adducts/10^6 nucleotides	
1	3.5	ND*
2	3.2	ND
3	1.2	ND
4	3.4	1.7
5	1.2	1.2
6	1.8	ND
7	–	ND
8	–	ND
9	1.3	NA

*ND, non detectable
NA, not assayed

One potential artifact of competitive ELISAs is nonspecific
inhibition of antibody binding by factors such as high or low salt
concentration and pH.   Thus, to further support these results,
samples were tested in a competitive ELISA using a monoclonal
antibody with specifity for unrelated 8-methoxypsoralen-DNA
adducts (8).   Since these patients had not been treated with
psoralen and there is little environmental exposure no psoralen
adducts should be present.   The level of inhibition of all samples
was less than 20% indicating there was no nonspecific inhibition of
antibody binding by the matrix of the DNA samples.   The difference
between tumor and nontumor tissue may not be significant because of
the small number of samples but will be further investigated in
ongoing studies.   In addition, because of the relatively high adduct

levels present in some samples, we have recently begun to utilize fluorescence methods to confirm adduct levels quantitated by ELISA. Initial studies with in vitro modified DNA indicate that fluorescence spectroscopy can readily detect adducts down to a level of $1/10^7$. We are currently developing immunoaffinity techniques to isolate AFB₁ modified DNA fragments from partially digested human DNA samples. By combining the specificity of the immunoaffinity technique with the sensitivity of fluorescence spectroscopy, we hope to obtain supportive data for the ELISA results. A similar approach has been used by others to confirm the presence of BPDE-I-DNA adducts in human placental DNA (34) and excised aflatoxin-DNA adducts in urine (see Groopman et al, this publication).

Immunohistochemical Detection of Aflatoxin B₁-DNA Adducts. The antibodies recognizing AFB₁-DNA are also being applied in immunohistochemical studies to localize adducts. Initial studies have utilized hepatocytes treated in vitro with 40uM AFB₁. Six hours after exposure, cells were fixed and treated with bicarbonate solution to convert the adducts to the stable ring opened form recognized by antibody 6A10. Staining with adduct specific antibody was followed by FITC labeled secondary antibodies. Figure 5 shows specific nuclear staining in treated cells but not in controls. This method is now being applied to the detection of adducts in human liver biopsies.

Multiple Adduct Analysis. Human exposure to environmental carcinogens usually occurs in the form of complex mixtures. To monitor exposure to these mixtures, we would ideally like to perform multiple ELISAs on DNA samples from the same individual to determine all adducts present. To circumvent the limited availability of white blood cell DNA, we have recently developed methods in which two different DNA adducts can be measured in a DNA sample with specific antibodies recognizing the individual adducts. Initial studies have been on DNA modified in vitro with BPDE-I and subsequently modified with 8-methoxypsoralen and UVA light. A mixture of the antibodies recognizing both adducts, each at the appropriate final dilution for the ELISA, was used. The competitive ELISA was carried out on serial dilutions of the modified DNA and mixed antibodies. This competitive mixture was first added to plates coated with BPDE-I-DNA and after a 90 min incubation, transferred to plates coated with 8-methoxypsoralen-DNA. Each plate was then incubated with alkaline phosphatase secondary antibody conjugate as in the standard assay. For both BPDE-I and 8-MOP-DNA, similar 50% inhibitions were found when only one adduct or both were present. These results suggest that it may be possible to make a cocktail of antisera to specific DNA adducts and, by sequential transfer to plates coated with the appropriate antigen, quantitate a number of different DNA adducts in a single sample. Since the adducts are not destroyed by incubation with antibodies in the ELISA, DNA can also be recovered from the competitive mixture on the microwell plate and repurified. The DNA can then be utilized for additional analysis by alternate methods such as postlabeling or fluorescence.

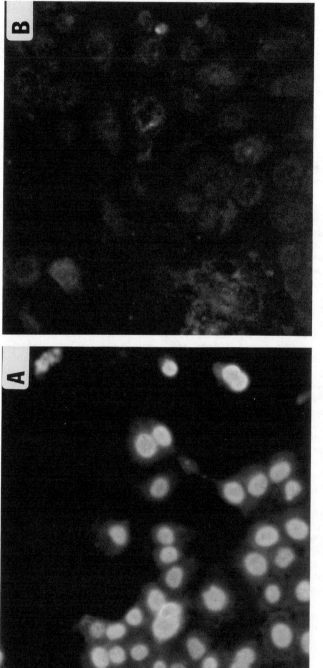

Figure 5. Indirect immunofluorescence staining of hepatocytes treated with 40uM aflatoxin B₁ (A); and control untreated cells (E). Slides were treated with 50mM sodium carbonate pH 9.5 after fixation, then with RNase, proteinase K and 50mM NaOH. Antibody 6A10 was used at a 1:2 dilution and goat anti-mouse IgG conjugated with fluorescein at 1:40.

Discussion

These studies demonstrate that immunologic methods have sufficient
sensitivity to monitor human exposure to environmental carcinogens.
Immunoassays also have the advantage of ease of application to large
number of samples making them ideal for epidemiologic studies.
However, human samples frequently have adducts levels near the
limits of sensitivity of these assays. In addition, because
competitive ELISAs measure general inhibition of antibody binding to
plates, artifacts can sometimes interfere with quantitation. For
these reasons, it is desirable to use other methods to confirm
adduct levels determined by ELISA. Unfortunately, alternate
methods, with the required sensitivity, are not always available.
The studies discussed above on PAH adducts in foundry workers are
examples of this approach. While the postlabeling assay may not be
measuring the identical adducts as the ELISA, the correlation of
both assays with exposure level and with each other provides support
for the relevance of the determinations. Similarly, for aflatoxin-
DNA adducts we are attempting to utilize fluorescence measurements
to confirm immunoassay data.

The current immunofluorescence method is limited to the
detection of adduct levels around $1/10^6$ nucleotides. Computer-
assisted video microscopy systems or the use of biotin-streptavidin
staining should further increase sensitivity. It may then be
possible to utilize these methods for adduct detection in human
samples from occupational or environmental exposures. Since adducts
can in theory be visualized in single cells, the small amount of
material obtained at biopsy could be utilized.

While methods for the determination of DNA adducts in humans
provides information about the biologically effective dose of a
carcinogen, and can therefore be used as a marker of exposure,
information about the relationship of these measurements to risk is
unknown. Future epidemiologic studies are needed to provide this
information.

Acknowledgments

Studies on placental DNA adducts in smokers and nonsmokers were
carried out in collaboration with R. Everson. Studies on coal tar
treated psoriasis patients were carried out in collaboration with D.
Warburton, M. Toor and V. DeLeo. The cooperation of S.W. Hsu and
D.S. Chen in the collection of samples from cancer patients in
Taiwan is also greatfully acknowledged. This work was supported by
grants from the National Institute of Health CA21111, OH02622 and a
gift from the Lucille P. Markey Charitable Trust

Literature Cited

1. Vainio, H.; Sorsa, M.; Hemminki, K. Amer. J. Ind. Med. 1983,
   4, 87-103.
2. Perera, F.P.; Weinstein, I.B. J. Chron. Dis. 1982, 35,
   581-600.
3. Yang, X.Y.; Santella, R.M. In Carcinogenic and Mutagenic
   Responses to Aromatic Amines and Nitroarenes; King, C.M.,
   Romano, L.J., and Schuetzle, D. Eds.; Elsevier: New York, 1987;
   pp 329-326.

4. Hsieh, L.L.; Hsu, S.W.; Chen, D.S.; Santella, R.M. Cancer Res. 1988, 48, 6328-6331.
5. Hsieh, L.L.; Jeffrey, A.M.; Santella, R.M. Carcinogenesis 1985, 6, 1289-1293.
6. Santella, R.M.; Lin, C.D.; Cleveland, W.L.; Weinstein, I.B. Carcinogenesis 1984, 5, 373-377.
7. Young, T.L.; Santella, R.M. Carcinogenesis 1988, 9, 589-592.
8. Santella, R.M.; Dharmaraja, N.; Gasparro, F.P.; Edelson, R.L. Nucleic Acids Res. 1985, 13, 2533-2544.
9. Miolo, G.; Stefanidis, M.; Santella, R.M.; Acqua, F.; Gasparro, F. Photochem. Photobiol. 1989, 3, 101-112.
10. Santella, R.M.; Lin, C.D.; Dharmaraja, N. Carcinogenesis 1986, 7, 441-444.
11. Jeffrey, A.M.; Kinoshita, T.; Santella, R.M.; Weinstein, I.B. In Carcinogenesis: Fundamental Mechanisms & Environmental Effects; Ts'o, P.O.P. and Gelboin, H. Eds.; D. Reidel Publ. Co.: Boston, 1980; p 565-582.
12. Poirier, M.C.; Santella, R.; Weinstein, I.B.; Grunberger, D.; Yuspa, S.H. Cancer Res. 1980, 40, 412-416.
13. Santella, R.M.; Gasparo, F.P.; Hsieh, L.L. Prog. Exper. Tumor Res. 1987, 31, 63-75.
14. Santella, R.M.; Weston, A.; Perera, F.P.; Trivers, G.T.; Harris, C.C.; Young, T.L.; Nguyen, D.; Lee, B.M.; Poirier, M.C. Carcinogenesis 1988, 9, 1265-1269.
15. Yang, X.Y.; DeLeo, V.; Santella, R.M. Cancer Res. 1987, 47, 2451-2455.
16. Perera, F.P.; Hemminiki, K.; Young, T.L.; Santella, R.M.; Brenner, D.; Kelly, G. Cancer Res. 1988, 48, 2288-2291.
17. IARC monographs on the evaluation of the carcinogenic risks of chemicals to humans: polynuclear aromatic compounds, Part 4; IARC: Lyon, 1985;.
18. Phillips, D.H.; Hemminki, K.; Alhonen, A.; Hewer, A.; Grover, P. L. Mutation Res. 1988, 204, 531-541.
19. Randerath, D.; Reddy, M.V.; Gupta, R.A.C. Proc. Natl. Acad. Sci. USA 1981, 78, 6126-6129.
20. Reddy, M.V.; Randerath, K. Carcinogenesis 1986, 7, 1543-1551.
21. Bickers, D.R. J. Invest. Derm. 1981, 77, 173-174.
22. Storer, J.S.; DeLeon, I.; Millikan, L.E.; Laseter, J.L.; Griffing, C. Arch. Derm. 1984, 120, 874-877.
23. Jongeneelen, F.J.; Anzion, R.B.M.; Leijdekkers, C.M.; Bos, R.P.; Henderson, P.T. Int. Arch. Occup. Environ. Hlth. 1985, 57, 47-55.
24. Clonfero, E.; Zordan, M.; Cottica, D.; Venier, P.; Pozzoli, L.; Cardin, E.L.; Sarto, F.; Levis, A.G. Carcinogenesis 1986, 7, 819-823.
25. Wheeler, L.A.; Saperstein, M.D.; Lowe, N.J. J. Invest. Derm. 1981, 77, 181-185.
26. Sato, F.; Zordan, M.; Tomanin, R.; Mazzotti, D.; Canova, A.; Cardin, E.L.; Bezze, G.; Levis, A.G. Carcinogenesis 1989, 10, 329-334.
27. Poirier, M.C.; Stanley, J.R.; Beckwith, J.B.; Weinstein, I.B.; Yuspa, S.H. Carcinogenesis 1982, 3, 345-348.

28. Everson, R.B.; Randerath, E.; Santella, R.M.; Cefalo, R.C.; Avitts, T.A.; Randerath, K. Science 1986, 231, 54-57.
29. Everson, R.B.; Randerath, E.; Santella, R.M.; Avitts, T.A.; Weinstein, I.B.; Randerath, K. J. Natl. Cancer Inst. 1988, 80, 567-576.
30. Perera, F.P.; Santella, R.M.; Brenner, D.; Poirier, M.C.; Munshi, A.A.; Fischman, H.K.; VanRyzin, J. J. Natl. Cancer Inst. 1987, 79, 449-456.
31. Perera, F.; Mayer, J.; Jaretzki, A.; Hearne, S.; Brenner D.; Young, T.L.; Fischman H.K.; Grimes, M.; Grantham S.; Tang, M.X.; Tsai W-Y.; Santella, R.M.. Cancer Res. 1989, 49, 4446-4451.
32. Paigen, B.; Ward, E.; Reilly, A.; Houten, L.; Gurtoo, H.L.; Minowada, J.; Steenland, K.; Havens, M.B.; Satori, P. Cancer Res. 1981, 41, 2757-2761.
33. Shugart, L. Anal. Biochem. 1986, 152, 365-369.
34. Weston, A.; Rowe, M.I.; Manchester, D.K.; Farmer, P.B.; Mann, D.L.; Harris, C.C. Carcinogenesis 1989, 10, 251-257.
35. Wallin, H.; Jeffrey, A.M.; Santella, R.M. Cancer Let. 1987, 35, 139-146.
36. Lee, B.M.; Santella, R.M. Carcinogenesis 1988, 9, 1773-1777.
37. Busby, W.F.; Wogan, G.N. In Chemical Carcinogens; Searle, C.D. Ed.; ACS: Washington, DC, 1984; pp 945-1136.
38. Wogan, G.N. In Hepatocellular Carcinoma; Okuda, K. and Peters, R.L. Eds.; John Wiley and Sons: New York, 1976; pp 25-42.
39. Croy, R.G.; Wogan, G.N. Cancer Res. 1981, 41, 197-203.

RECEIVED August 30, 1990

Chapter 21

Immunological Detection and Quantitation of Carcinogen–DNA Adducts

Mariko Tada[1], Misaki Kojima[1], Tomoyuki Shirai[2], Nobuyuki Ito[2], and Toshiteru Morita[3]

[1]Laboratory of Biochemistry, Aichi Cancer Center Research Institute, Chikusa-ku, Nagoya 464, Japan
[2]First Department of Pathology, Nagoya City University Medical School, Mizuho-ku, Nagoya 467, Japan
[3]College of General Education, Osaka University, Toyonaka 560, Japan

Antibodies against DNA modified by 4-nitroquinoline-1-oxide(4NQO), and 3,2'-dimethyl-4-aminobiphenyl(DMAB) were raised in rabbits. On application in enzyme linked immunosorbent assays (ELISA), these antibodies proved highly specific for the DNA-adducts which were used as the immunogens or structurally related forms. Sensitive immunoassays for the detection of the adducts isolated from tissues of rats exposed to carcinogens could therefore be established using competitive ELISA. Levels as low as 5-7 fmol adducts in 5 ug of DNA i.e. 2-3 adducts per 10^7 nucleotides could be detected. Demonstration of adduct formation in rat organs by immunoperoxidase staining revealed that the adducts were found preferentially in target tissues after exposure to DMAB, the staining intensity increasing dose dependently. Indirect immunofluorescence staining of human fibroblast (NSF) cells treated with 4NQO also indicated adducts to be localized specifically in the nuclei after treatment with 0.25 uM at mean lethal dose. These assays should prove useful for elucidation of the levels, localization and persistence of carcinogen carcinogen-DNA adducts in target organs in vivo.

The availability of methods for the direct measurement of specific DNA adducts should greatly facilitate studies on the biological significance of these lesions. In recent years, the immunoassay approach to detection of DNA damage has attracted interest because of the associated sensitivity and specificity, and the fact that DNA or other chemical adducts do not reqire radiolabeling. A large number of polyclonal and monoclonal antibodies have thus been elicited against carcinogen-DNA adducts(1-5) and used in highly sensitive immunoassays such as the enzyme-linked immunosorbent assay (ELISA) and radioimmunoassay (RIA) to quantitate adducts formed in DNA from animal tissues and cultured cells. With these assays, it has been possible to determine adduct formation in the fmol range. Moreover, immunohistochemical staining with these antibodies has allowed detection of the precise localization and examination of the kinetics of formation and removal of adducts at both single-cell and cell population levels. This information is of obvious significance for risk assessment and molecular epidemiology.

Our own research has concentrated on (i) detailed characterization of polyclonal antibodies against 4-nitroquinoline 1-oxide (4NQO)- modified DNA and 3,2'-dimethyl-4-aminobiphenyl(DMAB)-modified DNA, (ii) quantitation of adducts in biological samples and (iii) immunohistochemical studies of the adducts in tissues and cultured cells.

Experimental Methods

Polyclonal Antibodies: Carcinogen modified DNA was prepared as described previously (6,7). Modified DNA samples used as immunogens contained 18 pmol of 4NQO-adduct or 11.0 pmol of DMAB-adducts in 1 ug of calf thymus DNA, this is 6 4NQO adducts or 3.4 DMAB adducts per 10^3 nucleotides.

The modified DNA (0.25-0.5mg) was electrostatically complexed to equal amounts of methylated bovine serum albumin in phosphate buffered saline (PBS), and emulsified with complete Freund's adjuvant. Immunization of rabbits and purification of IgG fractions were performed as described previously (6,7).

Immunoassay (ELISA): Microtiter wells (Nunc Type II) coated with 10ng of highly modified DNA(1-6 adducts/10^3 nucleotides) were used for competitive ELISA. Antibodies were diluted in PBS containing 2 % FCS and mixed with equal volumes of serially diluted competitor. The mixtures were incubated in the wells for 60 min at $37^{\circ}C$, and the plates were then washed and incubated with peroxidase-conjugated goat anti-rabbit IgG for one hour at $37^{\circ}C$. Enzyme assays were carried out as described previously (6). The amount of inhibitor required for 50 % inhibition of antibody binding to immobilized antigen (IA_{50}) was regarded as a measure of the competitive ELISA sensitivity. **(Figure 1.)**

Immunohistochemical Staining

Rat Tissues: Immunohistochemical staining was carried out as described previously (14). Briefly, sections were treated with RNase A(100 ug/ml) at $37^{\circ}C$ for 2 hr and then with 2.5N HCl at room temperature for 45 min. Slides were incubated with 5 % normal goat serum for 20 min and then with anti-DMAB-DNA antibody for 60 min at $37^{\circ}C$. After treatment with 0.3 % H_2O_2 for 30 min at $37^{\circ}C$, binding of the antibody was visualized by the ABC method($37^{\circ}C$ for 30 min) with 0.05 % diaminobenzidine (for 10 min). The sections were also weakly counterstained with hematoxylin before dehydration through graded ethanols and xylene, and mounting.

Human Cells: Normal human skin fibroblast cells (NHSF) were grown as monolayer cultures on Lab-Tek 2 chamber-slides (Lab-Tek Products, Naperville, Ill) and treated with 2§10 x 10^7M 4NQO in Hank's solution. These cell samples were applied to immunofluoresence staining according to our method (6).

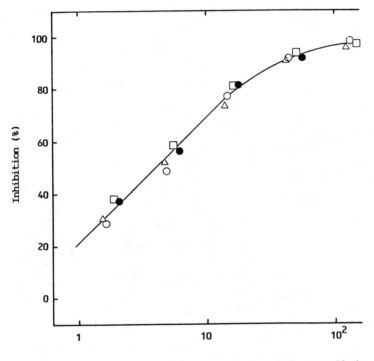

fmol DMAB-adduct

Fig. 1 Competitive inhibition   of antibody   binding to DMAB–DNA by
ELISA.   Each well was   coated with 10 ng of denaturated   DMAB–DNA (5
pmol adducts/ug DNA) and antibody   was 25 ng of IgG per well.   The
inhibiter contained   fmol   adducts in ug of DNA ; (●) 26, (○) 710,
(△) 5,140 and (□) 13,600, respectively.

Results

Characterization of Antibodies : Anti 4NQO−DNA−Antibody: The antibody affinity for 4NQO−DNA was studied in detail by competitive ELISA using DNA with large differences in modification levels. The results presented in Table I, clearly show that highly modified DNA is recognized more efficiently by the antibody than DNA having a lower degree of modification. For measurement of adducts in DNA from biological samples, standard modified DNA should be used as the competitor in the same modification range as the samples (1−10 adducts/10^6 nucleotides).

As the anti−4NQO−DNA antibody hardly reacted with nuclease P_1 digested nucleotide adducts, it may not in fact recognized the 4NQO adduct moiety in DNA, but rather may react with surrounding DNA structures. The antibody was also found to recognize DNA modified by 4HAQO derivatives such as 6−methyl−4−nitroquinoline 1−oxide (6−CH_3−4NQO) or N−methyl−4−nitroquinoline 1−oxide (N−CH_3−4NQO) with the same affinity as 4NQO−DNA. The adducts recognized by the antibody are presumably heat and alkaline stable, since the reactivity of 4NQO−DNA to antibodies did not decrease after heat treatment in PBS or 0.1N NaOH.

Table I. Competitive inhibition of polyclonal antibody binding to 4NQO DNA

Competitor	Modification adducts fmol/ug DNA	Amount of competitor (fmol) causing 50% inhibition of antibody binding
ss 4NQO−DNA[a]	20,000	5
	1,860	15
	356	60
	36.3	120
	21.0	120
	10.0	120
ds 4NQO−DNA	20,000	20
	1,860	460
	356	$>10^5$ [b]
ss AFF−DNA	25,000	$>10^5$
ss DMAB−DNA	13,600	$>10^5$
ss 6CH_3−4NQO−DNA	13,630	5
ss N−CH_3−4NQO−DNA	12,360	5

The microtiter plates were coated with 50 ng of ds 4NQO−DNA (20 pmol adducts/ug) and 25 ng of IgG were applied per well.
a) Modified DNAs were heated in PBS or 0.1N NaOH at 90°C for 4 min.
b) No inhibition detected at the highest concentration tested.

Anti DMAB−DNA Antibody: Specific for DMAB−DNA and 4−aminobiphenyl(AB)−modified DNA, this antibody was found to not cross-react with 4NQO−,4−aminoazoben derivative− or AAF−modified DNA (Table II). It is therefore concluded that the anti−DMAB−DNA antibody recognizes biphenyl rings bound to purine bases (7).

In contrast to anti-4NQO-DNA antibody, the IA_{50} values for heated denatured DNA were 7-8 fmol/assay, independent of modification levels and almost same affinity for enzymatic hydrolysates (mono- or oiligonucleotides). DMAB nucleotides are thought to be the antigenic determinants. (Table II)

The antibody has a greater affinity for highly modified native DMAB-DNA than for low modified   DNA.   Denaturation of DNA proved essential for competitive ELISA.

Table II. Competitive inhibition of antibody binding to DMAB-
DNA by various competitors

Competitor	Modification (fmol/ug)	Amount of competitor
ss DMAB-DNA[a]	13,600	7.0
	710	7.7
	26	8.0
ds DMAB-DNA	13,600	22.0
	710	44.0
	26	90.0
DMAB-DNA-digest[b]	13,600	30.0
ss AB-DNA[c]	866	4.0[d]
ss 4NQO-DNA	20,000	$>10^5$
ss AAF-DNA	25,000	$>10^5$
ss 2-MeO-AAB-DNA	15,600	$>10^5$
ss 3-MeO-AAB-DNA	11,300	$>10^5$

The microtiter plates were coated with 50 ng of ds DMAB-DNA
(12.3 pmol/ug) and 20 ng of IgG were applied per well.
a) Heated in PBS or 0.1 N NaOH at 90°C for 5 min.
b) DNaseI digested.
c) 4-Aminobiphenyl adducts were determined spectrophotometrically.
d) No inhibition detected at the highest concentration tested.

Quantitation of Adducts in DNA from Rat Tissues

4NQO: Donryu female rats (8 weeks old) were administered   an intravenous injection of $[^3H]$4HAQO at 20 mg per kg body weight. After one hour, the liver , pancreas, kidney, lung and uterus tissues were excised and DNA samples isolated by phenol extraction combined with protease and RNase A treatment.   The amounts of 4NQO adducts were determined by UV absorbance at 260 nm and by measurement of radioactivity.   The DNA samples were denatured by heating for 5 min at 90°C, and were assayed by competitive ELISA.

Competitive ELISA was carried out using highly modified   DNA (107 adducts/10^6 nucleotides)   and slightly modified DNA (1 - 3 adducts/10^6 nucleotides) as standards.   The results are shown in Table III and compared with values obtained from the radioactivity measurements.   Using the slightly modified   DNA (3 adducts/10^6 nucleotides) for   the standard inhibition curve, the values obtained were essentially the same as those gained from radioactivity measurements. Using the highly modified DNA (10^7 adducts/10^6 nucleotides) as the standard, the values were approximately 30% lower than radioactivity assay levels.   The pancreas has been found to be the main target organ after a single i.v injection of 4HAQO,   the highest amounts of adducts were also found in pancreatic tissue. Adducts were not present or only present at low levels in the liver nontarget organ (9). These results indicate a significant role of adducts formation in tumorigenesis.

Table III. Quantitation of 4NQO–adducts in DNA from rat tissues by radioactivity and competition ELISA

Tissue	Amount of [^{3}H] 4HAQO (fmol/ug DNA)		
	Radioactivity	ELISA (a)	(b)
Pancreas	392.0	373.3	274.0
Kidney	232.0	220.0	85.7
Stomach	221.0	206.7	–
Uterus	216.7	208.3	100.0
Lung	87.6	80.6	36.0
Liver	4.0	<7.0	–

Microtiter plates were coated with 50 ng of ds 4NQO–DNA (20 pmol/ug DNA) and 25 ng of IgG were applied per well. Modification levels of DNA used for standard curve generation were (a) 10–30 fmol/ug, 3–10/10^6 nucleotides (b) 350 fmol/ug, 107/10^6 nucleotides.

DMAB: F344 male rats (8 weeks old) were given a single s.c. injection of DMAB dissolved in corn oil at 50, 100 or 200 mg/kg body weight. After 24 hours, the rats were sacrificed and the liver, colon, intestine, kidney, pancreas and prostate tissues were removed and DNA samples were isolated as above.

Table IV. Quantitation of DMAB–adducts in DNAs from rat tissues exposed to DMAB

Tissue	Amount of DMAB adducts (fmol/ug DNA)		
Does (mg/kg)	50	100	200
Liver	47.0	66.7	117.3
Colon	7.6	32.3	73.6
Small intestine	5.7	10.2	21.3
Prostate (ventral)	19.6	51.0	91.3
(dorsolateral)	10.3	21.0	34.7

The percentage inhibition was used to calculate the numbers of adducts in the samples . (cf. Figure 1)

The DNA samples were heated in 0.02N NaOH at 70 °C for 10 min and then neutralized with 1N HCl and used in the competitive ELISA. Dose dependent increase in adduct formation in all the tissues was observed (Table IV). The amounts of adducts in liver and intestine at a dose of 200 mg/kg were 117 and 21.3 fmol in 1 ug of DNA , similar

results being obtained by Westra et al using a radioactivity assay(10).
Immunohistochemical Studies of DMAB Adducts in Rat Tissues Exposed to DMAB: Dose-dependent staining of liver nuclei was observed at 24 hr after administration of 50mg and 200mg of DMAB per kg body weight (Fig 2). Under the conditions of the immunohistochemical method applied, anti-DMAB-DNA antibody did not bind in detectable amounts to tissue sections of untreated rats and normal rabbit sera also did not react with tissues from DMAB-treated rats. The numbers of adducts determined by competitive ELISA, correlated well with the immunostaining intensity of nuclei in the treated rat tissues. With the immunohistochemical technique, a clear positive correlation between nuclear stain intensity and dose of carcinogen used was also observed. As little as 5 to 10 x10^4 adducts in one genome could be readily detected by immunostaining.
Immunofluorescence Studies on 4NQO Adducts in Human Fibroblast Cells: As illustrated in Fig 3, the antibody was found to bind specifically to the nuclei of the cells treated with 4NQO, with the exception of the nucleoli. The intensity of immunostaining fluorescence appeared to depend on the concentration of 4NQO administered to the cells. Weak fluorescent signal could be observed with 0.25 uM of mean lethal carcinogen dose. In this case, it was calculated that less than 10^4 adducts would be produced in one genome. In repair experiments, cells were treated for 60 min with 1x10^{-6}M 4NQO and, after removal of the carcinogen, then incubated with M-DEM medium containing 10% FCS for a further 16 hr. As shown in Fig 3, the intensity of nuclear fluorescence was diminished. These results demonstrated the possibility of visualizing repair processes in situ. Experiments to quantify adduct formation in situ by fluorescence or color intensity measurement are in progress.

Discussion

Polyclonal antibodies against 4NQO-, and DMAB-modified DNA were developed, both of which specifically recognized the carcinogen-DNA adducts used as the immunogenic antigens, while cross-reacting only with structurally related adducts. We have also recently elicited anti 2-methoxy- or 3-methoxy-4-aminoazobenzene modified DNA antibodies in rabbits. These antibodies are also highly specific to the modified DNAs used as immunogens, but cross-react with each other, and o-aminoazotoluol-modified DNA. However, they do not bind to DNAs modified by DMAB, 4NQO or AAF.
    The chemical structures of 4NQO and DMAB adducts have been well investigated. In 4NQO-DNA, 3-(deoxyguanosine-N^2-yl)aminoquinoline-1-oxide (4AQO), N-(deoxyguanosine-8-yl)-4AQO, 3-(deoxyadenosine-N^6-yl)-4AQO and one unstable deoxyguanosine adducts were recognized.(11-14) DMAB-DNA adducts were identified by Beland and kadlubar(15) and Flammang et al(16). as N-(deoxyguanosine-8-yl)-DMAB, 5-(deoxyguanosine-N^2-yl)-DMAB and N-(deoxyadenosine-8-yl)-DMAB.
    The antigenicities of 4NQO and DMAB adducts were not changed by heat treatment either in neutral or alkaline solution. Under alkaline conditions, guanine-C8 adducts are easily converted to imidazol-ring-opened products, which may also be antigenic determinants. Since the guanine N^2 and adenine N^6 adducts were also alkali stable, the possibility that these adducts are involved as antigenic determinants can not be excluded.(17,18)

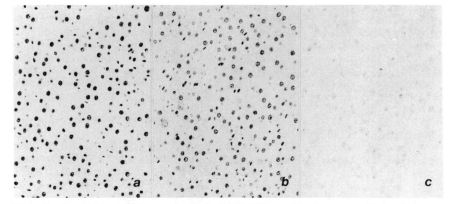

Fig. 3 Specific binding of anti–4NQO–DNA antibody to nuclei of human skin fibroblast cells demonstrated by immunofluorescence. Cells were exposed to (a) 2×10^{-6}M, (b) 1×10^{-6}M, (c) as b, but then incubated for 16 hr in fresh medium. (d) 5×10^{-7}M , (e) 2.5×10^{-7}M and (f) without 4NQO.

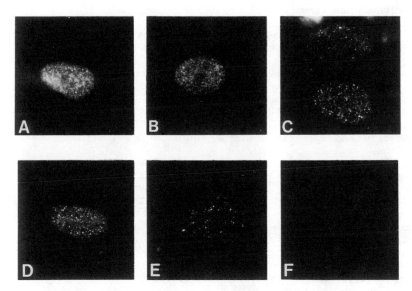

Fig. 2 Nuclear immunostaining of DMAB–DNA adducts in liver from the rat 24 hr after s.c. administration of DMAB. The dose of DMAB per kg of body weight (a) 200 mg, (b) 50 mg and (c) corn oil only.

The affinity of anti 4NQO–DNA–antibody for highly modified DNA was found to be higher than for DNA modified to a lesses degree, with as much as a 30 fold difference being observed between modification levels of 6 adducts/10^4 nucleotides and 2.4 adducts/10^5 nucleotides. As mentioned above, for quantitation of 4NQO adducts in DNA from biological samples, slightly modified DNA must be used for competitive immunoassay as the standard.(19,20)

In contrast, anti–DMAB–DNA antibody demonstrated the same affinity for both high and low modification levels of DNA after heat denaturation. As shown in Fig. 1, a single line standrad curve was obtained with various DMAB–DNA samples having different modification levels. Consequently the anti–DMAB–DNA antibody has an advantage over the anti–4NQO–DNA antibody for quantitate of a wide range of adduct levels in biological samples.

Immunohistochemical studies can give information on the formation and repair of adducts in single cells as well as clarifying the distribution and localization of adducts over different cell types in tissue sections. For immunohistochemical assays, however, it is necessary to denature DNA/or remove chromosomal proteins in order to allow the immune reagents access. Therefore optimal conditions for the acid treatment must be studied carefully, to minimize adduct instability. It remains to be established whether the presently applied methods are suitable for accurate quantification of adducts in cells.(6–8, 21)

Conclusion and Summary

Polyclonal antibodies against DNA modified by four kinds of chemical carcinogen, 4NQO, DMAB, 3- and 2-methoxy-4-AAB have been elicited in our laboratory.

The modified DNAs were presence by activation of N-hydroxy-compounds in the presence of seryl-AMP, and   DNAs containing 1-7 adducts per 10^3 nucleotides were obtained. It was found that modification levels of DNA of at least one per 10^3 nucleotides were necessery for suitable immunogencity. With highly modified DNAs very sensitive and specific antisera could be obtained from immunized rabbits.

Characterization of these antibodies was carried out by competitive ELISA. The difference in IA_{50} values between 60 and 2.4 adducts per 10^5 nucleotides was in the order of 30. Namely, the binding reaction was 30 fold greater with the highly modified DNA. Thus for quantification of DNA adducts of this type in biological samples, DNA with a low degree of modification must be used as the competitor standard.

Thus, using DNA modified at a low level (3 adduct/10^6 nucleotides), we have obtained results for adduct number almost the same as these gained by measurement of radioactivities in DNA from rats treated with labeled 4NQO.

In contrast to the anti 4NQO–DNA–antibody, the anti DMAB–DNA–antibody reacted with denatured DNA modified to widely differing extents with similar affinity. Moreover, anti DMAB–DNA– antibody could be shown to react with the same affinity to mono- or oligonucleotide–adducts. In this respect, the anti–DMAB–DNA antibody therefore has advantage for determination of adduct levels in biological samples.

Immunochemical studies should prove especially useful for non radiolabeled chemicals. As regards our anti azobenzen–DNA-antibodies, they could not be characterized in as great detail as for the anti–4NQO–DNA and anti–DMAB–DNA-antibodies, because radiolabeled compounds are not available, and the chemical nature of the relevant adducts has not yet been elucidated. However, these antibodies were found to react specifically with 4–aminoazobenzene–modified DNA and immunostaining was observed in the nuclei of hepatocytes of rats treated with 3–methoxyl–AAB.

Immunohistochemical studies have allowed detection of formation and removal of adducts in individual cells. However, prior to immunostaining, acid treatment is essential for antigen to have access and react with antibody and so the possibility of altration in adducts structure under such conditions require that methodological limitations be clearly defined.

In conclusion, we confirmed that immunochemical and immunohistochemical approaches are very useful tools for monitoring human exposure to environmental chemicals.

Acknowledgment

The authors wish to thank Dr. Y. Kawazoe and for preparing the N-hydroxy compounds and Mr. H. Aoki for excellent technical assistance This work was supported by Grants- in -Aid for Cancer Research from the Ministry of Education, Science and Culture, Japan.

Literature Cited

1. Stickland, P.T.; Boyle, J.M. In Prog. Nucleic Acid Res. Mol. Biol. 1984, p.1–48.
2. Sage, E.; Fuchs, R.P.P.; Leng, M. Biochemistry 1979, 18, 1328–1332.
3. Haugen, A.; Groopman, J.D.;Hsu, I–C.;Goodrich, G.R.; Wogen, G.N.; Harris, C.C. Proc.Natl.Acad. Sci. USA 1981, 78, 4124–4127.
4. Slor, H.; Mizusawa, H.; Neihart, N.; Kakefuda, T.; Day, 111, R.S.; Bustin, M. Cancer Res. 1981, 41, 3111–3117.
5. Heyting, C.; van der Laken, C.J.; Raamsdonk, W.; Pool, C.W. Cancer Res. 1983, 43, 2935–2941.
6. Morita, T.; Ikeda, S.; Minoura, Y.; Kojima, M.; Tada, M. Jpn. J. Cancer Res.(Gann) 1988, 79, 195–203.
7. Tada, M.; Aoki, H.; Minoura, Y.; Morita, T.; Shirai, T.; Yamada, H.; Ito, N. Carcinogenesis 1989, 10, 1379–1402.
8. Shirai, T.; Nakamura, A.; Fukushima, S.; Tada, M.; Morita, T. and Ito N. Carcinogenesis 1990, 11, 653–657.
9. Denda, A.; Mori, Y.; Yokose, Y.: Uchida, K.; Murata, Y.; Makino, T.; Tsutsumi, M.; Konishi, Y. Chem.–Biol.Interract. 1985, 56, 125–143.
10. Westra,J.G.; Flammang, T.J.; Fullerton, N.F.; Beland, F.A.; Weis, C.C.; Kadlubar, F.F. Carcinogenesis 1985 5, 37–41.
11. Tada, M.; Tada, M. Biochim. Biophys. Acta 1976, 454, 558–566.
12. Bailleul, B.; Galiegue, S.; Loucheux-Lefebvre, M.H. Cancer Res. 1981, 41, 4559–4565.
13. Galiegue-Zouitina, S.; Bailleul, B.; Ginot, Y–M.; Perly, B.; Loucheux-Lefebvre, M.H. Cancer Res. 1986, 46, 1858–1863.

14. Tada, M.; Kohda, K.H.; Kawazoe, Y. Gann, 1984, 75, 976–985.
15. Beland, F.A.; Kadlubar, F.F. Environ, Health Perspect. 1985, 62, 19–30.
16. Flammang, T.J.; Westra, J.G.; Kadkubar, F.F.; Beland, F.A. Carcinogenesis 1985, 6, 251–258.
17. Bailleul, B.; Galiegue-Zouitina, S.; Perly, B.; Loucheux-Lefebvre, M.H. Carcinogenesis 1985, 6, 319–322.
18. Roberts, D.W.; Benson, R.W.; Groopman, J.D.; Flammang, T.J. Nagel, W.A.; Moss, A.J.; Kadlubar, F.F. Cancer Res. 1988, 48, 6336–6342.
19. Santella, R.M.; Weston, A.; Perera, F.P.; Trivers, G.T.; Harris, C.C.; Young, T.L.; Nguyen, D.; Lee, B.M.; Poirier, M.C. Carcinogenesis 1988, 9, 1265–1269.
20. Kriek, E.; Van Schooten, F.J.; Hillebrand M.J.X.; Welling, M.C. In Methods for detecting DNA damaging agents in humans: Applications in cancer epidemology and prevention, (IARC Scientific publications No.89), 1988 Lyon, International Agency for Research on Cancer p.201–207.
21. Nakagawa, K.; Tada, M.; Morita, T.; Utsunomiya, J. and Ishikawa, T. J. Natl Cancer Inst 1988, 80, 419–425.

RECEIVED August 30, 1990

Chapter 22

Polycyclic Aromatic Hydrocarbon–DNA Adduct Load in Peripheral Blood Cells

Contribution of Multiple Exposure Sources

P. T. Strickland[1], N. Rothman[2], M. E. Baser[1], and M. C. Poirier[3]

[1]Department of Environmental Health Sciences and [2]Department of Epidemiology, School of Public Health, Johns Hopkins University, Baltimore, MD 21205
[3]Laboratory of Cellular Carcinogenesis and Tumor Promotion, National Cancer Institute, National Institutes of Health, Bethesda, MD 20892

Polycyclic aromatic hydrocarbons (PAH) are produced by combustion of organic materials, and humans are exposed to these compounds from a variety of sources. In a recent cross-sectional study of PAH exposure in residential fire fighters and controls, PAH-DNA adducts in peripheral nucleated blood cells were examined by immunoassay (ELISA) as potential markers of exposure. Mean adduct levels in caucasian participants (22 detectable/66 tested) increased with exposure to one or more sources of PAH: fire fighting, smoking, and/or char-broiled (CB) food consumption. The effect of dietary PAH on PAH-DNA adduct levels was further investigated in a controlled exposure study in which 4 volunteers consumed CB beef daily for 7 days. PAH-DNA adduct levels increased 3-fold and 6-fold above baseline levels and in two individuals and remained unchanged in two individuals during the feeding period. These results suggest that multiple sources of PAH contribute to the PAH-DNA adduct load in peripheral blood cells and confirm the importance of dietary contributions.

Polycyclic aromatic hydrocarbons (PAH) are introduced into the environment by a number of combustion processes, including heat and power generation, coke production, open refuse burning and motor vehicle emissions (1,2). Individuals can be exposed to PAHs from these environmental sources or from their occupation, diet or smoking habits.

In a recent study of exposure to combustion products in residential fire fighters (3,4), we measured several markers of genotoxic damage in nucleated blood cells as potential markers of exposure including sister chromatid exchanges and PAH-DNA adducts. The results suggest a contribution of occupation (fire fighting), diet (char-broiled (CB) food consumption) and cigarette smoking to individual levels of PAH-DNA adducts. In this report, we summarize these findings and present further evidence that dietary PAH contribute to PAH-DNA adduct levels in peripheral blood cells.

0097–6156/91/0451–0257$06.00/0

Multiple PAH Exposure in Fire Fighters

Blood samples (40 mls) were collected from 43 urban fire fighters
and 38 controls with similar demographics and smoking habits. A
detailed questionnaire was administered to participants to obtain
information on recent fire fighting activity, tobacco smoking (or
chewing), frequency of consumption of CB food and alcohol, and
other potential confounding factors (3). DNA was extracted from
the nucleated blood cell fraction (buffy coat) and analyzed for
PAH-DNA content by ELISA as previously described (5,6) using a
BPDE-DNA standard modified in the same range as the biological
samples (4.4 fmoles/ug DNA). The ELISA employed rabbit anti-BPDE-
DNA (antibody 33, diluted 1:70,000), goat-anti-rabbit IgG
conjugated to alkaline phosphatase (diluted 1:400) and p-nitro-
phenyl-phosphate as described previously (6). Since the antiserum
is capable of recognizing several PAH-DNA adducts (7), the results
are expressed as BPDE-DNA antigenicity and represent PAH-DNA
adduct formation (7,8). The lower limit of sensitivity of the
assay was 0.3 fmol adduct/ug DNA at 20% inhibition.
    PAH-DNA adducts were detectable ($\geq$ 0.3 fmole adduct/ug DNA)
in 35% (28/81) of the individuals tested. Smoking increased the
unadjusted risk for detectable PAH-DNA adducts (odds ratio = 2.4;
95% confidence interval = 0.96-6.26) (3). Caucasian fire fighters
(66/81) exhibited a higher risk for the presence of detectable
PAH-DNA adducts than caucasian controls after adjustment for CB
food consumption (odds ratio = 3.4; 95% confidence interval = 1.1-
10.5; p = 0.04) (3). Simple linear regression analysis indicated
that consumption of CB foods $\geq$3 times per month was associated
with increased PAH-DNA adduct levels (p = 0.04). Thus, the three
sources of exposure that contributed to PAH-DNA adduct levels in
the caucasian participants were fire fighting, tobacco smoking,
and CB food consumption (Figure 1, 2 and Refs 3,4). The percentage
of individuals with detectable adducts increased with the number
of exposure sources (Table I), but X^2 analysis for trend was not
significant.

Table I.   Percent Individuals With Detectable Levels
                of PAH-DNA Adducts

Number of Exposure Sources (FF,SM,CBF)	Number of Individuals (n)	Number with/without Detectable PAH-DNA Adducts with ($>$0.3 fmole/ug)	without ($<$0.3 fmole/ug)
0	10	2 (20%)	8
1	24	8 (33%)	16
2	25	9 (36%)	16
3	7	3 (43%)	4
	66	22	44

Analysis for Trend:
    (cutoff at $\geq$ 0.3 fmole/ug DNA) X^2 = 0.99   p = 0.32
    (cutoff at $\geq$ 0.75 fmole/ug DNA) X^2 = 2.68   p = 0.10
(Reproduced with permission from Ref 4. Copyright 1990, Oxford
University).

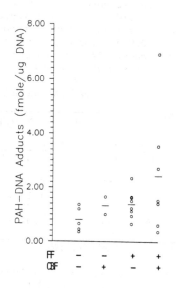

Figure 1.   Individual levels of PAH-DNA adducts in samples with detectable adducts stratified by fire fighting activity (FF) and charcoal-broiled food consumption (CBF). Mean adduct levels are indicated in each group. (Reproduced with permission from Ref 4. Copyright 1990, Oxford University).

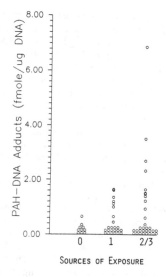

Figure 2.   Individual levels of PAH-DNA adducts stratified by number of exposure sources (fire fighting, smoking, or CB food consumption).

Mean adduct levels in the caucasian participants (22 detectable/
66 tested) increased with exposure to one or more of these three
sources of PAHs (Fig 2, Table II). Linear regression analysis
indicated an association between number of exposure sources and
mean adduct levels (p = 0.02). The major source of exposure
contributing to PAH-DNA adduct levels in the non-caucasian cohort
(6 detectable/15 tested) was smoking (data not shown).

Table II. Mean Levels of PAH-DNA Adducts

No. of Exposure Sources (FF,SM,CBF)	Mean Adduct Levels (fmole/ug DNA)	
	All Samples* (n = 66)	Detectable Samples (n = 22)
0	0.22 ± 0.16	0.50 ± 0.15
1	0.48 ± 0.53	1.13 ± 0.40
2	0.73 ± 0.11	1.75 ± 0.90
3	1.36 ± 2.48	2.97 ± 2.79

*Setting undetectable (<0.3 fmole/ug DNA) = 0.15 fmole/ug DNA

Linear regression analysis using number of exposures as
independent variable: r = 0.28, p = 0.023 (all samples);
r = 0.48, p = 0.025 (detectable samples).
(Reproduced with permission from Ref 4. Copyright 1990, Oxford
University).

Dietary PAH Exposure in Volunteers

A controlled dietary exposure study was conducted in order to test
the hypothesis that dietary sources of PAH can significantly
contribute to PAH-DNA adduct load (9). Four healthy caucasian
non-smoking male volunteers aged 33-39 adhered to a diet free of
CB foods for one month. During this period, three blood samples
(40 mls each) were collected for baseline measurements. Volunteers
then consumed CB ground beef daily for seven days (approximately
10 oz (280 gm) cooked weight per person per day). One individual
did not complete the planned schedule and ate CB beef on four days
only. Blood samples were collected immediately before eating CB
beef on the first day of feeding (day 1) and on days 2, 5, 8, 12
and 24-31. During the three weeks following feeding (days 8 to
31), volunteers refrained from eating any CB foods. Dietary
habits were monitored throughout the entire study by a daily diary
of food and beverage consumption.
     DNA was extracted from the nucleated blood cell fraction by a
modification of the method of Miller et al (10). PAH-DNA adduct
content was analyzed by ELISA as decribed above except that a
fluorescent enzyme substrate, 4-methylumbelliferyl phosphate, was
used in the final step as previously described (6). The lower
limit of detection was 0.04 fmole adduct/ug DNA.

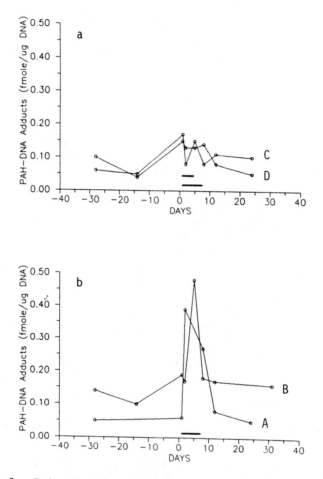

Figure 3. Individual levels of PAH-DNA adducts before, during and after CB beef consumption. Solid bar indicates period of daily CB beef intake. Volunteer D consumed beef for 4 days only. Volunteers A, B, and C consumed beef for 7 days. (Reproduced with permission from Ref 4. Copyright 1990, Oxford University).

Baseline levels of PAH-DNA adducts prior to CB beef consumption were between non-detectable and 0.19 fmole/ug DNA for all individuals (Figure 3a, 3b). During the week of CB beef consumption, two individuals exhibited an increase in PAH-DNA adduct level (Figure 3b). Volunteer A exhibited a 6-fold increase after one day of CB beef and declined to baseline levels 4 days after cessation of CB beef consumption. Volunteer B exhibited a 3-fold increase after 4 days of CB beef consumption and declined to baseline levels one day after cessation of CB beef consumption. PAH-DNA adduct level in volunteers C and D did not increase at any time during the study (Figure 3a). Volunteer D remained on the CB beef diet for four days only.

Discussion

These and other studies indicate that multiple sources of PAH contribute to the levels of PAH-DNA adducts measured in peripheral nucleated blood cells. The three sources of exposure identified as possibly contributing to PAH-DNA adduct levels were occupation, diet, and smoking. The occupation in this case was fire fighting which involves exposure to combustion products produced at the fire site and diesel and gasoline exhaust from fire trucks. The smoking exposure was cigarette or cigar smoking daily. The dietary exposure in the initial study was CB food consumption at least three times per month (about once per week).

Dietary exposure to PAH was examined in more detail in a controlled CB beef feeding experiment in which each individual served as his own control. Two individuals exhibited measurable increases in PAH-DNA adduct levels while the two other individuals did not respond. This difference could be due to interindividual differences in constitutive or induced physiological parameters such as absorption, digestion, metabolism, excretion and DNA adduct formation and repair (11). Conney et al (12) demonstrated the induction of phenacetin metabolizing enzymes in human volunteers fed CB beef for four days. Variation in the inducibility of PAH metabolizing enzymes in hepatocytes and peripheral blood cells in individuals fed CB beef may explain some of the observed interindividual differences in this study. Other studies (13,14) have demonstrated a 3-10 fold variation in the ability of cultured human lymphocytes or monocytes to form B(a)P-DNA adducts after treatment with B(a)P.

These results indicate that dietary sources of PAH can contribute to PAH-DNA adduct load in peripheral blood cells and should be considered when using PAH-DNA adducts as a marker of exposure to PAH from occupational or environmental sources. Furthermore, an understanding of the biochemical basis for the observed interindividual differences in PAH-DNA adduct levels should enhance the use of these adducts as markers in exposure assessment.

Acknowledgments

The authors thank C. Gentile, E. Bowman and E. Patterson for expert technical assistance. Research supported in part by DHHS grants ES03819, ES03841, and AR38884.

Literature Cited

1. Baum EJ. In Polycyclic Hydrocarbons and Cancer; Gelboin HV and Ts'o POP, Eds; Academic Press, New York: 1978, Vol. 1, p. 45-70.
2. Woo YT, Arcos JC. In Carcinogens in Industry and the Environment; Sontag JM, Ed; Dekker, New York: 1981, p. 167-281.
3. Liou SH, Jacobson-Kram D, Poirier MC, Nguyen D, Strickland PT, Tockman MS. Cancer Research 1989, 49:4929-4935.
4. Strickland PT, Liou SH, Poirier MC, Nguyen D, Tockman MS. In Biomonitoring and Carcinogen Risk Assessment; Garner C, Ed; Oxford University Press, Oxford (in press).
5. Lieberman MW, Poirier MC. Cancer Research 1973, 33: 2097-2103.
6. Santella RM, Weston A, Perera FP, et al. Carcinogenesis 1988, 9:1265-1269.
7. Weston A, Willey JC, Manchester DM, et al. In Methods for Detecting DNA Damaging Agents in Humans; Bartsch H, Hemminki K, O'Neill IK, Eds; IARC, Lyon: 1988, p. 181-189.
8. Perera FP, Santella RM, Brenner D, et al. JNCI 1987 79:449-456.
9. Rothman N, Poirier MC, Baser ME, Hansen JA, Gentile C, Bowman ED, Strickland PT. Carcinogenesis (in press).
10. Miller SA, Dykes DD, Polesky HF. Nucleic Acids Research 1988, 16:1215.
11. Harris CC. Carcinogenesis 1989, 10:1563-1566.
12. Conney AH, Pantuck EJ, Hsiao KC, Kuntzman R, Alvares AD, Kappas A. Fed Proc 1977, 36:1647-1652.
13. Nowak D, Schmidt-Preuss U, Jorres R, Liebke F, Rudiger HW. Int. J. Cancer 1988, 41:169-173.
14. Thompson CL, McCoy Z, Lambert JM, Andries MJ, Lucier GW. Cancer Research 1989, 49:6503-6511.

RECEIVED August 30, 1990

Chapter 23

Immunohistochemical Localization of Paraquat in Lungs and Brains

Masataka Nagao[1], Takehiko Takatori[1], Kazuaki Inoue[2], Mikio Shimizu[2], and Koichi Terazawa[1]

[1]Department of Legal Medicine, School of Medicine, Hokkaido University, Sapporo 060, Japan
[2]Department of Pathology, Hokkaido University Hospital, Sapporo 060, Japan

Immunohistochemistry was used to investigate the localization and dynamics of paraquat in lung and brain in paraquat-poisoned rats. Rats were sacrificed at 3 h, 12 h, 24 h, 3 days, 7 days and 10 days after the intravenous injection of paraquat (5 mg/kg). In lung tissues, paraquat was localized in walls of blood vessels and bronchiolar epithelial cells from 3 h to 10 days after the paraquat exposure. Furthermore, histiocytes containing paraquat were observed. Interstitial pulmonary fibrosis containing paraquat developed with time. These results indicate that histiocytes are the probable cause of the pulmonary fibrosis in paraquat-poisoned rats. On the other hand, in brain tissues, paraquat was localized only in capillary walls and glial cells but was not observed in nerve cells 10 days after the injection of paraquat, providing evidence that paraquat cannot pass through the blood-brain barrier.

0097–6156/91/0451–0264$06.00/0
© 1991 American Chemical Society

Paraquat (1,1'-dimethyl-4,4'-bipyridinium) is a widely used herbicide, and is one of major causes of mortality in human self-poisonings (1,2). Pulmonary fibrosis is one of the most harmful complications of paraquat poisoning. Although several studies concerning the mechanism of development of pulmonary fibrosis have been performed (1,3-6), there has not been a study on the localization and dynamics of paraquat in lung. Also, parkinsonism can be induced by 1-methyl-4-phenyl-1,2,3,6-tetrahydropyridine (MPTP), an analog of paraquat (7,8), and paraquat may also be a cause of parkinsonism. In this paper, we describe the localization and dynamics of paraquat in lung and brain tissues with immunohistochemical techniques using previously developed rabbit antisera (9).

Materials and Methods

Chemicals Paraquat dichloride (1,1'-dimethyl-4,4'-bipyridinium dichloride) was purchased from Aldrich Co., U.S.A. Stravigen (biotin-streptavidin amplified system (peroxidase)) was purchased from BioGenex Laboratories, U.S.A. 3-Amino-9-ethylcarbozole (AEC) and 3,3'-diaminobenzidine (DAB) were obtained from Sigma Chemical Company, U.S.A. All other reagents were purchased from Nakarai Chemical Co., Japan.

Animal Treatment Male Sprangue-Dawley rats (200-250g) were intravenously administered paraquat dichloride (5 mg/kg) dissolved in saline. They were sacrificed 3 h, 12 h, 24 h, 3 days, 7 days and 10 days after injection and the tissues (lung and brain) were removed.

Tissue Preparation The tissues were fixed in 0.1 M phosphate buffer (pH 7.4) containing 4 % paraformaldehyde for 6 h. After fixation, they were cut to a thickness of 2 mm and these slices were fixed again in the paraformaldehyde for 3 days; the slices were then dehydrated in upgrading series of ethanol, cleared in xylene, and embedded in paraffin. Sections of 3 μm in thickness were cut and used for the immunohistochemical analysis of paraquat.

Antibodies Against Paraquat The polyclonal antibodies against paraquat reported previously (9) were used. After these antisera were prepared a 1:10 dilution with 0.01 M phosphate-buffered saline (pH 7.4, PBS), they were absorbed with bovine serum albumin (BSA, 1 mg/ml) overnight at 4°C before use.

Immunohistochemical Procedures for Light Microscopy The immunohistochemical staining using a biotin-streptavidin-peroxidase complex was performed according to the

method as described previously (10). The sections
from lungs and brains were stained with AEC and with DAB,
respectively.

Histological   Staining   In   order   to   evaluate   the
morphological   changes   of   lung   and   brain,   a
hematoxylin-eosin   stain   was   performed.   A   Masson
trichrome stain was also used in order to evaluate the
collagen generation in lungs.

Results

Following paraquat exposure, infiltration of inflammatory
cells into lung tissue and fibrosis of interstitial lung
tissue developed with time (Figure 1).   Pathological
changes were not observed in brain tissues.
      In   all   immunohistochemical   sections   of   the
lungs   during   the   experiments,   antibody   binding   was
observed   in   walls   of   blood   vessels,   histiocytes   and
bronchiolar   epithelial   cells   (Figure   2).   It   was
also   observed   that   the   paraquat   localized   in   the
bronchiolar epithelial cells seemed to be secreted into
bronchiole (Figure 2-A).   On the other hand, the
paraquat-localization   in   brain   was   found   only
in   capillary   walls   and   glial   cells   (Figure
3-A).   However,   these   positive   reactions   were
completely inhibited when the antiserum was absorbed with
3 mg/ml paraquat   overnight at 4 °C before use (Figures
2-D and 3-B), and in control sections of lung and brain
antibody binding was not observed.

Discussion

The localization of paraquat in lung and brain of the
paraquat-exposed   rats   was   demonstrated   immunohisto-
chemically.   Since   paraquat   is   water-soluble,   it
would be removed by the fixation and staining processes
except for the intracellularly localized paraquat.   It
is   not   clear   whether   paraquat   is   localized
intracellularly   as   a   free   form,   or   bound   to   cellular
components.   However, the antiserum used in this study
selectively   recognizes   both   the   methyl   group   and
bipyridyl   ring   of   paraquat,   and   dose   not   bind   even
slightly to analogs of paraquat (9), indicating that the
antiserum   specificically   binds   to   the   intracellularly
localized   paraquat   which   keeps   the   original
structure.   The   further   investigations   for   the
intracellular   localization   of   paraquat   should   be
performed.
      In the lung 3 h after paraquat exposure, histiocytes
containing   paraquat   already   had   infiltrated   into
interstitial lung tissue and could be found there 10
days after the exposure.   Interstitial fibrosis was
also observed to increase with time.   Schoenberger et

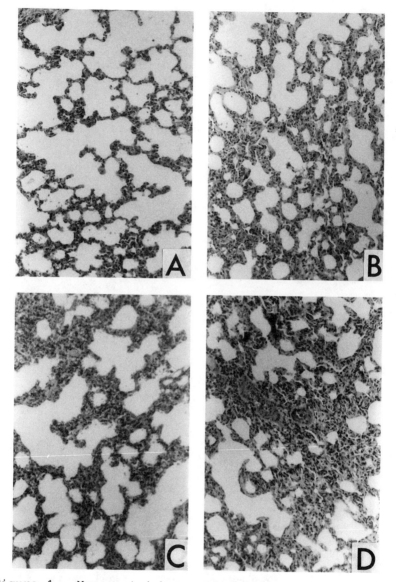

Figure 1. Masson trichrome stainings of lung; (x100)
3 h (A), 24 h (B), 3 days (C) and 10 days (D).

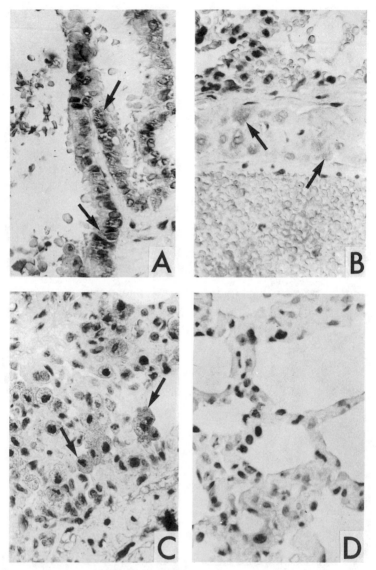

Figure   2.   Immunohistochemical   localization   of
paraquat in the lung tissue. (x400) 3 h (A), 3 days
(B), 10 days (C) and control (D).

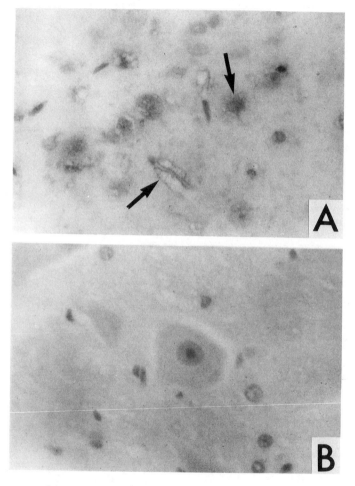

Figure 3. Immunohistochemical localization of paraquat in the brain tissue. (x400) 3 days (A) and control (B).

al. (3) reported that after paraquat exposure in vivo, alveolar macrophages produce fibronectin and release a fibroblast growth factor and Wong and Stevens (11) reported that after exposure in vitro, paraquat induces not only cytotoxicity in alveolar macrophages but also extracellular superoxide anion generation from alveolar macrophages.   From these findings it is likely that the development of the pulmonary fibrosis induced by paraquat is due to the release of the biologically active substances produced by macrophages, which are activated by phagocytizing herbicide damaged cells, in addition to the direct toxicity of paraquat.

The mechanism of migration of macrophages to the interstitial lung tissue is not clear.   However, it is possible that the increase in vascular permeability after the paraquat exposure in vivo (12) results in the production of a macrophage chemotactic factor from plasma proteins (13,14) at the interstitial lung tissue.

Webb (15) reported that in an autoradiographic study paraquat is localized in bronchiolar epithelium 24 hafter the paraquat exposure and implicates the bronchiolar epithelium as the site of paraquat uptake.   In our immunohistochemical study, we not only observed that paraquat was selectively localized in bronchiolar epithelial cells, but also observed that paraquat was secreted into the bronchiole.

In the brain, even 10 days after the paraquat exposure, paraquat was not found in nerve cells but observed only in glial cells and capillary walls.   These findings indicate that paraquat is unable to pass through the blood-brain-barrier, suggesting that paraquat itself is not a cause of Parkinson's disease (16).

Acknowledgments
This work was supported in part by the Shimabara Science Promotion Foundation.

Literature Cited

JOURNAL EXAMPLES:
1. Haley, T. J. Clin. Toxicol. 1979, 14, 1-46.
2. Carson D. J. L.; Carson E. D. Forens. Sci. 1976, 7, 151-60.
3. Schoenberger C. I.; Rennard S. I.; Bitterman P. B.; Fukuda Y.; Ferrans V. J.; Crystal R. C. Am, Rev. Respir. Dis. 1984, 129, 168-73.
4. Ilett K. R.; Stripp B.; Menard R. H.; Reid W. D.; Gillette J. R. Toxicol. Appl. Pharmacol. 1974, 28, 216-26.
5. Nerlich A, G.; Nerlich M. L.; Langer I.; Demling R. H. Exp. Mol. Pathol. 1984, 40, 311-19.
6. Yamaguchi M.; Takahashi T.; Togashi H.; Arai H.; Motomiya M. Chest 1986, 90, 251-7.

7. Langston J. W.; Ballard P. A.; Tetrud J. W.; Irwin I. Science 1983, 219, 979-80.
8. Burns R. S.; Chiueh C. C.; Markey S. P.; Ebert M. H.; Jacobowitz D. M.; Kopsin I. J. Proc. Natl. Acad. Sci. U.S.A. 1983, 80, 4546-50.
9. Nagao M.; Takatori T.; Terazawa K.; Wu B.; Wakasugi C.; Masui M.; Ikeda H. J. Forens. Sci. 1989, 34, 547-52.
10. Nagao M.; Takatori T.; Inoue K.; Shimizu M.; Terazawa K.; Akabane H. Toxicol. (in press)
11. Wong R. C.; Stevens J. B. J. Toxicol. Environ Health 1985, 15, 417-29.
12. Tanaka R.; Fujisawa S.; Kawamura K.; Harada M. J. Toxicol. Sci. 1983, 8, 147-59.
13. Honda M.; Hirashima M.; Hayashi H. Virchows Arch. B Cell Pathol. 1978, 27, 317-33.
14. Ishida M,; Honda M.; Hayashi H. Immunol. 1978, 35, 167-76.
15. Webb D. B. Br. J. exp. Pathol. 1980, 61, 217-21.
16. Koller W. C. Neurol. 1986, 36, 1147.

RECEIVED August 30, 1990

Chapter 24

Immunoassays for Molecular Dosimetry Studies with Vinyl Chloride and Ethylene Oxide

Michael J. Wraith, William P. Watson, and Alan S. Wright

Shell Research, Ltd., Sittingbourne Research Centre, Sittingbourne, Kent, ME9 8AG, United Kingdom

Specific, sensitive radioimmunoassays have been developed for detecting 1,N6-ethenodeoxyadenosine (EdA) and 3,N4-ethenodeoxycytidine (EdC), two adducts reported to occur in liver DNA of rats chronically exposed to vinyl chloride. Application of these assays in the analysis of liver DNA from rats exposed orally to five daily doses of 50 mg/kg vinyl chloride failed to clearly detect EdA and EdC. The levels of detection were 1 adduct in 3×10^8 nucleotides and 1 adduct in 6×10^8 nucleotides respectively. A novel immunochemical procedure has also been developed for monitoring human exposures to ethylene oxide based on the reaction of ethylene oxide with the amino group of the N-terminal valine residue of the α-chain of human hemoglobin. The method is designed to measure the extent of this reaction by determining the product in the form of the aducted N-terminal tryptic heptapeptide. The method has been employed in monitoring exposures of workers to ethylene oxide and has been validated by comparison with a gas chromatography-mass spectrometry procedure.

Despite the long-standing and diverse applications of immunochemistry in the clinical area, this technology is only now becoming widely exploited as a general tool in analytical and biochemistry, e.g. in environmental analysis. Initial studies in our laboratory were directed towards the analysis of the products of reactions between carcinogens and bio-macromolecules for applications in molecular dosimetry and biomonitoring programmes *(1)*. Subsequently, the focus has shifted to applications in environmental monitoring where the major attributes of the methods, i.e. speed and simplicity combined with high sensitivity and specificity, were needed to cope with the increasingly heavy demands. Particular emphasis was placed on enzyme-linked immunosorbent assays (ELISA) with a view to establishing practical field assays *(2)*.

These developments have been greatly facilitated by the development of monoclonal antibody technology which has expanded the horizons of immunoassay *(3)*. In particular this technology provides the potential for greater specificity and virtually unlimited supplies of antibodies which has allayed concerns about the long-term viability of the newly developed assays. The, so-called, non-traditional applications of immunoassay are therefore well-established *(4)* and have an important role in both applied and fundamental research.

0097–6156/91/0451–0272$06.00/0

Molecular dosimetry

It is now widely accepted that the identification and quantitation of the products of interactions between DNA and electrophilic chemicals (or their metabolites) is of key importance in the understanding of the mutagenic and carcinogenic processes. In particular, qualitative and quantitative determinations of DNA adducts formed during low-level exposures to chemicals are essential for rational prospective risk assessments *(5)*.

Problems arise when extrapolations are made from high experimental doses or concentrations, designed to give measurable effects in experimental species, to low concentrations occurring in the environmental or occupational situation. For example, it is uncertain whether the extrapolation from high doses to low doses should be based on a linear regression to zero dose or on some alternative dose-response relationship. Furthermore, translation of the risk estimates from the test (experimental) species to man are subject to error, the assessments being complicated by species differences among factors that determine the response to a given exposure or dose of the carcinogen *(6)*.

These interpretative problems led to the development of the 'target-dose' approach *(7)*, which seeks to improve the quality of risk assessment by providing a means of compensating for differences in the metabolism and metabolic disposition of carcinogens in the tissues of the test species and man. This approach, which is based on the estimation of the dose of ultimate carcinogen delivered to the critical cellular target (DNA), permits the investigation of the integrated operation of all the toxicokinetic and toxicodynamic factors that regulate tissue DNA dose.

A further important consideration is that human tissue DNA is generally inaccessible for experimental purposes and this led to the development of indirect methods of assessment of DNA doses *(8)*. Due to the electrophilic nature of mutagens and carcinogens they react with all types of nucleophilic centre, including those found in proteins. There are several examples where hemoglobin (Hb) is an appropriate dose monitor for DNA(9). Its advantages are that blood is readily and repeatedly obtainable from humans in useful quantities. Also the biological lifetime of the human erythrocyte (120 days) permits monitoring long after a particular exposure has ceased and provides high sensitivity by permitting adduct formation due to chronic, low-level exposure to accumulate.

The estimation of target dose, either directly at DNA or indirectly in proteins, is based on measurements of the amounts of the adducts formed by chemical reactions occurring between the ultimate carcinogen and DNA bases or amino acid residues. These adducts occur in tissues at extremely low concentrations, typically 1 adduct per 10^{6-10} bases or amino acids, and the technical demands on quantitative assays, without the use of radiolabelled carcinogens, are very high. In such instances, e.g. the monitoring of occupational exposure or conventional experimental carcinogenicity studies, the assay of adducts necessitates the development of extremely sensitive and specific procedures. In this respect immunochemical assays have made a significant contribution *(10,11)* and, because of their intrinsic selectivity, which often significantly reduces or obviates the need for preliminary purification steps, they are one of the principle methods of choice.

Vinyl Chloride Dosimetry Studies

Vinyl chloride (VC) has been known for many years to be a human carcinogen. It has therefore been the subject of intensive investigation and was chosen as a model for our molecular dosimetry studies in experimental animals. During 1980 work began on the development of radioimmunoassays (RIA) to quantitate VC-induced DNA adducts. The aim was to apply the RIAs in studies designed to investigate the quantitative relationships between exposures to low doses of VC, tissue DNA doses (including both target and non-target tissues) and hepatocarcinogenesis. At the commencement of the study, three VC-DNA adducts had been described in rats exposed to VC *in vivo (12,13,14)*. These comprised a major adduct,

N^7-(2-oxoethyl)guanine (I), and two minor (cyclic) adducts, 1,N^6-ethenodeoxyadenosine (II, EdA) and 3,N^4-ethenodeoxycytidine (III, EdC). During the course of the work evidence for a third cyclic adduct, N^2,3-ethenoguanine (IV), was reported *(15)*. (Figure 1).

Part of the original objectives of the study was the development of a RIA for the major VC-DNA adduct. However, N^7-alkylated deoxyguanosine adducts are chemically unstable and, although an immunogen was prepared, production of antibodies was not detected. Thus RIAs were developed for EdA and EdC.

Antibody Production. Immunogens were prepared by covalently linking 1,N^6-ethenoadenosine and 3,N^4-ethenocytidine to bovine albumin *(16)*. The use of the ribose forms of the adducts did not create a problem of specificity as the analytical samples were purified free of RNA. Ultimately, HPLC purification of EdA and EdC prior to analysis obviated all potential cross-reactivity problems. Rabbits were immunised by the multisite intradermal method *(17)* and high-titre antisera were generated.

Synthesis of Reference Standards and Radiolabelled Tracers. EdA and EdC were prepared *(18)* and purified by HPLC. In order to prepare [^{125}I]-radiolabelled tracers a strategy was devised to introduce a group into the adduct molecules, at a point remote from the antigenic determinant, which could be subsequently iodinated. In summary, the acetal derived from the reaction of anisaldehyde with the ribose moieties of adenosine or cytidine was demethylated with lithium t-butyl mercaptide to give phenolic derivatives. These were then allowed to react with chloroacetaldehyde *(18)* to give the modified etheno adducts. Iodination was accomplished using sodium [^{125}I]iodine and a solid-phase oxidising agent, 1,3,4,6-tetrachloro-3α-6α-diphenylglycoluril *(19)*.

Radioimmunoassay Parameters. The RIAs for EdA and EdC were developed by conventional approaches. The lower limit of detection for EdA was 1 pmol/ml which was equivalent to 0.05 pmol (3×10^{10} molecules) in the test sample volume. For EdC the lower limit of detection was 0.5 pmol/ml. Antibody affinities for both the EdA and EdC were estimated *(20)* to be in the order of 10^{-9}M.

Both antisera displayed high specificity for their respective antigen targets, with no cross-reactivity between these. Furthermore, cross-reactivity with normal deoxyribonucleosides, determined at the limits of solubility in the test system i.e. 3.6 μmol/ml, was also very low and suggested that the RIAs would be capable of detecting one adducted deoxyribonucleoside in the presence of approximately 10^5 non-adducted deoxyribonucleosides. This level of cross-reactivity indicated that the assays would be unable to meet the requirement of quantifying adducts in the presence of a molar excess of approximately 10^{10} deoxyribonucleosides i.e. 1 adduct per cell.

Assay sensitivity was therefore enhanced by concentrating and isolating the adducts by HPLC prior to RIA. Recoveries of 80% were typical. The removal of cross-reacting species maximised the assay sensitivities which were now limited only by the quantity of DNA available for analysis. The adopted procedure was based on a 5 mg DNA sample, equivalent to 9.25×10^{18} nucleotides. The lower limits of detection of the EdA and EdC RIAs were 3×10^{10} molecules and 1.5×10^{10} molecules respectively. Thus the EdA RIA was capable of detecting 1 adduct in 3×10^8 nucleotides and the EdC RIA 1 adduct in 6×10^8 nucleotides.

Vinyl Chloride Exposure Study. Rats were exposed to VC for 5 days (daily doses of 50 mg/kg in corn oil by oral intubation). DNA was isolated from the livers and hydrolysed enzymatically prior to HPLC.

Following collection of the appropriate fractions each was assayed by both RIAs. EdA was not detected. In the case of EdC values bordering on the lower limit of detection were obtained. Confirmation of this result would have required a 10-fold greater amount of DNA which was unavailable.

Conclusions. Specific and sensitive RIAs have been developed for the two minor adducts reported to occur in the liver DNA of rats exposed to VC. The results obtained in the experiment reported here were inconclusive and subsequent acute exposure studies performed in this laboratory and elsewhere *(14)* with ^{14}C-labelled VC failed to confirm the earlier findings *(13)*. However, a recent study *(21)* in which rats were exposed to 2000 ppm of VC for 10 days has reported measurements of the minor VC adducts by RIA.

The results observed in this range of VC exposure conditions supports the idea that measurable levels of the minor cyclic adducts are only found after long-term exposures or following exposures during rapid growth e.g. in neonates. Current views *(22)* are that the minor cyclic adducts are criticial to the promutagenic, and carcinogenic, activity of VC. However, there still remains a need to develop a simple quantitative method for the major N[7]-oxoethylguanine adduct as this will provide a more sensitive monitor, especially at low exposure doses.

Ethylene Oxide Exposure Monitoring

The utility of Hb as a DNA dose monitor was investigated, using ethylene oxide (EO) as the model carcinogen. Prior to this study, inhalation experiments in rats exposed to ^{14}C-EO had indicated a rapid absorption and equilibration of EO throughout the tissues *(23)*. The results of these studies were consistent with the view that, in the case of EO, Hb would be an effective tissue DNA dose monitor.

Peptide approach. A RIA was developed *(1)* for use in biomedical monitoring of EO exposure. The principal adducted amino acids formed in human Hb by reaction with EO are N-(2-hydroxyethyl)valine (α- and β-chain), N_1-(2-hydroxyethyl)histidine, N_3-(2-hydroxy-ethyl)histidine and S-(2-hydroxyethyl)cysteine. The possibility of developing antibodies against the adducted amino acids was judged to be low as their small molecular size offers only limited antigenicity. An alternative approach was to raise antibodies against a peptide from human Hb which contained an EO-adducted amino acid. The peptide selected was the N-terminal heptapeptide released from the α-chains of human Hb by the action of trypsin *(24)* (Figure 2).

The adducted heptapeptide and the unmodified analogue were synthesized chemically. Both peptides were analysed by high-performance liquid chromatography and were homogeneous on two different stationary phases. In addition, fast atom bombardment-mass spectrometry of both peptides gave $(M + H)^+$ ions consistent with the required molecular weights.

Radioimmunoassay development. The hydroxyethylated (HOEt) peptide was radioiodinated at the amino group of the C terminal lysine using a conjugation procedure *(25)*. This radioactive tracer was used in the optimized RIA and to monitor incorporation of the HOEt peptide during preparation of the immunogen. The HOEt peptide was coupled to horse albumin using 1-ethyl-3-(3-dimethylaminopropyl)carbodiimide, resulting in approximately 16 mol adducted peptide per mol immunogen. Four rabbits were immunized, and antisera from one animal (R103B9) demonstrated sufficiently low cross-reactivity when tested against the non-HOEt peptide, native human Hb and the peptides from trypsin-hydrolysed Hb, to be useful for the development of a RIA (see Figure 3). The cross-reactivity results indicated that it was possible to quantify the HOEt peptide in the presence of a 10^6-fold excess of the non-HOEt peptide. During development of the assay, attempts were made to analyse native Hb treated with EO; however, very low recoveries indicated that the antibody was capable of binding the HOEt peptide only after its release by trypsin hydrolysis. Additional assessments of specificity indicated that the antibody bound equally well to the equivalent HOEt peptide from rat Hb. This was probably due to the sequence homology of the first three amino acids of the α-chains of human and rat Hb (also rabbit, mouse and chimpanzee Hb). In contrast, the antibody did not bind the analogous propylene oxide adducted peptide, indicating high

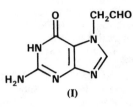

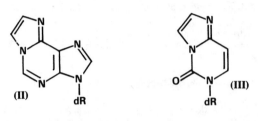

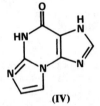

Figure 1. Vinyl chloride-DNA adducts

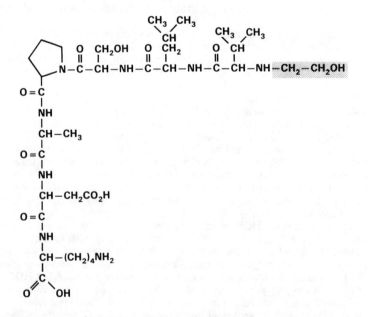

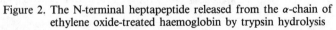

Figure 2. The N-terminal heptapeptide released from the α-chain of
ethylene oxide-treated haemoglobin by trypsin hydrolysis

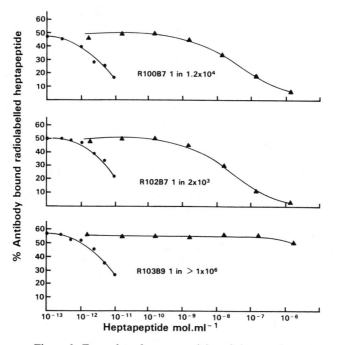

Figure 3. Examples of cross-reactivity of three antisera

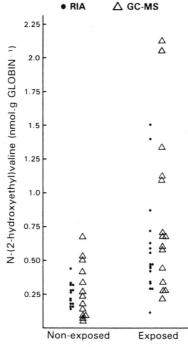

Figure 4. Levels of N-(2-hydroxyethyl)valine in the α-chain of Hb

specificity for the HOEt modification. The lower limit of detection of the optimized RIA was 25 fmol/50 μl sample, which in conjunction with the cross-reactivity data gave an overall sensitivity of 0.14 pmol HOEt peptide/g globin.

Subsequent to the RIA development an ELISA method for EO exposure monitoring, was developed. The ELISA method was more rapid and convenient than the RIA, retaining the advantages of immunochemical analysis but without the hazards of radioactivity.

Biomonitoring study. The RIA was validated in a study of hospital workers potentially exposed to EO. Samples of blood were obtained from a group of operatives employed in EO sterilization of medical equipment and supplies. Blood samples were also obtained from a group not involved in sterilization work. Test and control samples were analysed using the RIA procedure and were also analysed independently using a GC-MS method for N-(2-hydroxyethyl)valine *(26)* (Figure 4). Significant differences were found between potentially exposed workers and the control group. Background levels of hydroxyethylation were also found in the unexposed group, in agreement with earlier findings *(27)*. In the RIA, background levels of α-chain N-(2-hydroxyethyl) valine ranged from 0.14 – 0.44 nmol/g globin (mean 0.25; SD, 0.09; n = 14) in samples from the unexposed group. Samples from the potentially exposed operatives gave corresponding values ranging from 0.11 – 1.51 nmol/g globin (mean 0.58; SD, 0.37; n = 17). Corresponding data obtained by the GC-MS method were as follows: unexposed group 0.05 – 0.67 nmol/g globin (mean 0.27; SD, 0.2; n = 13), exposed group 0.21 – 2.11 nmol/g globin (mean 0.83; SD 0.61; n = 15). The independent GC-MS analysis thus gave results that were in very good agreement with the RIA data *(1)*.

Conclusions. A novel, sensitive and specific immunochemical biomonitoring method for EO exposure was developed. The RIA was validated against an existing GC-MS method, and the two widely differing analytical methods showed excellent agreement. An assumption made at the outset of this study was that the N-terminal valine residues of the α- and β-chains of Hb would display similar reactivities towards EO. This remains to be proven experimentally.

The observation of background levels of hydroxyethylation of Hb suggested the possible occurrence of the corresponding adducts as a 'background' in DNA. Potential sources of hydroxyethylating agents included cigarette smoke, engine exhausts, intestinal bacteria and lipid peroxidation. The origins and significance of background alkylations are under investigation *(28)*.

Final Remarks

In recent years the demands for trace analysis in toxicology and environmental monitoring have been steadily increasing. At the same time the costs of conventional methods of analysis have been escalating with continued advances in the sophistication of instruments. This has led to a search for more economic alternatives with performances comparable to existing technology. Immunoassay techniques match these requirements and have additional benefits which can fulfil the changing needs of the analyst. For example, although antibodies are usually very specific, it is possible, by accident or design, to produce antibodies of general specificity. This allows the development of generic immunoassays or immunoaffinity chromatography methods e.g. to quantify polyaromatic hydrocarbon adducts of DNA *(29)*. Applications of immunoassay outside the medical field are now widespread and fully established.

Acknowledgment

We thank our colleagues R. Davies, A.E. Crane, D. Potter, R.L. Ball, E. Akerman and D.W. Britton for their assistance and Mrs. B. Whitehead for typing the manuscript.

Literature Cited

1. Wraith, M.J., Watson, W.P., Eadsforth, C.V., van Sittert, N.J., Törnqvist, M. and Wright, A.S. in Methods for Detecting DNA Damaging Agents in Humans: Applications in Cancer Epidemiology and Prevention. Eds. Bartsch, H., Hemminki, K. and O'Neill, I.K. IARC Scientific Publications No. 89, Lyon 1988, p.271.
2. Wraith, M.J. and Britton, D.W. in Proceedings of the Brighton Crop Protection Conference, Pests and Diseases, BCPC Publication, 1988, Volume 1, p.131.
3. Köhler, G. and Milstein, C. *Nature* (London), 1975, *256*, 495.
4. Klausner, A. *Biotechnology*, 1987, *5*, 551.
5. Wright, A.S., Bradshaw, T.K. and Watson, W.P. in Methods for Detecting DNA Damaging Agents in Humans: Applications in Cancer Epidemiology and Prevention. Eds. Bartsch, H., Hemminki, K. and O'Neill, I.K. IARC Scientific Publications No. 89, Lyon 1988, p.237.
6. Wright, A.S. in The Pesticide Chemist and Modern Toxicology. Eds. Bardal, S.K., Marco, G.J., Golberg, L. and Leng, M.L. ACS Sympsoium series No. 160, Am. Chem. Soc., Washington, 1981, p.285.
7. Ehrenberg, L., Moustacchi, E. and Osterman-Golkar, S. *Mutat. Res.*, 1983, *123*, 121.
8. Osterman-Golkar, S., Ehrenberg, L., Segerback, D. and Hallstrom, I. *Mutat. Res.*, 1976, *34*, 1.
9. Watson, W.P. in Proceedings of Biomonitoring and Risk Assessment Meeting, Cambridge, July 1989, Eds. Garner, R.C., Farmer, P.B., Steel, G.T. and Wright, A.S. in press.
10. Lohman, P.H.M., Jansen, J.D. and Baan, R.A. in Monitoring Human Exposure to Carcinogenic and Mutagenic Agents. Eds. Berlin, A., Draper, M., Hemminki, K. and Vainio, H. IARC Scientific Publications No. 59, Lyon 1984, p.259.
11. Baan, R.A., Fichtinger-Schepman, A.M.J., Roza, L. and Van der Schans, G.P. *Arch. Toxicol.*, 1989, *Suppl. 13*, 66.
12. Osterman-Golkar, S., Holtmark, D., Segerback, D., Calleman, C.J., Gothe, R., Ehrenberg, L. and Wachtmeister, C.A. *Biochem. Biophys. Res. Comm.*, 1977, *76*, 259.
13. Green, T. and Hathway, D.E. *Chem. Biol. Interactions*, 1978, *22*, 211.
14. Laib, R.J., Gwinner, L.M. and Bolt, H.M. *Chem. Biol. Interactions*, 1981, *37*, 219.
15. Laib, R.J., Doerjer, G. and Bolt, H.M. *J. Cancer Res. clin. Oncol.*, 1985, *109*, A7.
16. Erlanger, B.F. and Beiser, S.M. *Proc. Natl. Acad. Sci.*, 1964, *52*, 68.
17. Vaitukaitis, J.L. in Methods in Enzymology. Eds. Langone, J.L. and van Vunakis, H., Vol. 73, Academic Press, 1981, p.46.
18. Barrio, J.R., Sechrist, J.A. and Leonard, N.J. *Biochem. Biophys. Res. Comm*, 1972, *46*, 597.
19. Fraker, P.J. and Speck, J.C. *Biochem. Biophys. Res. Comm.* 1978, *80*, 849.
20. Muller, R. and Rajewsky, M.F. *J. Cancer Res. Clin. Oncol.* 1981, *102*, 99.
21. Eberle, G., Barbin, A., Laib, R.J., Ciroussel, F., Thomale, J., Bartsch, H. and Rajewsky, M.F. *Carcinogenesis* 1989, *10*, 209.
22. Bolt, H.M. *Critical Reviews in Toxicology* 1988, *18*, 299.
23. Wright, A.S. in Developments in the Science and Practice of Toxicology. Eds. Hayes, A.W. Schnell, R.C. and Miya, T.S. Elsevier, Amsterdam, 1983, p.311.
24. Lehman, H. and Huntsman, R.G. *Man's haemoglobins*; North-Holland; Amsterdam, 1974.
25. Bolton, A.E. and Hunter, W.M. *Biochem. J.* 1973, *133*, 529.
26. Törnqvist, M., Mowrer, S., Jensen, S. and Ehrenberg, L. *Anal. Biochem.* 1986, *154*, 255.
27. Calleman, C.J. *Prog. clin. biol. Res.* 1986, *109B*, 261.
28. Törnqvist, M., Gustafsson, B., Kautiainen, A., Harms-Ringdahl, M., Granath, F. and Ehrenberg, L. *Carcinogenesis* 1989, *10*, 39.
29. Manchester, D.K., Weston, A., Choi, J.S., Trivers, G.E., Ferressey, P.V., Quintana, E., Farmer, P.B., Mann, D.L. and Harris, C.C. *Proc. Natl. Acad. Sci.*, 1988, *85*, 9243.

RECEIVED August 30, 1990

Chapter 25

Biological Response and Quantitation of Diethylstilbestrol via Enzyme-Linked Immunosorbent Assay

Paul Goldstein

Department of Biological Sciences, University of Texas, El Paso, TX 79968

Diethylstilbestrol (DES), a synthetic analog of estrogen, acts as a mitotic/meiotic inhibitor via the disruption of microtubules and severely affects gametogenesis. Treatment of cells with DES results in the production of aneuploid cells and gametes. To quantify the concentration of DES in tissues, cells or subcellular fractions, an anti-DES antibody was produced conjugated to keyhole limpet hemocyanin (KLH) and a modified enzyme-linked immunosorbent assay for DES was developed. The sensitivity of the assay was 10 ng/ml of DES which is in the same range as the RIA technique, but has the advantage of not using radioactive particles in the analysis. The high correlation coefficient (r^2=0.99) indicated a linear relationship between the concentration of DES in the fluid or tissue and the anti-DES antibody.

Diethylstilbestrol (DES), an artificial nonsteroidal estrogen, is recognized by the estradiol receptor (ER) with a high affinity (kD 10^{-9} M) and elicits the same cellular response as estrogen. Because DES is an analog of estrogen, it was used as a gynecological medication for women and as a feed additive for cattle and sheep (1). Studies have shown that intrauterine exposure to DES results in the development of adenocarcinomas with subsequent sterility of young female offspring (2). Activity of intermediates in the metabolic pathway are enhanced in the presence of oxidizing agents (3) which may account for the pleiotropic toxic effects manifested by elevated levels of DES. In addition to unscheduled DNA

0097–6156/91/0451–0280$06.00/0

synthesis (4), DES has been implicated in: increased or
induced sister-chromatid exchange (5), cell
transformation (6), increased chromosomal nondisjunction
(7), induction of aneuploidy and polyploidy (8), and
inhibition of cellular proliferation (9).

In addition to intranuclear effects, DES disrupts
mitotic and meiotic spindles resulting in errors in cell
division (10). It is believed that DES acts directly on
tubulin within 10 minutes of exposure since DES inhibits
the GTP-dependent self-assembly of tubulin (11) and
inhibits colchicine binding (12). The effect of DES is
similar to that of colchicine which also specifically
binds to tubulin and prevents the formation of
microtubules and spindles.

The mechanism of steroid action, which is limited to
specific target tissues, involves hormone binding to a
high affinity, low-capacity estrogen receptor
(ER)(13,14). The smallest active ER component is 30,000
daltons (15) and the nuclear concentration of ER depends
upon the kinetics of the interaction with the cytosolic
receptor (16), which is translocated to the nuclear
membrane (17). This increased binding of the
receptor-hormone complex to the cell nucleus (18), with
subsequent interaction with chromatin (19,20) results in
an altered transcriptional pattern (21,22). DES and many
of its oxidative metabolites (such as DIES- dienestrol)
also interact with high affinity with the ER and have
been shown to initiate high levels of mutagenicity and
carcinogenicity (23,14,24,25). Thus, the ER binds and
retains DES within the cell nucleus (14). Such an
interaction may be responsible for localizing DES to
target tissues and indicates the specificity of the
cellular response to DES. There have been no reports of
in vivo DES toxicity in sites which do not contain
estrogen receptors (14).

Since DES is still in use as a pharmaceutical drug
for humans and as a feed additive for livestock (in the
U.S.A., a recent case was discovered in Anthony, NM),
various techniques have been developed to quantitate the
amount of DES in both food and fluids. These include: 1)
High-pressure liquid chromatography (HPLC) , which
effectively measures DES to the pg level (26); 2)
radioimmunoassay (RIA), which is sensitive to the ng
level (27,28), and ELISA, which is also sensitive to the
ng level (as described in the current study). ELISA has
a distinctive advantage to the other techniques in that
it is non-radioactive, does not require expensive
equipment, interpretation of data is straight forward,
and it is rapid and in common use in numerous clinical
laboratories (29). In this study, a protocol is described
for the determination of DES in fluids (i.e. tissue
extractions, cell and subcellular preparations) and for
the production and characterization of highly specific

anti-DES antibodies. The ELISA is sensitive to 10 ng/ml
of DES in fluids, e.g. human and rabbit sera, cell
extracts.

MATERIALS AND METHODS

I. PREPARATION OF AN ANTI-DES ANTIBODY

Currently, anti-DES antibodies, all using a hapten
conjugated to bovine serum albumin (BSA), have been
produced from at least five different laboratories, as
follows: 1) Hoeschst #254 from Behring (Frankfurt,
Germany); 2) Laboratoire d'Hormonologie (Marloie,
Belgium); 3) Institut Pasteur (Paris, France); 4) Dr. B.
Hoffman and Dr. H. Meyer (Tech. Univ. Munchen, Freising,
Germany); and 5) Dr. A. Kambegawa (Teikyo Medical Univ.,
Japan). The variance in specificity between these
antibodies is distinctive although DES, with a molecular
weight of 248, has only a single epitope which should
ensure less variance between antibodies. DES itself is
non-immunogenic, therefore it must be conjugated to a
protein, such as BSA, to elicit an immune response.
However, specificity is compromised because of the
cross-species reactivity of BSA.

   The current study, describing the production of an
anti-DES antibody conjugated to keyhole limpet hemocyanin
(KLH), is a modification of Erlanger (30) and Nakamura
(28) and results in the production of a highly specific
antibody that is sensitive to a level of 10 ng/ml DES.
The specificity and sensitivity of the antibody were
verified using immunodot, ouchterlony and ELISA
techniques. The complete, detailed protocol for
production of the antibody is available upon request and
a brief version follows.

A. Preparation of the DES-4-CME-Me Intermediate

Methyl bromoacetate was added to DES that had been
dissolved in dimethyl sulfoxide (DMSO). Following this,
potassium carbonate was added and the mixture was stirred
for one hour at 50°C. After ether extraction, the ether
layer was evaporated over argon gas and the resulting
crystals were redissolved in chloroform and
chromatographed on thin layer plates. The middle of three
bands (the DES-4-CME-Me product; carboxymethylether) was
eluted by mixing in chloroform/methanol, centrifuged and
recrystallized over argon gas.

B. Saponification of the DES-4-CME-Me Intermediate

The DES-4-CME-Me crystals were redissolved in methanol.
NaOH was added and the mixture was incubated for one hour

at room temperature. The solution was then made acidic by adding HCl and the methanol was evaporated off using a Sorvall Speed Vac concentrator. The remaining solution was extracted with ether and evaporated to produce pure DES-4-CME to which the keyhole limpet hemocyanin was conjugated.

C. Conjugation of DES-4-CME to Keyhole Limpet Hemocyanin(KLH)

The DES-4-CME crystals were redissolved in dioxane to which tributylamine was added followed by isobutylchloroformate. After incubation at room temperature a yellow solution resulted. The DES-4-CME solution was added to the KLH solution (consisting of KLH, dH_2O, NaOH) and dialyzed overnight against dH_2O. The dialyzed emulsion was centrifuged and the supernatant was adjusted to pH 4.5 using HCl and readjusted to pH 6.5. The final solution was lyophilized and frozen.

D. Immunization of New Zealand White Rabbits

Rabbits were immunized with DES-4-CME-KLH, emulsified in Fruend's complete adjuvant, at two intramuscular sites. Secondary immunizations, consisting of DES-4-CME-KLH in Fruend's incomplete adjuvant were given at 21-28 days post primary immunization. The final boost was given 21-28 days post secondary immunization using the DES-4-CME-KLH in Fruend's incomplete adjuvant, the animal was sacrificed three days later and the serum frozen.

2. PROTOCOL FOR ELISA FOR DES

Diethylstilbestrol (DES) (Sigma),a small hapten, mw=248, will not effectively adhere to a microtiter plate. For this reason, the wells of the Immulon 2 plate were first coated with poly-l-lysine (10 ug/ml) followed by three washes of distilled water and then 100 ul of varying concentrations of DES (DES is dissolved in absolute ethanol) or sample were added to the wells. After 8 hrs. of evaporation at 4°C, a blocking agent (PBS/ 5% FCS)(PBS: phosphate buffered saline; FCS: fetal calf serum) was added, incubated overnight at 4°C, and the first wash consisted of PBS/Tween. Binding of the DES to the well was maximized to 46% with the Tween wash. Anti-DES antibody (rabbit IgG) was added at a concentration of 1:900, incubated at 37°C for one hour, washed with PBS, and followed with alkaline phosphatase conjugated anti-rabbit IgG (Sigma) at a concentration of 1:3000. This was incubated for one hour at 37°C, washed with PBS, and followed with p-nitrophenyl phosphate substrate (PNP)(1 mg/ml) in diethanolamine buffer (pH 9.6). After 45 minutes in darkness, an average positive

control was 2.9 +/- 0.4 at 405 nm. A test was considered
positive if the absorbance of the specimen was three
times that of the negative.

Controls:
1) Positive
   A)   To determine levels of DES in specimen, varying
          concentrations of the specimen were coated onto
          the microtiter wells, followed by the anti-DES
          antibody at a concentration of 1:900.
   B)   To determine the sensitivity of anti-DES antibody,
          3 ug/well of DES were coated onto the well, and
          the concentration of the antibody was varied from
          1:500 to 1:100000.

2) Negative
     Four separate negative controls are minimally
required for this test, during each run, to determine any
non-specific binding. The solutions are added as follows
at the appropriate step:
   1)   Buffer (no -DES) coating; Anti-DES antibody;
          anti-rabbit-IgG antibody; PNP
   2)   DES (3 ug/ml) coating; buffer (no anti-DES
          antibody); anti-rabbit IgG antibody; PNP
   3)   DES; Anti-DES antibody; buffer (no anti-rabbit
          IgG antibody); PNP
   4)   DES; anti-DES antibody; anti-rabbit IgG
          antibody; buffer (no PNP)
     Average absorbance values at 405 nm for these
negative controls were 0.05 +/- 0.02.
     In addition, untreated human and rabbit sera and
cell extracts were tested for any non-specific binding
inherent to these fluids. In all cases, the absorbance
values were similar to the negative controls.

Preparation of Standard Curve for DES determination

To prepare a standard curve, known ug/ml of DES are
coated onto the microtiter wells, ranging from 0.5 to 15
ug/well. The anti-DES antibody (1:900) is reacted to this
and the absorbance of each well is determined. A typical
result (which may differ for each laboratory) is as
follows (Fig. 3): (See Figures 1–5)

DES (ug/well	Optical Density (at 405 nm)
15	0.51
6	0.33
3	0.19
1.5	0.13
0.5	0.07

The region of this curve between DES concentrations of 15 to 0.5 g/well can be plotted as a linear function. In this example, the correlation coefficient is $r^2 = 0.98$ which demonstrates that the concentration of DES on the well is related to absorbance. Unknown quantities of DES in the fluid can then be determined by reading the absorbance of the test fluid and using the standard curve to identify levels of DES.

Determination of Sensitivity

Known quantities of DES (from 10,000 - 10 ng) were added to 1 ml of either human serum, rabbit serum or a phosphate buffer and added to microtiter wells that had been previously coated with anti-DES antibody (1:500). In this manner, it was determined that a concentration of 10 ng/ml of DES could be detected (Fig. 4) using this ELISA.

Results and Discussion

THE ELISA FOR DES

This heterogeneous ELISA for DES is, in its present form, noncompetitive (45). The single sandwich design of the ELISA (see Materials and Methods) is a necessity due to the sole epitope present on the small DES molecule. DES by itself will not elicit an immune response, thus, it must be conjugated to BSA or another substance (e.g. keyhole limpet hemocyanin) for the production of an antibody. The anti-DES antibody used in this study was produced against a carboxymethylether derivative of DES that was conjugated to KLH.

Dilutions of Antisera

Working dilutions of the antisera were established using a checkerboard on the microtiter plate. The working dilution was 1:900.

Non-specific binding of reagents

Negative controls, as listed in Materials and Methods including the use of untreated human and rabbit sera, were analyzed to check for non-specific binding of DES and antibodies. There was a significant problem of non-specific binding of the anti-DES antibody to the poly-1-lysine coated well (which was utilized to obtain an even coating of the DES to the well). This problem was removed by washing with PBS\Tween between the blocking step and the addition of the anti-DES antibody, followed by washes in straight PBS for the remainder of the ELISA. Non-specific binding was also reduced 50% by blocking

overnight at 4°C in PBS/5% FCS. The final absorbance of
0.02- 0.04 was considered acceptable as compared to
positive controls of 2.9 at 405 nm.

Sensitivity of the anti-DES antibody and the ELISA

To determine the percentage of binding of DES to the
microtiter well, pre-determined amounts of radioactive
DES in different fluids (100 % ethanol, PBS, and tissue
culture fluid) were coated onto the well. After
evaporation, followed by washes as outlined in the
Protocol, the average binding ratio was 46%, regardless
of the type of fluid the DES was in (e.g. started with
6900 counts of DES in the well and an average of 3150
counts remained after protocol).
        Demonstration of sensitivity of the anti-DES
antibody was shown in two different experiments. In the
first experiment, each well of the microtiter plate was
coated with 3 ug/well of DES and the amount of anti-DES
antibody in the system was varied from 1:500 to 1:100,000
(Fig. 2). The result was a linear graph with a
correlation coefficient (CC) of $r^2 = 0.98$. The average
for 10 tests was 0.95 with the CC ranging from 0.92 to
0.99. In the second experiment, varying amounts (0.5 to
15 ug) of DES were coated to the well, followed by
constant anti-DES concentration (1:900)(Fig. 3). The
linear graph also had a correlation coefficient of 0.98.
The average of the tests was 0.94 with a range of CC from
0.91 to 0.99. In both cases, there was no interaction
between FCS and the anti-DES antibody.
        Sensitivity of this ELISA was also determined by
adding known amounts of DES to 1 ml of either human
serum, rabbit serum or phosphate buffer. Antigen capture
was established by precoating the wells with anti-DES
antibody (1:500) and then reacting the "spiked" fluids.
Thus, this ELISA is sensitive to the 10 ng/ml level of
DES such that the positive absorbance is three times that
of the negative controls (Fig. 4).
        These experiments indicate that the anti-DES
antibody was highly discerning for the DES molecule and
that the concentration of DES can be determined using a
non-radioactive immunoassay. A unique application of
this ELISA has been in the determination of the effects
of DES on gametogenesis of the free-living nematode
Caenorhabditis elegans.

EFFECTS OF DES ON GAMETOGENESIS IN Caenorhabditis elegans.

The ELISA for DES was used to quantitate the amount of
DES that was absorbed to different body tissues in the
nematode Caenorhabditis elegans. The effects on

gametogenesis of various concentrations of DES were
characterized since even low levels of DES have an
immediate effect on cell division in C. elegans and most
organisms. For example, within 10 minutes of exposure to
low levels of DES, microtubules were dissociated and
showed a dose-dependent curve (11). In addition,
DES-induced DNA fragmentation (31), aneuploidy, nuclear
morphology alterations (32) and partial chromosome loss
resulted in induction of recessive spontaneous mutations
via pseudodominance (33).

In most organisms, DES concentration exceeding 100
uM are lethal (34), however, in C. elegans, there is a
low percentage of the population that remains viable even
under conditions of 400 uM DES (Fig. 1)(35). In the
cytoplasm, DES becomes bound to estrogen receptor (ER)
with a high binding affinity. Once the DES-cytosolic ER
is formed, the receptor protein undergoes a temperature
dependent activation and transformation and it is in this
transformed state that the DES-ER complex binds to DNA
and is retained within the cell nucleus (14). Thus, the
effects of DES on gene regulation are, in part, related
to the quantity of ER present and upon the kinetics of
the interaction of the receptor (16). The quantity of ER
present is also related to the amount of DES present (36)
and saturation can be achieved at relatively low levels.
Consequently, less than 100 uM DES can fully saturate the
ER, however, only a fraction of the total nuclear ER need
be associated with the nuclear matrix to stimulate the
biological response. Higher doses of DES, up to 10,000 uM
DES, did not promote additional elevation in
matrix-associated ER (37).

In C. elegans, the quantity of estrogen receptor has
not yet been determined, however, there is limited ER
available to bind to the DES and it is only this bound
DES that can be physiologically active (14,31). After
saturation of the ER, DES may bind to other proteins, but
at a lower avidity, yet this is inconsequential because
of the lack of any physiological response to non-bound
DES (38). This effect is clearly seen in C. elegans
whereby maximum organismal death occurs at 100 uM DES and
does not increase even as the concentration of DES
approaches 400 uM (Fig 1,5).

Summary and Conclusions

The effects of even minute quantities (5 ug/ml) of DES
are devastating to the process of cell division in
mitosis and meiosis (39) and the biologic effects, in
general, are linear on a dosage response curve. For this
reason, it is of great interest to accurately determine
extant levels of DES in tissues and fluids. This paper
describes the production of a highly specific
(non-reactive to DES-diproprionate and Hexoestrol) and

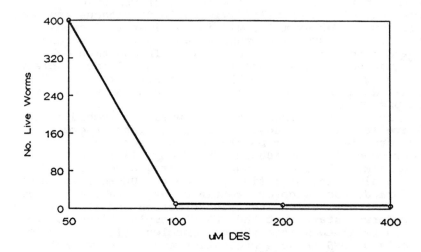

Figure 1. Response of C. elegans to various levels of DES. At 50uM DES, the numbers of worms alive after 5 days are within normal limits (>400). At concentrations of DES exceeding 100 uM, viability sharply decreases. However, the entire population does not die out and is sustained by a few (max. 10) individuals.

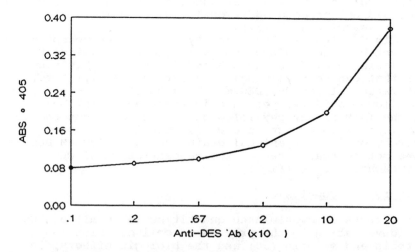

Figure 2. Constant quantities of DES were coated onto microtiter wells (3 ug/well) to which were added varying dilutions of anti-DES antibody (1:500- 1:100,000). The correlation coefficient of r^2 = 0.98 suggests that the relationship between DES and the anti-DES antibody is linear.

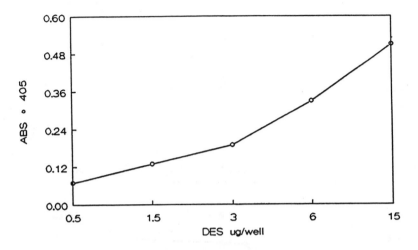

Figure 3. Varying the amount of DES coated onto a well (from 15 ug to 0.5 ug) also had a high correlation coefficient of r^2 = 0.98. This suggests that the amount of absorbance is relative to the amount of DES present in the fluid.

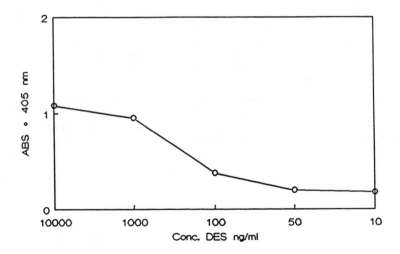

Figure 4. The sensitivity of the ELISA was determined by testing 1 ml samples of sera and buffers that had predetermined quantities of DES added to them. This ELISA is sensitive to 10 ng/ml as the absorbance at this level is still three times that of the negative controls.

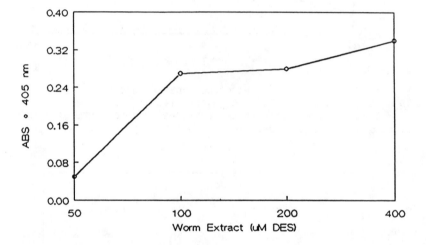

Figure 5. Worms incubated for 4 hrs. in varying concentrations of DES (50 - 400 uM) showed saturation of DES in cell-free extracts at a concentration of 100 uM. Analyses were via non-competitive ELISA and radioactive uptake.

sensitive anti-DES antibody using a keyhole limpet hemocyanin conjugate instead of more commonly used BSA conjugates. The ELISA for DES described in this study is the first for this small hapten and offers a distinct advantage over the RIA and HPLC techniques.

Acknowledgments

This work was supported by grants SO6 RR8012-17 and GM 08048 from the National Institute of Health. The technical and advisory assistance by Jose Aun, Lisa Magnano and Dr. D. Robert (Natl Center for Toxicol. Res.) was greatly appreciated.

LITERATURE CITED

1. Vorheer, H.; Messer, R.; Vorheer, U.; Jordan, S.; Kornfield, M. Biochem. Pharmacol. 1979, 28, 1865-1877.
2. Herbst, A.; Bern, H. Developmental effects of diethylstilbestrol DES) in pregnancy. 1981, New York, Thieme-Stratton, Inc.
3. Mehta, R.; von Borstel, R. Mutation Res. 1982, 92, 49-61.
4. Martin, C.; McDermid, A.; Garner, R. Cancer Res. 1978, 38, 2621-2627.
5. Rudiger, H.; Haenisch, F.; Metzler, M.; Desch, F.; Glatt, H.; Nature 1979, 281, 392-394.
6. Barrett, J.C.; Wong, A; McLachlan, J.A. Science 1981, 22, 1402-1404.
7. Tsutsui, T.; Maizumi, H.; McLachlan, J.; Barrett, J. Cancer Res. 1983, 43, 3814-3821.
8. Chrisman, C.; Hinkle, L.; Can. J. Genet. Cytol. 1974, 16, 831-835.
9. Ivett, J.; Tice, R. Environ. Mut. 1981, 3, 445-452.
10. Sawada, M.; Ishidate, M. Mutation Res. 1978, 57, 175-182.
11. Hartley-Asp, B.; Deinum, J.; Wallin, M. Mutation Res. 1985, 143, 231-235.
12. Sharp, D.; Parry, J. Carcinogenesis 1985, 6, 865-871.
13. Capony, F.; Rochefort, H. Mol. Cell Endocrinol. 1977, 8, 47-64.
14. Korach, K.; McLachlan, J. Arch. Toxicol., 1985, Suppl. 8, 33-42.
15. Lubahn, D.; McCarty, K.; McCarty, K. J. Biol. Chem. 1985, 260, 2515-2526.
16. Raynoud, J.; Bouton, M. In: Cytotoxic Estrogens in Hormone Receptive Tumors, 1988, Academic Press, NY
17. Raam, S.; Richardson, G; Bradley, F.; MacLaughlin, D.; Sun, L.; Frankel, F.; Cohen, J. Breast Cancer Res. Treat. 1983, 3, 179-199.

18.   Barrack, E.R. In: The Nuclear Envelope and the
        Nuclear Matrix, 1982, 259-269, A.R. Liss, NY
19.   Sheehan, D; Medlock, K.; Lyttle, C. J. Steroid
        Biochem. 1986, 25, 37-43.
20.   Garcia-Segura, L.; Olmos, G.; Tranque, P.;
        Aguilera, P.; Naftolin, A. J. Neurocytology 1987,
        16, 469-475.
21.   Jensen, E.; DeSombre, E. Science 1973, 182,
        126-134.
22.   DeSombre, E.R.; Mohla,S.; Jensen, E. J. Steroid
        Biochem. 1975, 6, 469-473.
23.   Bradford, P.A.; Olavesen, A.; Curtis, C.; Powell,
        G. Biochem. J. 1977, 164, 423-430.
24.   Metzler, M.; McLachlan, J. Biochem. Biophys. Res.
        Commun. 1978, 85, 874-884.
25.   Metzler, M. CRC Critical Reviews in Biochemistry,
        1981, 171-212.
26.   Gottschlich, R.; Metzler, M. Anal. Biochem. 1979,
        92, 199-202.
27.   Hoffman, B; Blietz, C.J. Anim. Sci. 1983, 57,
        239-246.
28.   Nakamura, K. Hiroshima J. Med. Sci. 1986, 35,
        325-338.
29.   Voller, A.; Bartlett, A.; Bidwell, D. J. Clin.
        Pathol. 1978, 1, 507-520.
30.   Erlanger, B.; Boref, F.; Beiser, S.; Lieberman, S.
        J. Biol. Chem. 1957, 228, 713-727.
31.   Sina, J.; Bean, C.; Dysart, G.; Taylor, V.;
        Bradley, M. Mutat. Res. 1983, 113, 357-391.
32.   Jones, K.; Pfaff, D.; McEwen, B. J. Comp. Neurol.
        1985, 239,255-266.
33.   McLachlan, J.; Wong, A.; Degen, G.; Barrett, J.
        Cancer Res. 1982, 42, 3040-3045
34.   Wheeler, W.; Cherry, L.; Downs, T.; Hsu, T.
        Mutation Res. 1986, 171, 31-41.
35.   Goldstein, P.; Watts, M.; Aun, J. Cytobios 1988,
        58, 7-17.
36.   Medlock, K.; Sheehan, D.; Nelson, C.; Branham, W.
        J. Steroid Biochem. 1988, 29, 527-532.
37.   Simmen, R; Means, A.; Clark, J. Endocrinology
        1984, 115, 1197-1202.
38.   Sheehan, D.; Young, M. Endocrinology 1979, 104,
        1442-1446.
39.   Goldstein, P. Mutation Research 1986 174, 99-107.

RECEIVED August 30, 1990

Chapter 26

Preparation and Characterization of Mouse Antibodies Against Hemoglobins Modified by Styrene Oxide

Robert A. Haas, Doris Hollander, and Mitchell Rosner

Air and Industrial Hygiene Laboratory, California Department of Health Services, Berkeley, CA 94704

Antibodies were raised which recognize styrene-7,8-oxide modified human hemoglobin. The covalent binding of styrene-7,8-oxide with human hemoglobin was measured at varying styrene oxide concentrations. The same reaction with Swiss-Webster mouse hemoglobin was carried out and resulted in a heterogeneous mixture of chemically-modified hemoglobins plus unreacted hemoglobin which was used without further purification to immunize BALB/c mice. Antibodies produced by this method cross-reacted with styrene-oxide modified human hemoglobin and demonstrated a preference for the chemically-modified hemoglobins. Each serum sample also cross-reacted to a slight but measurable extent with unmodified human hemoglobin but no cross-reactivity was observed using unmodified mouse hemoglobin. Antibody recognition of human hemoglobin required that the level of styrene oxide modification be greater than 0.03 styrene oxide residues per hemoglobin tetramer.

The analysis of chemically-altered hemoglobin (Hb) as a indicator of chemical exposure was pioneered by Ehrenberg and colleagues ([1]). Hemminki ([2]) demonstrated that the styrene metabolite, styrene-7,8-oxide (SO) (1,2-epoxy-ethylbenzene, CAS 96-09-3) became covalently bound to serum proteins and hemoglobin when incubated with human blood. This chemically-modified hemoglobin should be immunologically distinct from unmodified hemoglobin. Thus antibodies that recognize the modified Hb could be raised and utilized in an immunoassay to detect such molecules in blood samples from exposed individuals. This report describes the development of such antibodies in mice and their binding properties to styrene oxide-modified hemoglobins from mice and humans.

0097–6156/91/0451–0293$06.00/0

Materials and Methods

Reactions of styrene oxide (SO) with whole blood *in vitro* were performed using freshly-drawn human blood. To determine SO incorporation into blood, [^{14}C]styrene oxide (28 mC$_i$/mmol, Amersham, Arlington, IL) in 100% ethanol was added to 0.9 ml whole human blood plus 0.1 ml sterile saline (0.9% NaCl) to final concentrations of styrene oxide of 3.6 μM, 36 μM, 357 μM, and 1.8 mM and incubated at 37° for 2 h. The reaction mixtures were centrifuged at 3000 x g for 5 min and the pellets washed three times with 0.7 ml 0.9% saline. The red blood cell pellets were lysed by addition of an equal volume of deionized water, the debris removed by centrifugation, and globin prepared by the acidic acetone method (3). SO incorporation was then determined by liquid scintillation counting of the redissolved globin pellet and expressed as nmol SO per ml whole blood assuming [Hb]$_{whole\ blood}$=15 mg/ml (4). Mouse and human hemoglobin were prepared by hypotonic lysis of red blood cells, removal of stroma, and ammonium sulfate precipitation (50% saturation) of contaminating proteins. The hemoglobin-containing supernatant was dialyzed against 0.01 M Bis-Tris (pH 7.0) and used within one week. For preparation of immunogen and competitors, styrene oxide-modified hemoglobin was prepared by addition of an appropriate volume of a 5 M solution of [^{14}C] styrene oxide (50 DPM/nmol) to a solution of 310 μM human or mouse hemoglobin in 10 mM Bis-Tris buffer (pH 7.4) to yield the final desired concentration of styrene oxide. Ethanol (10% v:v) was added and the reaction mixture vortexed to aid in solubilizing the styrene oxide. The reaction solutions were incubated in the dark for 24 hours at ambient temperature. The styrene oxide/hemoglobin reaction mixtures were then chromatographed twice on 150 μm Sephadex G-25 size-exclusion spun columns to remove the unreacted styrene oxide. The excluded volume containing the hemoglobin was dialyzed against three changes of water for 36 hours. Aliquots of the hemoglobin were analyzed for radioactivity by liquid scintillation counting and for hemoglobin content by reaction with Drabkin's reagent to measure cyanmethemoglobin at 540 nm. The extent of modification expressed as nmol styrene oxide /nmol hemoglobin tetramer was calculated.

To raise antibodies against SO-Hb, BALB/c mice were immunized intradermally on days 0, 7, and 21 with SO-modified mouse hemoglobin (1-2 SO residues/Hb tetramer) with 100 μg conjugate in 2X RIBI (Ribi Immuno-Chemical Res. Inc., Hamilton, MT) adjuvant. On day 28, the mice were bled via the tail vein and the sera screened against SO-modified mouse and human hemoglobin and unmodified human hemoglobin.

Ascites production was induced by intraperitoneal injection of ATCC sarcoma TG180 (200 μl, $\approx$10^5 cells).

A competitive enzyme-linked immunosorbent assay (ELISA) using SO-modified Hb as competitor was used to characterize the antibodies. Solutions for assay were prepared by addition of equal volumes of a 1:100 dilution of the polyclonal ascites to an equal volume of the hemoglobin samples (at various dilutions) to yield the desired concentration of competitor. The control well contained only a 1:200 dilution of the ascites fluid. These solutions were prepared 12-16 h before the assay and stored at 4°. Immulon 2 flat bottom 96 well microtitration plates (Dynatech Laboratories, Chantilly, VA) were coated with 0.5 μg of the desired hemoglobin solution in 100 μl phosphate buffered saline (pH 7.2) at 4° for 16 h. The wells were aspirated, washed 3 times with 0.05% Tween 20 phosphate buffered saline solution (PBS-Tween), and subsequently filled to capacity with a 1% bovine serum albumin in PBS-Tween (v:v) and incubated for 0.5 h at room temperature to block any uncoated sites in the microplate wells. After discarding the blocking solution, the antibody/competitor solutions were added to the appropriate wells. After 2 hours the wells were aspirated and washed 3 times with the PBS-Tween solution. To each well was added 100 μl of a 1:1000 dilution of goat antimouse IgG alkaline phosphatase conjugate (Sigma Chemical Co. St. Louis, MO). Following incubation for 2 h at 20°, the wells were washed 3 times with PBS-Tween before applying the alkaline phosphatase substrate (p-nitrophenyl phosphate, 4 mg/ml) solution in 0.1% $MgCl_2$ (w:v) diethanolamine buffer (pH 9.8). The microplates were read at 405 nm either kinetically or after stopping the reactions at 1 hour with 50 μl of 0.1 M EDTA.

Serum titers were determined using plates coated as described above. Serum was obtained by tail bleeding.

Results and Discussion

The covalent binding of styrene oxide to human hemoglobin when SO is incubated with whole blood or washed red blood cells is shown in Figure 1. Over the concentration range of 3 μM-1.8 mM SO, the amount of SO in washed red blood cells is approximately proportional to its initial concentration. In whole blood, the linear range of SO incorporation into hemoglobin is [SO]=30 μM-1.8 mM. The ratio of SO/Hb tetramer ranges from 0.0006-0.6 SO/Hb tetramer. Much higher levels of modification may be achieved by lysing the erythrocytes and removing cell debris (5). When this semi-purified human hemoglobin is reacted with SO at concentrations up to 50 mM, the styrene oxide reacts in a linearly-related, dose-dependent manner similar to that seen for the whole blood reaction. Concentrations of SO above 50 mM do not lead to additional modification but to extensive denaturation and precipitation of the protein. It appears that saturation of soluble Hb occurs when about five SO residues/Hb tetramer are achieved (unpublished observations). Semi-purified mouse hemoglobin was also reacted with SO to prepare immunogens. Mouse hemoglobin (0.29 mM) from Swiss-Webster mice reacts with SO (50 mM) yielding 1.2 S0 residues/Hb tetramer. This preparation was used to immunize BALB/c mice. The serum titers of two such immune mice are shown in Figure 2. Three different immobilized antigens were used: (a) unmodified human Hb; (b) SO-modified human Hb; and (c) SO-modified murine Hb (the immunogen). The antisera from both mice showed no cross-reactivity with unmodified human hemoglobin. The antisera from mouse #5 cross-reacts with modified human Hb

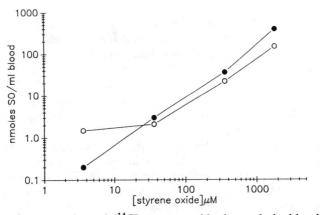

Figure 1. Incorporation of [^{14}C]-styrene oxide into whole blood (open circles) and washed red blood cells (filled circles) as a function of initial styrene oxide concentration.

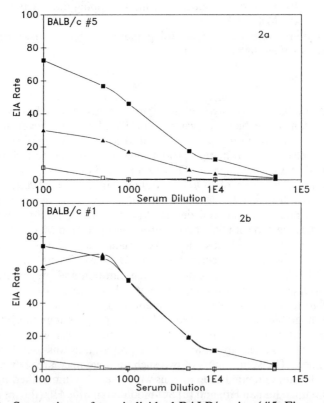

Figure 2. Serum titers of two individual BALB/c mice (#5, Figure 2a and #1, Figure 2b) immunized with styrene oxide-modified Swiss-Webster mouse hemoglobin. The antisera is titered against the immunogen, SO-modified S-W mouse Hb, (filled squares), styrene oxide-modified human hemoglobin (filled triangles), and unmodified human hemoglobin (open squares).

to about half the extent as the immunogen. Mouse #1 antisera demonstrates approximately 100% cross-reactivity with SO-modified human hemoglobin.

Injection of a sarcoma cell line into the intraperitoneal cavity of these immune mice provided a polyclonal ascites fluid. The use of antibodies in the ascites fluid in competitive ELISA experiments using SO-modified human Hb at different levels of modification is summarized in Figure 3. In all cases the immobilized antigen was 0.5 μg of a high level (3.6 SO/Hb) modified human Hb preparation. As shown in Figure 3, the degree of inhibition is related to the extent of modification. The Hb with the lowest level of modification tested, 0.03 styrene oxide residues per tetramer, produced no inhibition even when as much as 500 μg of the modified hemoglobin was incubated with antibody (data not shown). Higher levels of competitor could not be used because inhibition of binding occurred even with unmodified hemoglobin as competitor at concentrations greater than 100μg/well. This can be seen by the nearly 30% inhibition that resulted using unmodified hemoglobin at 500 μg/well.

Inhibition relative to unmodified Hb is first seen at levels of modification of 0.15 SO/Hb. It may be that the lack of inhibition at lower levels of modification is the result of too few binding sites at very low levels of modification, or there may be something qualitatively different about hemoglobin modified at different levels. The latter possibility is unlikely since peptide mapping of SO-modified hemoglobin from whole blood reactions indicates that there is no single preferred site of reaction and that the pattern of modification on several different peptides is qualitatively similar regardless of whether the hemoglobin is within the red blood cell or not (manuscript in preparation). The ascites fluid can be used for screening of samples where the modification levels are high enough to permit detection, eg. in animal studies employing relatively high SO doses. There is no significant cross-reaction of these antibodies with unmodified murine hemoglobin and thus these preparations can be used for *in vivo* mouse experiments, now in progress.

Summary and Conclusions

Mouse antibodies against SO-modified hemoglobin were raised using a heterogeneous immunogen consisting of chemically-modified mouse hemoglobin molecules in a mixture that contained many different modified sites (5) as well as a large proportion of unmodified molecules. It is shown that BALB/c mice can mount an immune response to the SO-modified hemoglobin of Swiss-Webster mice. Although there are amino acids differences between the β-chains of these strains (6, 7), the chemical modification of hemoglobin provides the immunodominant epitopes as evidenced by a much higher serum response to the modified hemoglobin. In addition, antibodies raised against styrene oxide-modified mouse hemoglobin also bind to similarly modified human hemoglobin. These mouse antibodies have a "threshold" of 0.15 SO residues per human Hb tetramer below which no binding is observed.

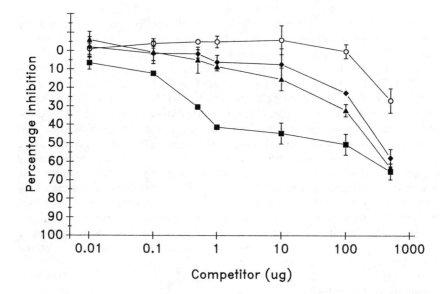

Figure 3. Competitive enzyme-linked immunosorbent assay. Immobilized antigen is styrene oxide-modified human hemoglobin (3.6 SO/Hb tetramer), competitors are human hemoglobins modified at different levels: no modification (open circles), 0.15 SO/Hb tetramer (filled diamonds), 0.4 SO/Hb tetramer (filled triangles), and 3.6 SO/Hb tetramer (filled squares).

Acknowledgments

We thank Jean Grassman for many helpful discussions and C. Peter Flessel for a critical review of the manuscript. Support by NIEHS grant ES04705 is greatly appreciated.

Literature Cited

1. Tornqvist, M.; Mowrer, J.; Jensen, S.; Ehrenberg, L. Analytical Biochemistry 1986, 154, 255-266.

2. Hemminki, K. Arch. Toxicol. Suppl. 1986, 9, 287-290.

3. Ascoli, F.; Fanelli, M. R. R.; Antonini, E. In Meth. Enz.; Antonini, E., Rossi-Bernardi, L., Chiancone, E., Eds.; Academic: New York, 1981; Vol. 76, p 72.

4. Pereira, M. A.; Chang, L. W. Chem.-Biol Interactions 1981, 33, 301-305.

5. Kaur, S.; Hollander, D.; Haas, R.; Burlingame, A. M. J. Biol. Chem. 1989, 264, 16981-12984.

6. Popp, R. A. Biochim. Biophys. Acta 1973, 303, 52-60.

7. *ibid.* 61-67.

RECEIVED August 30, 1990

Chapter 27

Detection of Cisplatin–DNA Adducts in Humans

Miriam C. Poirier[1], Shalina Gupta-Burt[1], Charles L. Litterst[2], and Eddie Reed[1]

[1]National Cancer Institute and [2]National Institute of Arthritis and Infectious Diseases, National Institutes of Health, Bethesda, MD 20892

An ELISA, which measures cisplatin-DNA intrastrand adducts, and atomic absorbance spectrometry, which measures total platinum bound to DNA, have been used to quantify DNA modification in samples from patients receiving platinum drug-based therapy and rats in which the treatment of human cancer patients has been modeled. Adducts measured in blood cell DNA samples from cancer patients have correlated with dose and chemotherapeutic efficacy. Human tissue DNA adducts have a widespread distribution, and long-term adduct persistence (> 1 year) has been observed in many organs including tumor and target sites for drug toxicity.

In recent years a large number of polyclonal and monoclonal antisera have been produced against adducts or modified DNA samples of a variety of chemical classes, including methylating and ethylating agents, aromatic amines, polycyclic aromatic hydrocarbons (PAH), aflatoxins, psoralens and platinum-ammine complexes (1,2). These antisera have been used to establish highly-sensitive quantitative immunoassays and have been adapted for immunohistochemistry, and electron microscopy.

   The determination of carcinogen-DNA adduct levels by quantitative immunologic procedures has certain advantages over other techniques (1). The sensitivity is frequently better than that obtained with radiolabeled carcinogens. Carcinogen-DNA adduct antisera generally do not react with structurally-dissimilar adducts of the same carcinogen, the carcinogen alone, unmodified nucleosides or unmodified DNA. They are therefore highly-specific probes for chemical bound to DNA. Immunologic assays are rapid, highly reproducible, and can be used in many situations where the cost of a radiolabeled compound would be prohibitive; for example, long-term chronic administration of a carcinogen. In addition, because these techniques are highly sensitive they have been successfully applied to the determination of DNA adducts in tissues of humans exposed to chemicals (2).

   Cisplatin is a chemotherapeutic agent responsible for the cure of testicular cancer, and in widespread use for ovarian, head and neck and lung cancers. Because it is a rodent carcinogen (3) and causes the formation of DNA adducts it was of interest for us to investigate DNA adducts formed by this drug in tissues of human cancer patients receiving platinum drug chemotherapy. Efforts have been made to correlate the extent of cisplatin-DNA adduct formation with dose, toxicity and response of the disease to chemotherapy.

The Use of Enzyme-Linked Immunosorbent Assay (ELISA) and Atomic Absorbance Spectrometry (AAS) for Determination of Cisplatin-DNA Adducts.

DNA modified with the chemotherapeutic agent cisplatin (cis-diammine-dichloroplatinum [II]) has been used to elicit antisera specific for the intrastrand bidentate N7-d(GpG)- and N7-d(ApG)-diammineplatinum adducts (4). These adducts comprise a major portion (80%) of platinum bound to DNA in biological samples from cells, animals or humans receiving platinum-drug exposure (5,6). The antiserum recognizes modified DNA samples and is not specific for the individual adducts alone, unmodified DNA or unmodified nucleotides (4). It has been demonstrated (7) that the ELISA, as performed with a calf thymus DNA standard modified 1-4%, similar to the immunogen DNA, substantially underestimates the total intrastrand adducts in a biological DNA sample. This is because the three-dimensional conformation in a DNA modified in the range of several adducts per 100 nucleotides is significantly different from a biological sample DNA modified in the range of one adduct in a million nucleotides. There is, however, good correlation between biologically-relevant events such, as dose response and disease response, and DNA adducts determined by the ELISA. Thus the assay measures DNA damage in an internally-consistent fashion, even though the values do not reflect the total number of adducts.

In order to obtain quantitative evaluation of cisplatin bound to DNA, we now assay all human samples by both ELISA and AAS. Because at least 80% of the total platinum bound to DNA in a biological sample is in the form of intrastrand adducts, AAS gives values which are close to the desired quantitation. AAS with Zeeman background correction has sufficient sensitivity to measure cisplatin in biological samples (8). In order to compare results with the two methods, 44 samples of rat tissue were assayed by both ELISA and AAS, and when data from the two methods were plotted simultaneously (Figure 1), the correlation observed was essentially linear. The actual numbers generated by the AAS were approximately 500 times higher than those obtained by ELISA, as shown in the Figure.

Monitoring of Cisplatin-DNA Adducts in Blood Cell DNA of Cancer Patients

The ELISA, described above, has been utilized to measure DNA adducts in nucleated blood cell DNA and tissue DNA of cancer patients receiving platinum drug-based chemotherapy. Advantages of this approach include the opportunity to obtain human samples from unexposed controls, and the potential to observe a dose-response for adduct formation since precise dose information is available. In addition, it was anticipated that if adducts could be correlated with biological effects such as disease response or toxicity, any manipulations which would enhance adduct formation might actually improve therapy.

Blood cell DNA samples were obtained from testicular and ovarian cancer patients receiving platinum drug-based therapy during 5 day drug infusions at the NIH Clinical Center (9), and assayed for adducts by cisplatin-DNA ELISA. These studies demonstrated the first dose-response for DNA adduct formation in humans (Figure 2), and samples from unexposed controls were consistently negative. About half of the treated patients experienced a dose-related increase in adducts with subsequent monthly cycles of chemotherapy, and the other half did not have measurable adducts even after very high total cumulative doses (9). This suggests that there may be biochemical or metabolic factors which govern interindividual variability in DNA adduct processing.

When adducts were correlated with disease response in 72 ovarian and poor-prognosis testicular cancer patients (Figure 3), a trend was observed in which the complete responders had the highest mean adduct levels and DNA from many of the non-responders did not contain measurable adducts (10,11). The trend analysis, which compared mean adduct levels for complete responders, partial responders and non-responders was significant with a P value of 0.03 (10). These data support the contention that the formation of high levels of the cisplatin-DNA intrastrand adducts, as measured by the ELISA, is associated with successful chemotherapy, and that those individuals who fail to form measurable adducts have a high rate of therapy failure. In an extension of these studies (12) a series of prognostic variables, including DNA adduct formation, Karnofsky status, total cumulative platinum dose, stage of disease, bulk of disease at initiation of therapy and histologic type and grade, was correlated with disease response by univariate and multivariate analysis. In an analysis of data from 24 ovarian cancer patients the only variable to be significantly associated with disease response was cisplatin-DNA adduct formation measured by the ELISA.

Monitoring of Cisplatin-DNA Adducts in Tissue DNA of Cancer Patients

Even though blood cell DNA adducts correlate with platinum-drug dose in some individuals, and are associated with a positive response to therapy, experiments with blood cell DNA do not indicate the extent of adduct formation in the target tumor tissue or in organs which undergo toxicity. In order to investigate DNA adduct formation in tumor and other human tissues, adducts were measured in tumor biopsies and in tumor tissues obtained at autopsy. In one study, biopsies of cervical tumor were taken before and 24 hr after treatment with carboplatin (7). In addition, blood was drawn from the same individuals before, 24 hr after and 8 days after chemotherapy. DNA adducts, measured by ELISA (Table I), were similar in tumor and in blood 24 hr after the drug infusion, and continued to be elevated in blood 8 days later. The results demonstrated that adduct levels in blood cell DNA were reflective of those in cervical tumor DNA after a single carboplatin dose.

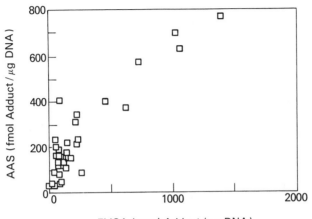

ELISA (amol Adduct / μg DNA)

Figure 1. Comparison of cisplatin–DNA adduct values determined in 44 samples of rat kidney DNA by ELISA using an anti-cisplatin–DNA antiserum, and by AAS.

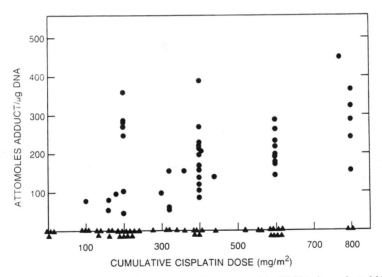

Figure 2. Dose response for cisplatin–DNA adducts, assayed by ELISA, in nucleated blood cell DNA of 77 testicular and ovarian cancer patients receiving their first course of cisplatin-based therapy on 21 or 28 day cycles. Adduct levels are plotted as a function of total cumulative cisplatin dose, with 45 positive samples (0) and 47 negative samples (Δ).

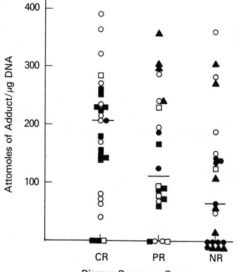

Figure 3. Correlation between cisplatin–DNA adduct level in nucleated blood DNA (ordinate) and disease response for 55 ovarian and 17 poor prognosis testicular cancer patients receiving 6 different single-agent or combination treatment protocols (different symbols) (10,11). Disease response is classified as complete response (CR), an absence of detectable disease; partial response (PR), a greater thatn 50% remission; and no-response (NR), a less thabn 50% remission.

Table I. Carboplatin-DNA adducts (attomol adduct/ug DNA) in cervical tumor and nucleated blood cell DNA of patients.

| Patient | Dose[b] | Adduct Levels Before/After Carboplatin Therapy[a] | | | | |
| | | Tumor of Cervix | | Nucleated Blood Cells | | |
		Before	24hr	Before	24hr	8days
A	399	0	97	0	160	137
B	393	0	250	0	83	127
C	226	0	none	0	51	none

[a] = attomol of adduct/ug DNA; [b] = mg/m^2

SOURCE: Reproduced with permission from ref.7.

Table II. Cisplatin-DNA adducts measured by ELISA (attomol adduct/ug DNA) and AAS (femtomol adduct/ug DNA) in human tissues obtained at autopsy.

Tissue	Patient 2 ELISA	AAS	Patient 4 ELISA	AAS	Patient 6 ELISA	AAS	Patient 7 ELISA	AAS
Tumor	-	-	58	0	176	0	73	11
Bone Marrow	0	0	77	0	-	-	45	14
Lymph Node	143	28	0	14	-	-	-	-
Spleen	343	0	-	0	-	-	283	53
Kidney	511	41	50	6	315	204	122	38
Liver	457	23	45	29	342	90	96	198
Peripheral Nerve	-	0	0	0	-	62	4	19
Brain, Gray	143	0	62	2	-	-	-	-
Brain, White	306	0		112	-	-	-	-

"-" indicates absence of tissue sample
"0" indicates adduct levels too low to measure

Measurements of cisplatin-DNA adducts in autopsy tissues of cancer patients have demonstrated a widespread adduct distribution and long-term persistence in many tissues. Thus, it has been possible to compare adduct levels in many organs with those in the target tumor. Cisplatin-DNA adducts have been measured in multiple tissues obtained at autopsy from 9 individuals who received platinum drug-based therapy (Table II). Samples from all but 2 patients were assayed by both ELISA and AAS. Evidence obtained to date suggests that adducts are formed in many organs of the body, and that the levels observed are similar in all organs including tumor. In addition, many of the individuals studied had acheived their latest chemotherapy 4-15 months prior to death, and still had measurable adducts. This demonstrates that cisplatin-DNA adducts are highly persistent.

Summary and Conclusions

These studies were initially established to validate the use of immunoassays for carcinogen-modified DNA with human samples. The original goals were to define a dose-response for DNA adduct formation in humans under controlled conditions, and to explore a possible relationship between adduct formation and observable biological consequences such as tumor remission. In the area of assay development it has become apparent that the immunoassays have limitations and that corroborative methods should be employed where possible. The human studies have established a dose response for the 50% of blood cell DNA samples that are positive. In addition, patients with the highest blood cell DNA adduct levels were those most likely to enter into remission, while those with unmeasurable adducts responded poorly. This finding opens the possibility that the cisplatin-DNA ELISA may be useful in predicting the response of an individual patient to therapy. Adduct determinations in human tissues have shown that platinum drug bound to DNA has a broad physiological distribution, and is present in tumor as well as organs which are targets for drug toxicity. In addition, DNA adducts are highly persistent in human tissues for many months after treatment. It is likely that the widespread distribution and long-term persistence of drug-induced DNA damage may be related to chemo-therapeutic efficacy, toxicity and a potential for second malignancy, all of which are associated with the use of the diammineplatinum compounds for human cancer chemotherapy.

Literature Cited

1. Poirier, M. C. Monitoring Human Exposure to Carcinogens: Analytical and Ethical Considerations; Skipper, P. L., Koshier, F., Groopman, J. D., Eds.; Telford Press: Caldwell, in press, 1989.
2. Poirier, M. C.; Beland, F. A. Prog Exp Tumor Res 1987, 31, 1-10.
3. Hennings, H.; Shores, R. A.; Poirier, M. C.; Reed, E.; Tarone, R. E.; Yuspa, S. H. JNCI, in press, 1990.
4. Poirier, M. C.; Lippard, S. J.; Zwelling, L. S.; Ushay, H. M.; Kerrigan, D.; Thill, C. C.; Santella, R. M.; Grunberger, D.; Yuspa, S. H. Proc Natl Acad Sci USA 1982, 79, 6443-6447.

5. Fichtinger-Schepman, A. M. J.; van Oosterom, A. T.; Lohman, P. H. M.; Berends, F. Cancer Res 1987, 47, 3000-3004.
6. Dijt, F. J.; Fichtinger-Schepman, A. M. J.; Berends, F.; Reedijk, J. Cancer Res 1988, 48, 6058-6062.
7. Poirier, M. C.; Egorin, M.; Fichtinger-Schepman, A. M. J.; Reed, E. Methods for Detecting DNA Damaging Agents in Humans: Application in Cancer Epidemiology and Prevention; Bartsch, H., Hemminki, K., O'Neill, I. K., Eds.; IARC Sci Publications, Vol. 89, 1988; p 313-320.
8. Reed, E.; Sauerhoff, S.; Poirier, M. C. Atomic Spectroscopy 1988, 9, 93-95.
9. Reed, E.; Yuspa, S. H.; Zwelling, L. A.; Ozols, R. F.; Poirier, M. C. J Clin Invest 1986, 77, 545-50.
10. Reed, E.; Ozols, R. F.; Tarone, R.; Yuspa, S. H.; Poirier, M. C. Proc Natl Acad Sci U S A 1987, 84, 5024-8.
11. Reed, E.; Ozols, R. F.; Tarone, R.; Yuspa, S. H.; Poirier, M. C. Carcinogenesis 1988, 9, 1909-11.
12. Reed, E.; Ostchega, Y.; Steinberg, S. M.; Yuspa, S. H.; Young, R. C.; Ozols, R. F.; Poirier, M. C. Cancer Res, in press, 1990.

RECEIVED August 30, 1990

Chapter 28

Detection of O^4-Alkylthymine in Human Liver DNA

Molecular Epidemiological Study on Human Cancer

Nam-ho Huh[1], Chieko Moriyama[1], Masahiko S. Satoh[1], Junji Shiga[2], and Toshio Kuroki[1]

[1]Department of Cancer Cell Research, Institute of Medical Science, University of Tokyo, 4-6-1, Shirokanedai, Minato-ku, Tokyo 108, Japan
[2]Department of Pathology, Faculty of Medicine, University of Tokyo, 7-3-1, Hongo, Bunkyo-ku, Tokyo 113, Japan

Quantitation of premutagenic DNA structural modifications is among the most reliable markers for the assessment of possible human exposure to environmental carcinogens. Considering the ubiquitous distribution of alkylating N-nitroso compounds and their capability to induce malignant tumors in a wide variety of animal tissues, we quantitated O^4-ethylthymine in human liver DNA. Among 33 cases analysed, O^4-ethylthymine was detected in 30 cases and the mean O^4-ethylthymine level of 19 cancer cases was significantly higher than that of 11 non-cancerous cases. Further, when we screened 5 human liver DNA samples for some other premutagenic O-alkylated DNA adducts, O^4-methylthymine was detected in 3 cases, but none of the cases showed any detectable O^6-methyl- or O^6-ethylguanine levels. These results indicate that humans are actually exposed to alkylating N-nitroso compounds and O^4-alkylthymine may be a suitable marker for monitoring the exposure because of its stable nature in mammalian cells. For more extensive studies in this direction, we have established a highly sensitive and specific method to detect O^4-ethylthymine by a combination of pre-fraction by HPLC, ^{32}P-postlabelling, and immunoprecipitation with monoclonal antibody (PREPI), thus increasing the sensitivity about 30 fold compared to our previous method without sacrifying its extremely high specificity.

Chemical carcinogenesis is a complex multistep process involved in the induction of a large proportion of human cancers. Among the most important factors in this process are structural modifications of DNA by chemical carcinogens and their subsequent repair, the

---

Abbreviations: O^6-EtG, O^6-ethylguanine; O^4-EtT, O^4-ethylthymine; O^4-EtdThd, O^4-ethyl-2'-deoxythymidine; O^4-Et-3'-TMP, O^4-ethyl-2'-deoxythymidine-3'-monophosphate; O^6-MetG, O^6-methylguanine; O^4-MetT, O^4-methylthymidine; T, thymidine; PREPI, A detection method by pre fractionation, postlabeling and immunoprecipitation.

0097–6156/91/0451–0308$06.00/0
© 1991 American Chemical Society

efficiency of which differs widely depending on the type of DNA adducts, cells, tissues, and animal species (1-3). In this regard, alkylating N-nitroso compounds are most well studied, and a number of specific DNA modifications including O^6-alkylguanine and O^4-alkylthymine are identified as premutagenic and/or pretrans-formational lesions (2, 4-10). In a transplacental carcinogenesis with a pulse treatment of N-ethyl-N-nitrosourea in rats, Goth and Rajewsky showed that persistance of O^6-ethylguanine (O^6-EtG) in cellular DNA may be responsible for the particular susceptibility of brain cells to malignant conversion (2). This notion was further confirmed by an in vitro study, where transformation frequency was compared in a set of clonal fibroblasts with different repair capacities for O^6-EtG (Thomale et al., manuscript in preparation). The repair-proficient variants were far more resistant to malignant conversion by N-ethylnitrosourea, but not by benz(a)pyrene-7,8-diol-9,10-epoxide, than the repair-deficient clone. On the other hand, continuous feeding of rats with diethylnitrosamine lead to accumulation of O^4-ethylthymine (O^4-EtT) in target hepatocytes, while O^6-EtG content in the DNA remained at about 50 fold lower level than O^4-EtT because of rapid enzymatic elimination (10).

Detection of such critical modifications in DNA derived from human tissues may be a reliable marker for monitoring human exposure to environmental alkylating agents, and contribute to possible identification of high-risk group or individuals who may develop cancer in the future. Since humans are exposed chronically to low concentration of diverse environmental carcinogens, DNA adducts in human tissues are expected to be a complex mixture of different modifications at extremely low levels. Highly sensitive and specific methods are, therefore, essential for precise quantitation of DNA adducts in human materials. In this paper, we describe a study on the detection of O^4-EtT in human liver DNA using HPLC fractionation and ratioimmunoassay (RIA), and the development of a detection method for O^4-EtT with improved sensitivity and specificity.

Table I. $\underline{O}^4$-EtT in human liver DNA

Cases	No. of cases analysed	No. of cases above detection limit[a]	$\underline{O}^4$-EtT/T (x 10^8) Mean+SD[b]	Range
I Liver cancer	13	12	39.9+40.2	3.6-113
II Cancer in other organs	8	7	54.3+74.0	3.4-206
III Nonmalignancy	12	11	11.7+6.5	3.4-25.8

[a]Detection limit for $\underline{O}^4$-EtT/T, 3×10^{-8}

[b]Difference between I and III: $\underline{p} < 0.05$; difference between I + II and III: $\underline{p} < 0.05$

Detection of O^4-Ethylthymine and Some Other Alkylated DNA Adducts in Human Liver DNA

In the first series of molecular epidemiological studies (11), we decided to detect O^4-EtT in human liver DNA, because O^4-EtT is chemically and biologically stable (10,12,13), relevant to mutation and/or transformation (7-10). In addition, liver shows very low cell proliferation (thus minimizing dilution of DNA adducts by DNA replication) and is most active in metabolic activation of chemical carcinogens. DNA was isolated from 20-50 g of liver tissue obtained at autopsy by a conventional phenol extraction, and hydrolysed enzymatically to nucleosides. O^4-ethyl-2'-deoxythymidine (O^4-EtdThd) in the DNA hydrolysates was fractionated by reverse phase HPLC system and quantitated by competitive RIA using anti-(O^4-EtdThd) monoclonal antibody ER-01 (14). Detection limit of the assay was 3×10^{-8} as a molar ratio of O^4-EtT to thymidine (T) when 20 mg of DNA was analysed. Table I shows the summary of the result obtained from 13 liver cancer cases, 6 cases with other cancer than liver, and 12 cases with non-cancerous diseases. All cases except one in each group showed detectable O^4-EtdThd content. The adduct levels were significantly higher in the liver cancer group and non-liver cancer cases as compared to the non-malignant control group ($p < 0.05$). This result indicates that humans are actually exposed to ethylating agents and that O^4-EtT accumulated in cellular DNA might be involved in the induction of human cancer.

Encouraged with the above result, we are now extending the study to detect some other premutagenic alkylated adducts as well, including O^4-methylthymidine (O^4-MetT), O^6-methylguanine (O^6-MetG), and O^6-EtG. The preliminary result was shown in Figure 1. About 5 mg of liver DNA hydrolysates in each case was fractionated by the reverse phase HPLC system and the fractions were screened for the 4 alkylated nucleosides using corresponding monoclonal antibodies (detailed conditions will be reported elsewhere). Among 4 cases analysed, O^4-MetT was detected in 3 cases, O^4-EtT in 2 cases, but no cases contained detectable amount of O^6-MetG or O^6-EtG. This again indicates that persistence of adducts in cellular DNA largely depends on efficiency of removal, and not much on the initial formation rate in case of chronic exposure, and that the O^4-alkyl modification of thymidine is a better marker for monitoring human exposure to methylating or ethylating agents.

Table II. Comparison of the methods to quantitate O^4-EtT with respect to absolute and relative sensitivity

Methods	Absolute sensitivity	DNA	Relative sensitivity
Immuno-Slot-Blot	3 fmol	25 μg	1.5×10^{-7}
HPLC + RIA	600 fmol	20 mg	3×10^{-8}
PREPI	20 fmol	20 mg	1×10^{-9}
		1 mg	2×10^{-8}

Sensitivity and Specificity of Detection Methods

Humans are chronically exposed to a diversity of chemical substances at low concentrations. Highly sensitive and specific methods, therefore, are a prerequisite of molecular epidemiological studies. The term "sensitivity" in the present context possibly implies two different meanings, i.e., absolute and relative sensitivity. The absolute sensitivity may be defined as minimum amount of DNA adducts to be quantitated precisely by a given method, while the relative sensitivity is the lowest relative modification level to be determined in cellular DNA (Table II). When the absolute sensitivity is fixed, the relative sensitivity depends on the amount of DNA which can be analysed in a system. To identify and quantitate DNA adducts in human materials, we need an assay with very high absolute as well as relative sensitivity.

Structural modification of human DNA by environmental carcinogens is also considered to vary widely and, therefore, we need a highly specific method as well. In contrast to some experimental situations, a detection method whose specificity comes solely from antibody (e.g., ELISA, Immuno-Slot-Blot (15), or RIA without prefractionation) is not adequate. For example, when we screened each fraction of HPLC elutes of human liver DNA hydrolysates with respect to the binding to antibody ER-01, we observed two peaks in addition to O^4-EtdThd (see reference 11). If we had quantitated by simple RIA without prefractionation by HPLC, we might have falsely interpreted the result as a higher value. In Figure 1, we also notice some peaks reactive with anti-O^6-ethyl-2'-deoxyguanosine antibody at the aberrant position. Thus, combined use of several analytical methods with a different molecular basis for specificity is essential for reliable assessment of DNA modification levels in human materials.

PREPI: A New Detection Method of O^4-EtT by the Combination of HPLC-Prefractionation, ^{32}P-Postlabelling, and Immunoprecipitation

In the previous study where we quantitated O^4-EtT using the combination of prefractionation by HPLC and competitive RIA, the absolute detection limit was 600 fmol (11). For the analysis of 3×10^{-8} level of O^4-EtT/T, we needed 20 mg of DNA isolated from 20 - 50 g of liver tissues. This large DNA requirement precluded the more extensive studies using biopsy specimens or peripheral blood cell DNA. We have developed, therefore, a more sensitive and specific method to quantitate O^4-EtT. The detailed procedure will be described elsewhere (Moriyama, C. et al., Manuscript in preparation). Briefly, the assay is composed of enzymatic hydrolysis of DNA to nucleosides-3'-monophosphate, fractionation by HPLC, 5'-phosphorylation of O^4-ethyl-2'-deoxythymidine-3'-monophosphate (O^4-Et-3'-TMP) with ^{32}P-gamma ATP and polynucleotide kinase (16), and immunoprecipitation using monoclonal antibody after 3'-dephosphorylation. Figure 2 shows the standard curve of O^4-Et-3'-TMP obtained by the method. Radioactivity of the final precipitates increased linearly from 5 to 20 fmol of O^4-Et-3'-TMP. Therefore, 20 fmol of O^4-EtT in sample DNA will be enough for precise determination in duplicates after serial dilution. Thus the absolute sensitivity was increased about 30 fold from 600 to 20 fmol

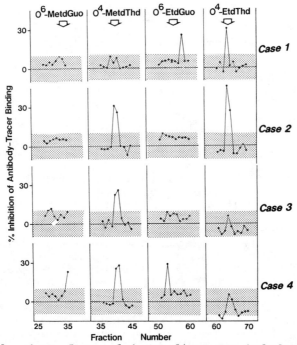

Figure 1. About 5 mg of human liver DNA hydrolysates was fractionated by reverse phase HPLC system. Each fraction was screened with respect to the reactivity with specific monoclonal antibodies against O^6-MetG, O^6-EtG, O^4-MetT or O^4-Et-T. The arrows indicate the position of the corresponding authentic standards.

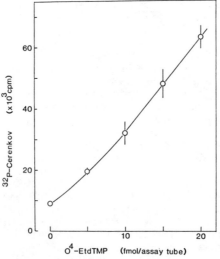

Figure 2. A standard curve of O^4-Et-3'-TMP obtained by PREPI (Prefractionation, Postlabeling and Immunoprecipitation). The ordinate means radioactivity determined by Cerenkov-counting. The abscissa is the amount of O^4-Et-3'-TMP in an assay tube.

of O^4-EtT. The specificity of the method is also very high because each step contributes to the increase of specificity depending on different molecular basis. We are now further optimizing conditions and also applying the method to actual human samples.

Acknowledgments

This work is supported in part by Grant-in-Aid for Special Project Research, Cancer-Bioscience, from the Ministry of Education, Science and Culture, Japan. We thank Dr. Manfred F. Rajewsky, Universität Essen, Federal Republic of Germany, for kind collaboration and providing the antibodies and tracers.

Literature Cited

1. Pegg, A.E. In Biochemical Basis of Chemical Carcinogenesis; Greim, H.; Jung, R.; Kramer, M.; Marquardt, H.; Oesch, F., Eds.; Raven Press: New York, 1984; pp 265-274.
2. Goth, R.; Rajewsky, M. F. Proc. Natl. Acad. Sci. USA 1974, 71, 639-43.
3. Hall, J.; Brésil, H.; Montesano, R. Carcinogenesis 1985, 6, 209-11.
4. Singer, B.; Kusmierek, J. J. Annu. Res. Biochem. 1982, 52, 655-93.
5. Pegg, A. E. In Reviews in Biochemical Toxicology; Hodgson, E.; Bend, J. R.; Philpot, R. M., Eds.; Elsevier/North-Holland Biomedical Press: Amsterdam, 1983; Vol. 5, pp 83-133.
6. Saffhill, R. Chem.-Biol. Interact. 1985, 53, 121-30.
7. Singer, B.; Spengler, S.J.; Fraenkel-Conrat, H.; Kusmierek, J. T. Proc. Natl. Acad. Sci. USA 1986, 83, 28-32.
8. Singer, B. Cancer Res. 1986, 46, 4879-4885.
9. Dyroff, M. C.; Richardson, F. C.; Popp, J. A.; Bedell, M. A.; Swenberg, J. A. Carcinogenesis 1986, 7, 241-46.
10. Swenberg, J.A.; Dyroff, M. C.; Bedell, M. A.; Popp, J. A.; Huh, N.; Kirstein, U.; Rajewsky, M. F. Proc. Natl. Acad. Sci. USA 1984, 81, 1692-95.
11. Huh, N.; Satoh, M. S.; Shiga, J.; Rajewsky, M. F.; Kuroki, T. Cancer Res. 1989, 49, 93-97.
12. Müller, R.; Rajewsky, M. F. Z. Naturforsch. 1983, 38c, 1023-29.
13. Huh, N.; Rajewsky, M. F. Carcinogenesis 1986, 7, 435-39.
14. Adamkiewicz, J.; Eberle, G.; Huh, N.; Nehls, P.; Rajewsky, M. F. Environ. Health Perspect. 1985, 62, 49-56.
15. Nehls, P.; Adamkiewicz, J.; Rajewsky, M. F. J. Cancer Res. Clin. Oncol. 1984, 108, 23-29.
16. Gupta, R. C.; Reddy, M. V.; Randerath, K. Carcinogenesis 1982, 3, 1081-92.

RECEIVED August 30, 1990

Chapter 29

Sensitive Immunochemical Assays for Monitoring Acetaminophen Toxicity in Humans

Dean W. Roberts[1], Robert W. Benson[1], Neil R. Pumford[1,3], David W. Potter[1,4], Henrik E. Poulsen[2], and Jack A. Hinson[1]

[1]Division of Biochemical Toxicology, National Center for Toxicological Research, Jefferson, AR 72079–9502
[2]Department of Pharmacology, University of Copenhagen and Department of Medicine A, Rigshospitalet DK-2100 Copenhagen, Denmark

Acetaminophen (paracetamol) is a commonly used analgesic which is hepatotoxic at high doses in humans and in laboratory animals. Toxicity is believed to be mediated by the reactive metabolite N-acetyl-p-benzoquinone imine which binds to protein thiols as 3-(cystein-S-yl)acetaminophen adducts. Ultrasensitive immunoassays for 3-(cystein-S-yl)acetaminophen derivatives were developed and extensively characterized. Using these assays the formation of this adduct in protein has been correlated with the development of the hepatotoxicity in mice and humans. In mice, adduct levels in the liver reached maximal levels at 2-4 hours and then exhibited a marked decrease which was inversely correlated with parallel elevations in serum adducts and serum levels of the liver-specific transaminase ALT. This suggested that the serum adducts were of hepatic origin and could be monitored as a biomarker of acetaminophen toxicity. Analysis of serum samples from acetaminophen overdose patients demonstrated a positive correlation between immunochemically detectable serum adducts and hepatotoxicity.

Acetaminophen (APAP, N-acetyl-p-aminophenol, paracetamol) is a widely used over-the-counter analgesic. At therapeutic doses it is a safe drug. However, at high doses it may produce severe hepatic necrosis and has also been reported in some individuals to be nephrotoxic (1-3). Available evidence indicates that acetaminophen hepatotoxicity is not a result of the parent compound but is mediated by a reactive metabolite N-acetyl-p-benzoquinone imine (NAPQI). This metabolite is the two-electron oxidation product of acetaminophen and is formed by the microsomal cytochrome P-450 mixed function oxidase system (4-8). Following a therapeutic dose of acetaminophen the reactive metabolite is detoxified

---

[3]Current address: National Institutes of Health, Bethesda, MD 20892
[4]Current address: Rohm and Haas Company, Spring House, PA 19477

by reaction with the cysteine-containing tripeptide glutathione (GSH) to form 3-(glutathion-S-yl)acetaminophen. Following an acetaminophen overdose, hepatic GSH levels are depleted and covalent binding of reactive metabolite to cellular proteins correlates with the development of the hepatotoxicity (9,10). Covalent binding is believed to be primarily a reaction of NAPQI with cysteinyl sulfhydryl groups to produce the corresponding 3-(cystein-S-yl)acetaminophen (3-Cys-A)-protein adduct (11,12).

We perceived the need for sensitive assays that do not rely on the use of radioisotopes or extensive analytical methodology and that could accurately detect protein-bound acetaminophen in biological fluids in the presence of unbound acetaminophen. To this end, we recently developed sensitive avidin biotin-amplified ELISA (A-B ELISA) and particle concentration fluorescence immunoassays (PCFIA) which use antiserum specific for the major acetaminophen-protein adduct associated with toxicity (13-16). These assays are new tools to study the relation between formation of the 3-Cys-A protein adduct and acetaminophen-induced toxicity. In this report we review how these assays were developed, validated in laboratory animals, and used to quantify 3-Cys-A protein adduct formation in human acetaminophen overdose patients.

Development of Immunoassays for Protein-bound Acetaminophen

Previous assays for acetaminophen covalently bound to protein required the use of radiolabeled acetaminophen which was detected after extensive solvent extraction of the protein to remove unbound radioactivity and subsequent quantification of radiolabel associated with the protein (10). Since it was shown that acetaminophen-binding to protein is primarily via cysteine residues (3-Cys-A) (11,12), an immunogen was synthesized which contained the acetaminophen-cysteine adduct. Synthetic NAPQI was allowed to react with N-acetylcysteine to produce 3-(N-acetyl-cystein-S-yl)acetaminophen (NAC-acetaminophen) which was purified by HPLC. The conjugate was subsequently coupled to an immunogenic carrier protein, keyhole limpet hemocyanin (KLH), using 1-(3-dimethyl-aminopropyl)-3-ethyl-carbodiimide hydrochloride as a coupling reagent. Rabbits were immunized with the resulting KLH-NAC-acetaminophen (13).

An antigen was also synthesized for use as a solid-phase coating antigen in the A-B ELISA and as particle-bound antigen in the PCFIA. The requirement was for an immobilized antigen that contained acetaminophen bound to cysteinyl groups in protein as occurs in acetaminophen toxicity. Metallothionein was selected because it contained a high molar content of free cysteine sulfhydryl groups and it was thought that it would react directly with synthetic NAPQI to yield the relevant 3-Cys-A adduct. The synthetic scheme used to prepare the immunogen and solid phase antigen is presented in Figure 1 (13).

To select the antiserum best suited for detection of 3-Cys-A adducts in the presence of free acetaminophen, the relative inhibitory potencies of

NAC-acetaminophen and acetaminophen were compared for nine antisera in competitive A-B ELISA. For all responding rabbits, NAC-acetaminophen was a more efficient inhibitor than free acetaminophen. The relative efficiency of NAC-acetaminophen and acetaminophen to inhibit binding of antibody to solid phase metallothionein-acetaminophen are presented in Figure 2. Inhibition curves using polyclonal rabbit anti 3-Cys-A from one of the rabbits at a dilution of 1:4308, and NAC-acetaminophen and acetaminophen as inhibitors, demonstrated that NAC-acetaminophen was detected with 3.8 orders of magnitude greater sensitivity than the free drug.

Immunochemical Quantification of Acetaminophen Adducts

In initial work to characterize the epitope and to assay samples from acetaminophen-dosed animals, the competitive A-B ELISA was utilized (14). In subsequent work the assay was modified to adapt it to a competitive PCFIA format (15). The PCFIA has advantages over ELISA including: a covalently coupled solid phase, shorter incubation times, the availability of internal standards, and a more flexible assay format. It utilizes a fluorimeter and specially designed assay plates (Baxter Healthcare Corp., Mundelein, IL). Solid phase antigen for PCFIA was prepared by coupling metallothioin-acetaminophen to amino-substituted polystyrene beads using N-succinimidyl 3-(2-pyridyldithio)propionate as a coupling reagent (Figure 3). In both assays, a limiting amount of rabbit anti 3-Cys-A antibody was incubated with either 3-(N-acetyl-L-cystein-S-yl)acetaminophen standard or an unknown sample (mouse liver fraction, serum, or structurally related inhibitor) and then with solid phase acetaminophen-derivatized metallothionein. Detection of rabbit anti 3-Cys-A antibody bound to solid phase metallothioin-acetaminophen assay antigen was accomplished using avidin-biotin-horseradish peroxidase amplification and substrate conversion in the A-B ELISA, and using fluorescein isothiocyanate conjugated second antibody in the PCFIA.

To evaluate the amount of acetaminophen bound to proteins, utilizing the competitive A-B ELISA or PCFIA, inhibition by unknown samples was compared with an assay standard prepared by derivatizing protein with NAPQI. For some experiments, 3-(N-acetyl-L-cystein-S-yl)acetaminophen was used as an assay standard, in which case, the values obtained were corrected for differences in the relative inhibitory potency of 3-(N-acetyl-L-cystein-S-yl)acetaminophen and 3-Cys-A protein adduct (120 fmol/well and 2300 fmol/well, respectively) (14). After dialysis, unknown samples were diluted to a final concentration of approximately 4 μg protein/assay well, assayed in duplicate, and expressed as nmoles of 3-Cys-A per mg of protein. The ELISA and PCFIA were shown to have similar limits of detection (20 pmole/mg protein) and to recognize the same epitope as demonstrated by similar relative inhibitory potencies for N-acetylcysteine-acetaminophen, acetaminophen-bound

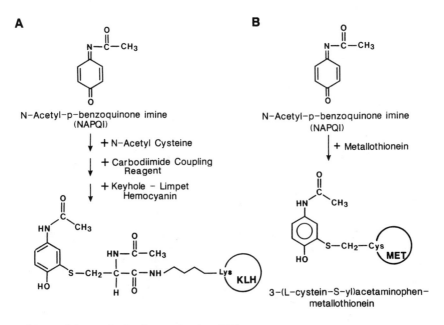

3-(N-acetyl-L-cystein-S-yl)acetaminophen-KLH

Figure 1. A. Preparation of 3-Cys-A-KLH immunogen. B. Preparation of metallothionein-acetaminophen assay antigen. Detailed methods for the syntheses are described in 13 (Reproduced from Ref. 13).

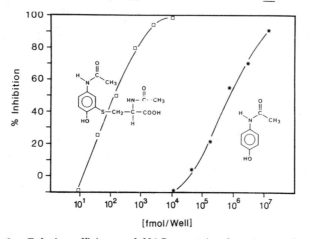

Figure 2. Relative efficiency of NAC-acetaminophen (□) and acetaminophen (*) in the competitive A-B ELISA. The relative efficiency of NAC-acetaminophen and acetaminophen to compete for a limited amount of rabbit anti KLH-NAC-acetaminophen antibodies were determined in the presence of excess metallothionein-acetaminophen adsorbed in wells of 96-well polystyrene assay plates. The 50% inhibitory concentration was 110 fmole/well for NAC-acetaminophen and 687,000 fmole/well for acetaminophen (Reproduced from Ref. 13).

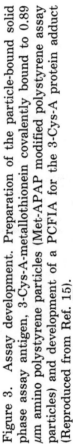

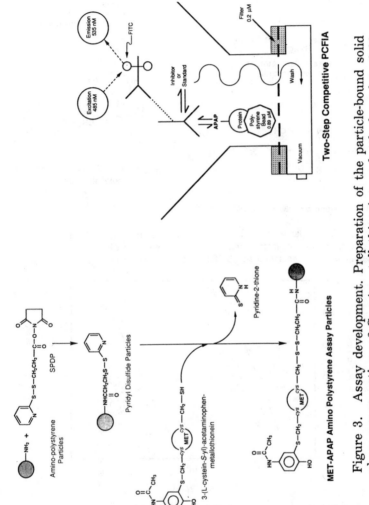

Figure 3. Assay development. Preparation of the particle-bound solid phase assay antigen, 3-Cys-A-metallothionein covalently bound to 0.89 μm amino polystyrene particles (Met-APAP modified polystyrene assay particles) and development of a PCFIA for the 3-Cys-A protein adduct (Reproduced from Ref. 15).

glutathione S-transferase, and acetaminophen in both assays. Serum samples and liver fractions containing 3-Cys-A protein adducts assayed by ELISA and PCFIA produced similar results (r= 0.89) (15).

Epitope Characterization

One of the primary reasons for development of an immunoassay that recognized acetaminophen bound to protein was to examine acetaminophen-toxicity in human overdose patients. It was therefore essential to fully characterize the epitope recognized in the assay and to evaluate the degree to which metabolites, structural analogs or other analgesics might cross react. To determine the nature of the epitope and to quantify the relative importance of specific substituent groups, twenty structurally related compounds were evaluated as inhibitors in the competitive immunoassay. These data are presented in Table I. and Figure 4 (14).

The most effective inhibitor was 3-(N-acetyl-L-cystein-S-yl)acetaminophen which had an observed 50% inhibitory concentration of 120 + S.D. 30 fmol/well (n=19). Approximately 6,200-fold higher concentrations of unbound acetaminophen and 5.2 x 10^6-fold higher concentrations of N-acetyl-L-cysteine were required for comparable inhibition. It was demonstrated with acetaminophen analogs, that the hydroxyl group and the N-acetyl moiety of acetaminophen were important in epitope recognition. A 5,000-fold decrease in detection was observed when the analog did not contain the hydroxyl group or when the N-acetyl moiety was replaced with a hydroxyl substituent. Recognition by antibody was also dependent upon the stereochemistry of the analogs. The 50% inhibitory concentration for 3-(L-cystein-S-yl)acetaminophen was 2,300 fmol/well, whereas a 25-fold higher concentration of 3-(D-cystein-S-yl)acetaminophen was required for 50% inhibition. Although 3-(glutathion-S-yl)acetaminophen was an efficient inhibitor at very low concentrations, the 50% inhibitory concentration for GSH was 2.3 x 10^9 and for S-methylglutathione was 2.6 x 10^9 fmol/well. Other metabolites of acetaminophen were poor inhibitors in the immunoassay. 3-Hydroxyacetaminophen was nearly 2-fold less efficient than acetaminophen and over 10,000-fold less efficient than 3-(N-acetyl-L-cystein-S-yl)acetaminophen. The acetaminophen sulfate and acetaminophen glucuronide were ineffective competitive inhibitors at concentrations below 10^6 fmol/well. Other analgesics such as aspirin and phenacetin were not inhibitory even at high concentrations (Table 1) (14). Collectively, these data indicate that primary antibody specificity involves antigenic determinants found on acetaminophen bound covalently via carbon 3 to the sulfur of cysteine residues, that are not found on protein alone or on free acetaminophen (Figure 4).

This antiserum has also been used to detect acetaminophen-protein adducts in Western blots of serum and liver fractions from acetaminophen-dosed animals (16) and to localize the 3-Cys-A adduct in target tissues (17 and 18, Bucci et al. this volume).

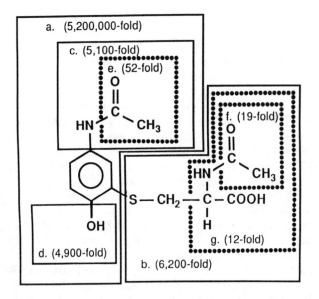

Figure 4. Effect of acetaminophen-conjugate substituents on antibody
recognition in the acetaminophen-adduct immunoassay. The decrease in
inhibition due to particular substituents is shown. These values were
calculated by comparing the ability of the following acetaminophen-
conjugate analogs to competitively inhibit antibody binding to solid-
phase acetaminophen-bound metallothionein. a: N-acetyl-L-cysteine compared to 3-(N-acetyl-L-cystein-S-yl)acetaminophen. b: acetaminophen
compared to 3-(N-acetyl-L-cystein-S-yl)acetaminophen. c: 2-(N-acetyl-L-
cystein-S-yl)hydroquinone compared to 3-(N-acetyl-L-cystein-S-yl)-
acetaminophen. d: 3-(methylthio)acetanilide compared to 3-(methylthio)acetaminophen. e: 2-(L-cystein-S-yl)-p-aminophenol compared to 3-
(L-cystein-S-ylacetaminophen. f: 3-(L-cystein-S-yl)acetaminophen compared to 3-(N-acetyl-L-cystein-S-yl)acetaminophen. g: 3-(methylthio)acetaminophen   compared   to   3-(N-acetyl-L-cystein-S-yl)-
acetaminophen (Reproduced from Ref. 14).

Table I. Competitive Inhibition of Acetaminophen Analogs in the Acetaminophen-Adduct Assay

Inhibitor	50% Inhibition (fmol/well)
3-(N-acetyl-L-cystein-S-yl)acetaminophen	120
3-(glutathion-S-yl)acetaminophen	300
3-(methylthio)acetaminophen	1,400
3-(L-cystein-S-yl)acetaminophen	2,300
3-(diglutathion-S-yl)diacetaminophen	22,000
3-(D-cystein-S-yl)acetaminophen	57,000
3-(glutathion-S-yl)diacetaminophen	75,000
2-(L-cystein-S-yl)-4-aminophenol	120,000
acetaminophen dimer	160,000
2-(N-acetyl-L-cystein-S-yl)hydroquinone	610,000
acetaminophen	740,000
3-hydroxyacetaminophen	1.3×10^6
3-(methylthio)acetanilide	6.9×10^6
N-acetyl-L-cysteine	6.2×10^8
glutathione	2.3×10^9
S-methylglutathione	2.6×10^9
acetaminophen sulfate	$>1.0 \times 10^6$[a]
acetaminophen glucuronide	$>1.0 \times 10^6$[a]
phenacetin	$>1.0 \times 10^6$[a]
aspirin	$>1.0 \times 10^6$[a]

[a]Inhibition was not detected at this concentration.

Effect of Substitution Level

Since our intent was to use the competitive A-B ELISA to quantify 3-Cys-A adducts formed in biological samples at unknown and perhaps variable levels of protein modification, experiments were conducted to determine the effect of adduct substitution level on quantification. Standards of known substitution level were prepared by derivatizing 9,000 g liver supernatant with various concentrations of [3H]NAPQI. After extensive dialysis to remove noncovalently bound materials, protein concentrations were determined and the substitution level of each standard was determined by scintillation counting. These synthetic standards, which ranged from 0.5 to 30 nmol 3-(cystein-S-yl)[3H]acetaminophen per mg protein, were analyzed in the competitive immunoassay. When the data were plotted with percent inhibition as a function of protein concentration, the results show an ordered family of inhibition curves where the most highly substituted proteins were the most efficient inhibitors, and the least substituted protein was the least efficient inhibitor (Figure 5, Panel A). When the same inhibition data were plotted as a function of covalently bound 3-(cystein-S-yl)[^{3}H]acetaminophen, the family of curves were superimposed indicating that under these conditions the competitive A-B ELISA accurately quantifies acetaminophen covalently bound to protein regardless of substitution level (Figure 5, Panel B).

Acetaminophen-induced Hepatotoxicity in Mice

To determine the relationship between the formation of 3-Cys-A adducts in protein and the development of hepatotoxicity, dose response and time course experiments were conducted in male B6C3F1 mice. Using the A-B ELISA specific for 3-Cys-A adducts, we quantified the formation of this adduct in liver and serum protein of mice dosed with acetaminophen. Serum levels of alanine aminotransferase (ALT) were monitored as an index of hepatotoxicity. Administration of acetaminophen at doses of 50, 100, 200, 300, 400, and 500 mg/kg to mice resulted in an increase in serum levels of liver-specific transaminase (evidence of hepatotoxicity) at four hours in the 300, 400, and 500 mg/kg treatment groups only. The formation of 3-Cys-A adducts in liver protein was not observed in the groups receiving 50, 100, and 200 mg/kg doses, but was observed in the groups receiving doses above 300 mg/kg of acetaminophen. Levels of liver adduct were higher in animals receiving the higher doses. 3-Cys-A protein adducts were also observed in serum of mice receiving hepatotoxic doses of acetaminophen. This was an unexpected result. In the time course study, 3-Cys-A adducts in the liver protein reached maximal levels two hours after a 400 mg/kg dose of acetaminophen. By twelve hours the levels decreased to approximately ten percent of the maximal level. In contrast, 3-Cys-A adducts in serum protein were delayed, reaching a sustained maximum six to twelve hours after dosing. The correlation between the appearance of serum aminotransferase and 3-Cys-A adducts in serum protein and the temporal correlation between the decrease in 3-Cys-A adducts in liver protein and the appearance of adducts in serum protein are consistent with a hepatic origin of the 3-Cys-A adducts detected in serum protein (Figure 6). We thus hypothesized that the adducts appearing in serum were of hepatic origin, derived from injured hepatocytes during the development of drug-induced hepatotoxicity and postulated that serum 3-Cys-A protein adducts are a specific biomarker that can be used to study acetaminophen hepatotoxicity (19). Subsequently, the hepatic origin of the 3-Cys-A protein adducts in serum was further confirmed by comparison of adducts detected in SDS-PAGE immunoblots of serum and hepatic protein of B6C3F1 mice at various times after acetaminophen dosing (16). More than 15 proteins containing 3-Cys-A adducts were detected in the liver 10,000 g supernatant. The most prominent protein containing 3-Cys-A adducts in the hepatic 10,000 g supernatant had a relative molecular mass (M$_r$) of 55 kDa.

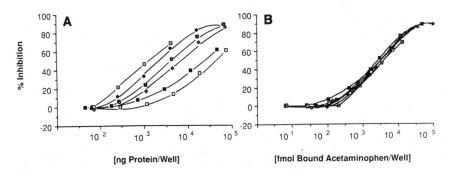

Figure 5.   Effect of substitution level on quantitation in the competitive A-B ELISA. The substitution levels based on the nmoles of [³H] NAPQI per mg of protein were (□) 30, (♦) 10, (◘) 5, (◊) 3, (■) 1, (□) 0.5. In panel A acetaminophen-protein adducts were expressed in terms of protein concentration and in panel B, the same inhibition data was plotted as a function of covalently bound acetaminophen equivalents based on radioactivity (Reproduced from Ref. 19).

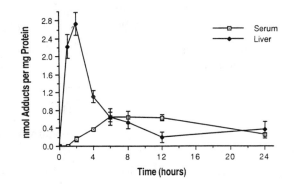

Figure 6.   Time course for 3-Cys-A adduct formation in liver and serum proteins. B6C3F1 male mice which had been fasted overnight were administered acetaminophen (400 mg/kg). The procedures for the preparation of the protein samples and the competitive A-B ELISA for quantitation of the acetaminophen are described in (19). The data points are mean + SEM. All data on acetaminophen adduct formation in serum from 2 to 24 hours are significantly different from control values. The acetaminophen adduct formation in liver protein from one hour to eight hours is significantly different ($P < 0.05$) from control values; however, the acetaminophen adduct formation at the 12 and 24 hour time points are not significantly different from the untreated controls (Reproduced from Ref. 19).

Serum proteins containing 3-Cys-A adducts had molecular masses similar to those found in the liver 10,000 g supernatant (55, 87 and approximately 102 kDa). Collectively, these data indicated that liver adducts were released into the serum following lysis of hepatocytes (16,19).

Detection of Acetaminophen-Protein Adducts in Humans

To determine if acetaminophen toxicity in humans was mediated by a similar mechanism as reported for experimental animals, we looked for the occurrence of 3-Cys-A protein adducts in plasma samples from patients who had taken an overdose of acetaminophen. Plasma was obtained from 30 patients presenting with acetaminophen poisoning at Rigshospitalet, Copenhagen. Plasma samples were stored at -20°C and subsequently shipped frozen to the National Center for Toxicological Research for immunochemical quantification of 3-Cys-A adducts in protein using the competitive PCFIA (20). The patient histories indicated that all patients, except one presenting 58 hours after overdose, were immediately treated with NAC (antidote, NAC) i.v. 300 mg/kg body weight. Plasma obtained at admission was assayed for serum ALT activity and acetaminophen concentration. Of the 30 patients, eleven were at high risk for developing severe liver damage according to Prescott's classification (a nomogram relating risk as a function of time and plasma paracetamol concentration; 21); five patients (Group I) were treated with NAC within 8 hours after overdose, five patients were treated later than

Table II. Concordance between Hepatotoxicity and Acetaminophen-Protein Adducts

Group [n]	I[5]	II[6]	III[3]	IV[16]
Risk Factor#	high	high	moderate	low
Time to NAC* (hours)	4 (1-7)	41 (13-74)	10 (4-15)	10 (2-56)
ALT* (I.U. x 1000)	0.02 (0.01-0.05)	8.4 (4.4-14)	0.04 (0.01-0.08)	0.02 (0.01-0.06)
Plasma 3-Cys-A* (nmol/mg protein)	0	1.9 (0.1-4.1)	0	0▽ (0-0.2)
Plasma acetaminophen*(μM)	1.8 (0.09-2.8)	0.4 (0-1.5)	1.1 (0.2-1.8)	0.6 (0.4-1.3)

Risk of severe liver damage according to Prescot (21).
* Values are the median (range) at admission.
▽ One patient had a value of 0.192.
Adapted from Hinson et al. (20).

8 hours and one was untreated (Group II). Three patients were at moderate risk for developing hepatotoxicity (Group III), and 16 patients were at low risk for developing hepatotoxicity (Group IV) (Table II). All patients that had liver damage, as indicated by elevated plasma ALT, had immunochemically detectable 3-Cys-A adducts in their plasma. These were the patients at high risk of severe liver damage that did not receive antidotal NAC treatment within 8 hours after ingesting acetaminophen (Group II). In contrast, adducts were not found in the plasma of patients who did not show evidence of hepatotoxicity. The relationship between plasma ALT and levels of 3-Cys-A protein adducts at the time of admission is the first direct evidence of a mechanism involving 3-Cys-A adducts in acetaminophen induced liver toxicity in man.

Summary

Immunological approaches were developed to study the the relationship between the binding of acetaminophen to protein and the development of acetaminophen induced hepatotoxicity. Knowledge of the toxic reactive metabolite formed during acetaminophen metabolism and the structure of the resultant 3-Cys-A adduct in protein, suggested the synthesis of a corresponding cysteine-acetaminophen derivative for use as an immunogen. Competitive A-B ELISA and PCFIA were developed and the epitope recognized was extensively characterized. These immunoassays constituted new tools which were used to establish the relationship between the formation of the 3-Cys-A protein adduct in liver and serum and the pathogenesis of acetaminophen toxicity in mice. These tools made it possible to test the hypothesis of an identical mechanism for acetaminophen toxicity in man. Our finding that 3-Cys-A adducts occur in plasma from patients with acetaminophen overdose (20) is preliminary evidence in support of this hypothesis. The close correlation between serum transaminase levels and serum 3-Cys-A adducts is consistent with a hepatic origin for this adduct. In future work this characterized antiserum, coupled with sensitive immunochemical assays and modern protein technology, will provide experimental approaches for the identification and characterization of the protein structures damaged by the acetaminophen metabolite. Such knowledge is needed to understand the processes that ultimately lead to acetaminophen-induced cellular necrosis. This insight is a prerequisite for the development of improved treatment strategies for patients with acetaminophen overdose, and hopefully can be extended to improve understanding of cellular damage from other arylating agents.

Literature Cited

1. Hinson, J.A. Biochemical toxicology of acetaminophen. Rev. Biochem. Toxicol. 1980, 2, 103-129.
2. Black, M. Acetaminophen hepatotoxicity. Ann. Rev. Med. 1984, 35, 577-593.

3. Nelson, S.D. Metabolic activation and drug toxicity. J. Med. Chem. 1982, 25, 753-765.
4. Mitchell, J.R., Jollow, D.J., Potter, W.Z., Davis, D.C., Gillette, J.R. and Brodie, B.B. J. Pharmacol. Exp. Ther. 187, 1973, 185-194.
5. Potter, W.Z., Davis, D.C., Mitchell, J.R., Jollow, D.J., Gillette, J.R. and Brodie, B.B. J. Pharmacol. Exp. Ther. 1973, 187, 203-210.
6. Jollow, D.J., Thorgeirsson, S.S., Potter, W.Z., Hashimoto, M., and Mitchell, J.R. Pharmacology 1974, 12, 251-271.
7. Dahlin, D.C., Miwa, G.T., Lu, A.Y.H. and Nelson, S.D. Proc. Natl. Acad. Sci. U.S.A. 1984, 81, 1327-1331.
8. Potter, D.W. and Hinson, J.A. J. Biol. Chem. 1987, 262, 966-973.
9. Mitchell, J.R., Jollow, D.J., Potter, W.Z., Gillette, J.R. and Brodie, B.B. J. Pharmacol. Exp. Ther. 1973b, 187, 211-217.
10. Jollow, D.J., Mitchell, J.R., Potter, W.Z., Davis, D.C., Gillette, J.R. and Brodie, B.B. J. Pharmacol. Exp. Ther. 1973, 187, 195-202.
11. Hoffmann, K.J., Streeter, A.J., Axworthy, D.B. and Baillie, T.A. Chem.-Biol. Interact. 1985a, 53, 155-172.
12. Hoffmann, K.J., Streeter, A.J., Axworthy, D.B. and Baillie, T.A. Mol. Pharmacol. 1985b, 27, 566-573.
13. Roberts, D.W., Pumford, N.R., Potter, D.W., Benson, R.W., Hinson, J.A. J. Pharmacol Exp. Ther. 1987, 241, 527-533.
14. Potter, D.W., Pumford, N.R., Hinson, J.A., Benson, R.W., Roberts, D.W. J. Pharmacol. Exp. Ther. 1989, 248, 182-189.
15. Benson, R.W., Pumford, N.R., McRae, T.A., Hinson, J.A., and Roberts, D.W. Toxicologist 1989, 9 47.
16. Pumford, N.P., Hinson, J.A., Benson, R.W., and Roberts, D.W. Toxicol. Appl. Pharmacol. 1990, In Press.
17. Roberts, D.W., Hinson, J.A., Benson, R.W., Pumford, N.R., Warbritton, A.R., Crowell, J.A., and Bucci, T.J. Toxicologist 1989, 9, 47.
18. Bucci, T.J., Warbritton, A.R., T.J., Hinson, J.A. and Roberts, D.W. Immunoassays for Monitoring Human Exposure to Toxic Chemicals; American Chemical Society: Washington DC. This volume.
19. Pumford, N.R., Hinson, J.A., Potter, D.W., Rowland, K.L., Benson, R.W., Roberts, D.W. J. Pharmacol. Exp. Ther. 1989, 248, 190-196.
20. Hinson, J.A., Roberts, D.W., Benson, R.W., Dalhoff, K., Loft, S., and Poulsen, H.E. The Lancet 1990, 335, 732.
21. Prescott, L.F., Illingworth, R.N., Critchley, J.A.J.H., Stewart, M.J., Adam, R.D., Proudfoot, A.T. Br. Med. J. 1979, 2, 1097-1100.

RECEIVED August 30, 1990

Chapter 30

Immunohistochemical Detection of Antigenic Biomarkers in Microwave-Fixed Target Tissues

Acetaminophen–Protein Adducts

Thomas J. Bucci[1], Alan R. Warbritton[1], Jack A. Hinson[2], and Dean W. Roberts[2]

[1]Pathology Associates, Inc., Jefferson, AR 72079–9502
[2]Division of Biochemical Toxicology, National Center for Toxicological Research, Jefferson, AR 72079–9502

Immunohistochemical localization of adducts can be accomplished using antiserum developed and characterized for quantitative immunoassays. The 3-(cystein-S-yl)acetaminophen protein adduct is associated with the toxicity of the prototype hepatotoxin, acetaminophen. Immunohistochemical localization of this adduct is described to illustrate the technique as an adjunct to other methods in the assessment of toxicity. This method provides direct correlation between presence of adduct and morphologic evidence of cell injury. Microwave irradiation was pioneered as a fixation method to simultaneously preserve tissue structure and adduct antigenicity.

Many chemicals that are active as toxicants or carcinogens have electrophilic properties or are metabolically converted to electrophiles which react with nucleophilic centers in nucleic acids and proteins to form covalent adducts. Covalent binding of a reactive metabolite to essential cell constituents is frequently a critical event that leads to toxicity. The complex formed by reactive metabolite binding to cell macromolecules (adduct) constitutes a novel molecular species in the organism, and it may contain unique antigenic determinants or "neoantigens". Immunological assays that recognize these neoantigens can be used to study the role of the macromolecular adduct in toxicity.

Once an important neoantigen is known, antisera raised against an immunogen that contains the critical epitope may be the basis of a variety of immunoassays. Since the parent compound is usually a hapten which, by itself, is too small to be an immunogen, appropriate immunogenic conjugates must be isolated or synthesized. Antisera raised against macromolecular conjugates of the parent compound or a structurally related metabolite may recognize 1) the parent compound 2) the modified macromolecule, or 3) a unique adduct. The criteria used to select an antiserum or monoclonal antibody clone are dictated by the objectives of the intended application. Thus, the use of highly characterized antisera is essential to meaningful interpretation.

In many cases, antibodies developed to detect and quantify these antigenic adducts (and other antigenic target molecules) are also suitable for use in parallel immunohistochemical studies. These studies can provide both spatial and temporal correlation of adduct localization with cell injury and thus provide information on pathogenesis that cannot be obtained by other means. The integrity of information gained from immunohistochemical localization of antigens is critically dependent on knowledge of the antigenic determinants that the primary antibody will bind.

As part of the localization procedure, the antigen-bound primary antibody is detected by a system that permits direct observation of the site of antibody binding. A number of detection systems are available to enhance sensitivity and to visualize primary antibody bound in situ to target antigen. Most of them use either a radiolabeled second antibody and visualization by autoradiography, or an enzyme, fluorescent, or electron dense probe coupled to a linking second antibody, protein A, or avidin-biotin amplification system (1,2).

Immunohistochemistry provides structural specificity that is absent in autoradiography of radiolabeled compounds and further affords the opportunity to study human tissue without having first to administer a radioactive compound. In addition, immunohistochemical techniques have the advantage that many antigens are preserved by solutions used for histologic fixation, e.g., formaldehyde. Thus, tissues preserved for other purposes can be studied retrospectively to correlate histopathology, adduct localization and other indices of injury. In principal, the method lends itself well to identification of biomarkers of toxicant exposure in the context of "molecular epidemiology". Although a variety of fixation techniques are available, no single system provides optimal preservation of morphology of all tissues and all target antigens simultaneously. Thus there is a spectrum of effectiveness in achieving good preservation of both structure and antigenicity, including several highly satisfactory techniques (3-6).

For prospective studies, we have successfully pioneered the simultaneous preservation of adduct antigenicity and tissue structure using fixation by microwave irradiation (7). Microwave irradiation had been demonstrated to provide superior preservation of antigens used in diagnostic pathology. For 22 of 23 antigens in nine tissues in comparison with formaldehyde, it yielded immunostaining that was more intense and more extensive while providing excellent preservation of cytomorphologic detail (8). By causing excitation of water molecules, the microwave energy effects controlled heating and coagulative stabilization of protein.

Localization of Drug-Protein Adducts in Animals Dosed with the Prototype Hepatotoxin, Acetaminophen

Immunologic techniques were used to study acetaminophen toxicity in mice to elucidate mechanisms of liver toxicity. The hepatotoxicity of acetaminophen is mediated by a reactive metabolite, N-acetyl-p-benzoquinone imine (NAPQI). The metabolite binds to protein as 3-(cystein-S-yl)acetaminophen (3-Cys-A) and the amount of binding correlates with toxicity. This covalent binding, and in particular the 3-Cys-A adduct, is the most reliable biomarker of acetaminophen toxicity (9-12).

Previous reports described: an ELISA specific for 3-Cys-A protein adducts that does not rely on radiolabeled samples (13); characterization of the relevant epitope (14); and quantification of 3-Cys-A adducts in liver and serum as a function of hepatotoxicity (9). This highly characterized antiserum specific for the acetaminophen-protein adduct, 3-Cys-A, was used to correlate the immunohistochemical localization of 3-Cys-A protein adducts with the development of cell injury, in microwave-fixed liver.

Approach. Male B6C3F1 mice were fasted overnight and dosed with saline (control) or acetaminophen (400 mg/kg) intraperitoneally the following morning. Mice were killed at various times after dosing and serum and tissues were collected for clinical chemistry and adduct analysis as described previously (9). The alanine aminotransferase (ALT) activity in serum was determined as an index of hepatotoxicity (15). Total hepatic glutathione (GSH) levels were determined, according to the method of Tietze (16), as an indicator of the degree to which reactive metabolite had depleted this constitutive detoxification mechanism (11).

The polyclonal antiserum used for these studies was raised in rabbits immunized with 3-(N-acetylcystein-S-yl)acetaminophen-keyhole limpet hemocyanin and has specificity for antigenic determinants which are not found on the parent compound or on host protein, but are found on adducts in which acetaminophen is bound via carbon 3 to the sulfur of cysteine in proteins (13,14, Roberts et al., this volume).

Comparison of Fixation Techniques. For microwave fixation, fresh tissues were trimmed to approximately 2 mm thickness, placed in plastic cassettes, and held in ice cold saline (30-120 minutes) until irradiation in a 700-watt oven set to hold at 60°C for 2 minutes (7). Microwave-fixed tissues were stored overnight in 70% ethanol at room temperature, processed routinely and then embedded in paraffin, sectioned, deparaffinized, and rehydrated using standard techniques. Other pieces of liver were similarly trimmed, then fixed by immersion in either 3.7% formaldehyde in neutral phosphate buffer (48 hours), 2% glutaraldehyde in phosphate buffer (2 hours), Bouin's solution (24 hours), or B-5 solution (4 hours). These tissues were similarly embedded in paraffin then sectioned routinely. In other studies, specimens of lung and kidney were fixed by microwave irradiation with the same schedule as liver.

The five fixation procedures were evaluated systematically to compare preservation of adduct antigenicity in immunostained sections and cell structure in hematoxylin and eosin (H&E)-stained sections (Table I).

Table I.   Fixation Outcome: Simultaneous Preservation of Adduct
            Antigenicity and Tissue Structure

Fixative	Structural Preservation	Antigen Preservation	Comments
Glutaraldehyde 2%	3*	3*	tissue brittle, immunostain weak
B-5 solution	3	3.5	tissue brittle, immunostain weak
Bouin's solution	4	0	antigen not preserved
Formaldehyde 3.7%	4.5	4	immunostain and structure acceptable, some cell shrinkage
Microwave irradiation	5	5	immunostain intense, uniform, boundaries sharp; structure excellent

*subjective scale 1-5; 5 = maximum score

Immunohistochemical Staining. Based on the comparison of fixation techniques, microwave irradiation was used for all subsequent evaluations. Acetaminophen-protein adducts were detected in tissue sections with a modification of the unlabeled antibody enzyme method described by Sternberger (1). Drug-protein antigens were stained by sequential incubations with 1) rabbit anti 3-Cys-A (diluted 1:350 in PBS containing 1% fetal bovine serum, 20 minutes), 2) sheep anti-rabbit linking antiserum, 20 minutes, 3) rabbit peroxidase-antiperoxidase (PAP), 20 minutes, and 4) 3,3'-diaminobenzidine (DAB) substrate (containing 0.03% H_2O_2, 20 minutes). All incubations were at room temperature and slides were washed twice with PBS between incubations. Sections were counterstained with Mayer's hematoxylin. These reagents for immunohistochemical detection of rabbit anti 3-Cys-A were components of an immunoperoxidase staining kit (Cambridge Research Laboratory, Cambridge, MA). The relative content of 3-Cys-A protein adduct in liver cells was determined by the deposition of the microscopically visible brown DAB reaction product at the sites of antibody binding. The amount of reaction product was proportional to the quantity of adduct. The subjective estimates of adduct quantity localized in histologic sections of liver were corroborated by quantitative immunoassay of adduct in liver homogenates (Roberts et al., this volume).

Correlation of Adduct Localization and Toxicity. Saline-treated controls had no hepatocyte injury and no adduct formation. Figure 1 is an H&E-stained section of liver from a saline-treated mouse (normal control) and Figure 2 is a (negative) immunostain for 3-Cys-A adduct, from the same animal. In controls in which normal pre-immune serum was substituted for anti 3-Cys-A antibody, no immunohistochemical reaction product was present in sections from adduct-positive animals (Figure 5).

Acetaminophen-related hepatotoxicity was inferred from the morphologic changes in sections of liver fixed by microwave irradiation, processed routinely, and stained with H&E using standard techniques. Figure 3 depicts an H&E-stained section of liver from a mouse killed three hours after 400 mg/kg acetaminophen. The centrilobular necrosis of all lobules with sparing of periportal areas is consistent with previous descriptions of acetaminophen-induced hepatotoxicity at the light microscopic (10-11), and ultrastructural (17) levels.

Three hours after a dose of 400 mg/kg acetaminophen the area immunostained for adduct with anti 3-Cys-A antibody (Figure 4), coincided exactly with the necrotic zone evident in the H&E-stained replicate section from the same tissue block (Figure 3). In time-course studies, the adduct was present in centrilobular hepatocytes as early as 15 minutes after administration of acetaminophen, with progressive concentric enlargement of the immunostained area, with increasing intensity of stain, to a maximum at two to four hours. Thereafter, the intensity of stain decreased in animals examined at intervals to 24 hours, as hepatocytes disintegrated. Figure 6 is an immunostained liver from a mouse killed seven hours after 400 mg/kg acetaminophen. Less intense adduct staining was evident, compared with the 3-hour sample depicted in Figure 4.

The quantity of adduct in hepatocytes, as revealed by immunostaining, correlated well with the adduct content of the 10,000 x g liver supernate from the same liver, measured by quantitative immunoassay. Further, in dose-response studies, adduct accumulation in hepatocytes followed depletion of hepatocellular GSH. By one hour after a 400 mg/kg dose, GSH was reduced by >90%, and the adduct was abundant (7).

Whereas immunostaining revealed adduct as early as 15 minutes after a 400 mg/kg dose, morphologic evidence of hepatic injury, as determined by light microscopy, was absent until one hour after dosing. At that time, cloudy swelling of centrilobular hepatocytes and "piece-meal" necrosis of individual cells were evident. By two to three hours, however, there was massive centrilobular necrosis, and significant hepatotoxicity was also indicated by an increase in ALT in serum (7). Available evidence indicates that the 3-Cys-A adducts that appear in serum are

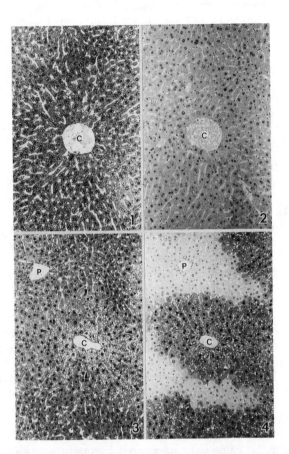

Figures 1-4. (Figure 1) Normal liver. Saline-treated control mouse. H&E
stain. C = central vein. (Figure 2) Normal liver. Saline-treated control
mouse. Anti 3-Cys-A immunostain for acetaminophen adduct is negative.
C = central vein. (Figure 3) Mouse liver 3 hours after 400 mg/kg acetaminophen. H&E stain. Hepatocytes in centrilobular region (within
arrows) are swollen and vacuolated. Many are anuclear. Most are necrotic. The periportal region appears normal. C = central vein, P = portal
vein. (Figure 4) Same liver as Figure 3. Anti 3-Cys-A immunostain for
acetaminophen adduct is strongly positive in centrilobular hepatocytes
(dark cells), and negative in periportal hepatocytes. C = central vein, P
= portal vein.

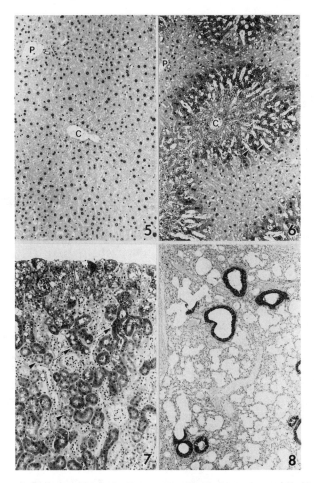

Figures 5-8.   (Figure 5) Same liver as Figure 4. Immunostain control. Preimmune serum was substituted for the omitted primary antibody to verify the specificity of the immunostain to reveal 3-Cys-A. No reaction product is present, despite the presence of adduct (compare with Figure 4). (Figure 6) Mouse liver 7 hours after 400 mg/kg acetaminophen. Anti 3-Cys-A immunostain for acetaminophen adduct is positive in centrilobular hepatocytes (dark cells). Adduct has leached from lysed necrotic hepatocytes nearest the central vein (C). The centrilobular hepatocytes more distal to the central vein are also necrotic, but still retain adduct. Sinusoids (arrows) are dilated. P = portal vein. (Figure 7) Mouse kidney 3 hours after 400 mg/kg acetaminophen. Anti 3-Cys-A immunostain for acetaminophen adduct is positive (dark cells) in some proximal (long arrows) and some distal (short arrows) convoluted tubules. (Figure 8) Mouse lung 3 hours after 400 mg/kg acetaminophen. Anti 3-Cys-A immunostain for acetaminophen adduct is positive in most epithelial cells (dark cells) lining the bronchioles. The remainder of the lung is negative.

leached from damaged hepatocytes (9). By eight hours nearly all centrilobular hepatocytes were lysed, and the intervening sinusoids were dilated. The progressive lysis of necrotic cells that was evident in H&E-
stained sections correlated temporally with increases in serum of both
ALT and adducts (data not shown), liberated from the disintegrating
hepatocytes (9). At 24 hours, there was a moderate inflammatory response at the junction of the necrotic centrilobular zone with the surviv-
ing periportal hepatocytes, that included prominent phagocytosis by
macrophages. The macrophages also contained adduct which was presumably scavenged from hepatocyte debris. Brisk mitotic activity and
regeneration of hepatocytes were also present. The necrotic centrilobular
region collapsed and was remodeled over a three to five-day period.

The uniform centrilobular distribution of adduct and of necrosis
coincided with the reported hepatic distribution of the ethanol-inducible
cytochrome P-450 enzyme previously shown to be the primary enzyme
responsible for conversion of acetaminophen to NAPQI (18,19). These
adduct localization studies further corroborate the role of that enzyme
system in production of the reactive metabolite responsible for the adduct formation. In addition, the data support the concept that NAPQI
reacts with protein at the site of its formation and does not diffuse from
centrilobular hepatocytes to other cells such as periportal hepatocytes.

This system was also used to demonstrate 3-Cys-A adducts in the
mouse kidney and lung. Three hours after 400 mg/kg acetaminophen,
adduct was present in most proximal and some distal convoluted tubules
in the kidney (Figure 7). In the lung, adducts were restricted to epithelial
cells lining the bronchi and bronchioles, sites previously reported to be
the target of acetaminophen-induced necrosis (20) and the location of the
ethanol-inducible P-450 (19) (Figure 8).

Conclusion

Using a drug-protein adduct as a biomarker of cell injury, we have shown
that immunohistochemistry can be a potent technique to complement
other immunochemical analyses to reveal in situ the molecular insult
associated with pathologic change at the subcellular level. Well-
characterized antibody is essential, as are well preserved tissue structures and adduct antigenicity. Corroborative evidence afforded by other
techniques, both quantitative and temporal, provides ample validation of
the localization procedure. Immunohistochemistry is subjectively semi-
quantitative. Careful attention to procedural details such as section
thickness, incubation periods, and standardized methods improve reproducibility and quantitation. Automated image analysis systems using
densitometry or photometry can provide further enhancement. These
methods can be employed ultrastructurally as well, to provide greater
detail regarding the localization of antigenic markers.

Once validated, immunohistochemical techniques can be used in
retrospective studies, taking advantage of tissues collected for other pur-

poses. For example, they can be used after the fact to identify possible toxic agents in human tissues. Immunohistochemical surveys provide a potential approach to environmental monitoring in which sentinel animals or plants could be collected and evaluated for presence of adducts that were biomarkers of exposure.

Localization of a specific adduct in tissue provides a visual fixed point of reference from which further questions may be posed regarding other correlates of injury: altered structure or function and quantification of adduct. In the case of acetaminophen adducts in the liver, we showed that morphologic evidence of cell injury (histology), coincided consistently in time and location with quantity of localized adduct. Functional measurements revealed that clinical evidence of hepatic dysfunction also were proportional to histologic evidence of localized adduct and of injury. The concordance between adduct visualized in tissue sections and adduct measured in liver homogenates by quantitative immunoassay was temporally consistent as the adduct accumulated and also later as the adduct leaked from damaged cells and appeared in serum (7). Other correlations that strengthened the postulated mechanism of acetaminophen hepatotoxicity were the depletion of hepatic GSH antecedent to cell injury, and the coincident distribution of localized adduct, tissue injury, and ethanol-inducible P-450.

Literature Cited

1. Sternberger, L.A. Immunocytochemistry; 3rd ed.; John Wiley: New York, 1986; Chapter 3.
2. Rogers, A.W. Techniques of Autoradiography; Elsevier: Amsterdam, 1979.
3. Falini, B. Arch. Pathol. Lab. Med. 1983, 107, 105-17.
4. Pettigrew, N.M. Arch. Pathol. Lab. Med. 1989, 113, 641-4.
5. Gerrits, P.O.; van Goor, H. J. Histotechnol. 1988, 11, 243-6.
6. Rickert, R.R.; Maliniak, R.M. Arch. Pathol. Lab. Med. 1989, 113, 673-9.
7. Roberts, D.W.; Hinson, J.A.; Benson, R.W.; Pumford, N.R.; Warbritton, A.R.; Crowell, J.A.; Bucci, T.J. Toxicologist 1989, 9, 47.
8. Leong, A. S.Y.; Milie, J.; Duncis, C.G. J. Pathol. 1985, 156, 275-89.
9. Pumford, N.R.; Hinson, J.A.; Potter, D.W.; Rowland, K.L.; Benson, R.W.; Roberts, D.W. J. Pharmacol. Exp. Ther. 1989, 248, 190-6.
10. Mitchell, J.R.; Jollow, D.J.; Potter, W.Z.; Gillette, J.R.; Brodie, B.B. J. Pharmacol. Exp. Ther. 1973, 187, 211-7.
11. Jollow, D.J.; Mitchell, J.R.; Potter, W.Z.; Davis, D.C.; Gillette, J.R.; Brodie, B.B. J. Pharmacol. Exp. Ther. 1973, 187 195-202.
12. Gillette, J.R. Biochem. Pharmacol. 1974, 23, 2785-94.
13. Roberts, D.W.; Pumford, N.R.; Potter, D.W.; Benson, R.W.; Hinson, J.A. J. Pharmacol. Exp. Ther. 1987, 241, 527-33.

14. Potter, D.W.; Pumford, N.R.; Hinson, J.A.; Benson, R.W.; Roberts, D.W. J. Pharmacol. Exp. Ther. 1989, 248, 182-9.
15. Popper, H. In The Liver: Biology and Pathobiology; Arias, I.M., Jakoby, W.B., Popper, H., Schachter, D., and Shafritz, D.A., Eds., Raven Press, New York, 1988; p 1087-1103
16. Tietze, F. Anal. Biochem. 1969, 27, 502-22.
17. Placke, M.E.; Ginsberg, G.L.; Wyand, D.S.; Cohen, S. Toxicol. Pathol. 1987, 15, 431-8.
18. Anderson, L.M.; Ward, J.M.; Park, S.S.; Rice, J.M. Path. Immunopathol. Res. 1989, 8, 61-94.
19. Okey, A.B. Pharmacol. Ther. 1990, 45, 241-298.
20. Jeffery, E.H.; Haschek, W.N. Toxicol. Appl. Pharmacol. 1988, 93, 452-61.

RECEIVED August 30, 1990

APPENDIX 1
ENVIRONMENTAL MONITORING

Readers are refered to the journal Food and Agricultural Immunology where a majority of the papers deal with food monitoring by immunoassay.

English Language Review Articles

Aherne, G. W. Immunoassays in environmental analysis. Anal. Proc. (London), 24: 140-1. 1987.

Aherne, G. W. Immunoassays in the analysis of water. Int. J. Environ. Anal. Chem., 21: 79-88. 1985.

Aherne, G. W.; English, J.; Marks, V. The role of immunoassay in the analysis of microcontaminants in water samples. Ecotoxicol. Environ. Saf., 9: 79-83. 1985.

Allen, J. C., Immunoassays in food analysis. BNF Nutr. Bull., 11: 46-54. 1986.

Allen, J. C.; Smith, C. J. Enzyme-linked immunoassay kits for routine food analysis. Trends Biotechnol., 5: 193-9. 1987.

Cheung, P. Y. K.; Gee, S. J.; Hammock, B. D. Pesticide immunoassay as a biotechnology. ACS Symp. Ser., 362(Impact Chem. Biotechnol.), 217-29. 1988.

Coleman, J. W. Enzyme-linked immunoassays for detection of anti-drug antibodies. Rev. Immunoassay Technol., 2: 159-73. 1988.

Finglas, P. M.; Morgan, M. R. A. The determination of vitamins in food by biospecific analysis. Int. Ind. Biotechnol., 8: 9-12. 1988.

Goh, K. S.; Hernandez, J.; Powell, S. J.; Greene; C. D. Atrazine soil residue analysis by enzyme immunoassay: solvent effect and extraction efficiency. Bull. Environ. Contam. Toxicol., 45: 208-14. 1988.

Hammock, B. D.; Mumma, R. O. Potential of immunochemical technology for pesticide analysis. In: Pestic. Anal. Method., J. Harvey, Jr.; G. Zweig, Eds. American Chemical Society, Washington, DC, pp. 321-52. 1980.

Hammock, B. D.; Gee, S. J.; Cheung, P. Y. K.; Miyamoto, T.; Goodrow, M. H.; Van Emon, J.; Seiber, J. N. Utility of immunoassay in pesticide trace analysis. In: Pestic. Sci. Biotechnol., R. Greenhalgh; T. R. Roberts, Eds. Blackwell: Oxford, UK. 1987.

Harrison, R. O.; Gee, S. J.; Hammock, B. D. Immunochemical methods of pesticide residue analysis. ACS Symp. Ser., 379(Biotechnol. Crop Prot.), 316-30. 1988.

Herman, B. W. Immunoassay of pesticides. In: Immunological Techniques in Insect Biology. L. I. Gilbert; T. A. Miller, Eds. Springer-Verlag, New York, NY, pp. 135-80. 1988.

Hitchcock, C. H. S. Immunoassay techniques for food analysis. Anal. Proc. (London), 24: 146-7. 1987.

Jung, F.; Gee, S. J.; Harrison, R. O.; Goodrow, M. H.; Karu, A. E.; Braun, A. L.; Li, Q. X.; Hammock, B. D. Use of immunochemical techniques for the analysis of pesticides. Pestic. Sci., 26: 303-17. 1989.

Morgan, M. R. A. Newer techniques in food analysis - immunoassays and their application to small molecules. J. Assoc. Public Anal., 23: 59-63. 1985.

Morris, B. A.; Clifford, M. N.; Jackman, R.; Immunoassays for Veterinary and Food Analysis. (Elsevier Appl. Science: London, UK), 392 pp. 1988.

Mumma, R. O.; Brady, J. F. Immunological assays for agrochemicals. In: Pestic. Sci. Biotechnol., R. Greenhalgh; T. R. Roberts, Eds. Blackwell: Oxford, UK. 1987.

Newsome, W. H. Potential and advantages of immunochemical methods for analysis of foods. J. - Assoc. Off. Anal. Chem., 69: 919-23. 1986.

Van Emon, J. M.; Seiber, J. N.; Hammock, B. D. Immunoassay techniques for pesticide analysis. Anal. Methods Pestic. Plant Growth Regul., 17(Adv. Anal. Tech.), 217-63. 1989.

Vanderlaan, M.; Van Emon, J.; Watkins, B.; Stanker, L. H. Monoclonal antibodies for the detection of trace chemicals. In: Pesticide Science and Biotechnology R. Greenlaigh, T.R. Roberts Eds., Blackwell Scientific Publications, London, pp 597-602. 1987.

Vanderlaan, M.; Watkins, B.; Stanker, L. Environmental monitoring by immunoassay. Environmental Science and Technology. 22: 247-54. 1988.

Wraith, M. J.; Britton, D. W. Immunochemical methods for pesticide residue analysis. Brighton Crop Prot. Conf.-Pests Dis., (1), 131-7. 1988.

Immunosensors and Other Novel Detection Schemes

Bright, F.V.; Betts, T. A.; Litwiler, K.S. Regenerable fiber-optic-based immunosensor. Anal. Chem., 62: 1065-69. 1990.

Elling, W.; Huber, S. J.; Bankstahl, B.; Hock, B. Atmospheric transport of atrazine: a simple device for its detection. Environ. Pollut., 48: 77-82. 1987.

Guilbault, G. G.; Ngeh-Ngwainbi, J. Use of protein coatings on piezoelectric crystals for assay of gaseous pollutants. Biotec, 2(Biosens. Environ. Biotechnol.), 17-22. 1988.

Mascini, M.; Moscone, D.; Palleschi, G.; Pilloton, R. In-line determination of metabolites and milk components with electrochemical biosensors. Anal. Chem. Acta, 213: 101-11. 1988.

Northrup, M. A.; Stanker, L. H.; Vanderlaan, M.; Watkins, B. E. Development and characterization of a fiber optic immuno-biosensor. NATO ASI Ser., Ser. C, 280(Spectrosc. Inorg. Bioact.), 229-41, 1989.

Ohlson, S.; Lundblad, A.; Zopf, D. Novel approach to affinity chromatography using "weak" monoclonal antibodies. Anal. Biochem., 169: 204-8. 1988.

Russell, A. J.; Trudel, L. J.; Skipper, P. L.; Groopman, J. D.; Tannenbaum, S. R.; Klibanov, A. M. Antibody-antigen binding in organic solvents. Biochem. Biophys. Res. Commun. 158: 80-5. 1989.

Vo D. T.; Nolan, T.; Cheng, Y. F.; Sepaniak, M. J.; Alarie, J. P. Phase-resolved fiber-optics fluoroimmunosensor. Appl. Spectrosc., 44: 128-32. 1990.

Wimpy, T. H.. An enzyme immunoassay for milk progesterone: development of a flow-through technique. 119 pp. Avail. Univ. Microfilms Int., Order No. DA8902824 From: Diss. Abstr. Int. B 1989, 49: 4618-19. 1988.

Specific Residue Immunoassays

Afghan, B. K.; Carron, J.; Goulden, P. D.; Lawrence, J.; Leger, D.; Onuska, F.; Sherry, J.; Wilkinson, R. Recent advances in ultratrace analysis of dioxins and related halogenated hydrocarbons. Can. J. Chem., 65: 1086-97. 1987.

Banerjee, B.D. Development of an enzyme-linked immunosorbent assay for the quantification of DDA (2,2-bis(p-chlorophenyl) acetic acid) in urine. Bull. Environ. Contam. Toxicol., 38: 798-804. 1987.

Brady, J. F.; Fleeker, J. R.; Wilson, R. A.; Mumma, R. O. Enzyme immunoassay for aldicarb. ACS Symp. Ser., 382(Biol. Monit. Pestic. Exposure: Meas., Estim., Risk Reduct.), 262-84. 1989.

Brimfield, A. A.; Lenz, D. E.; Graham, C.; Hunter, Jr., K. W. Mouse monoclonal antibodies against paraoxon: potential reagents for immunoassay with constant immunochemical characteristics. J. Agric. Food Chem., 33: 1237-42. 1985.

Bushway, R. J.; Perkins, B.; Savage, S. A.; Lekousi, S. J.; Ferguson, B. S. Determination of atrazine residues in water and soil by enzyme immunoassay. Bull. Environ. Contam. Toxicol., 40: 647-54. 1988.

Bushway, R. J.; Perkins, B.; Savage, S. A.; Lekousi, S. L.; Ferguson, B. S. Determination of atrazine residues in food by enzyme immunoassay. Bull. Environ. Contam. Toxicol., 42: 899-904. 1989.

Cheung, P. Y. K.; Hammock, B. D. Monitoring Bacillus thuringiensis in the environment with enzyme-linked immunosorbent assay. ACS Symp. Ser., 379(Biotechnol. Crop Prot.), 359-72. 1988.

Collier, T. L.; ApSimon, J. W.; Sherry, J. P. Development of an improved hapten for use in the radioimmunoassay of dioxins. Chemosphere, 20: 301-8. 1990.

Dixon-Holland, D. E.; Katz, S. E. Direct competitive enzyme-linked immunosorbent assay for sulfamethazine residues in milk. J. Assoc. Off. Anal. Chem., 72: 447-50. 1989.

Elsasser, T. H.; Munns, R. K.; Shimoda, W. Methodological considerations for penicillin radioimmunoassay. J. Immunoassay, 8: 73-96. 1987.

Ercegovich, C. D.; Vallejo, R. P.; Gettig, R. R.; Woods, L.; Bogus, E. R.; Mumma, R. O. Development of a radioimmunoassay for parathion. J. Agric. Food Chem., 29: 559-63. 1981.

Fickling, S. A.; Hampton, S. M.; Teale, D.; Middleton, B. A.; Marks, V. Development of an enzyme-linked immunosorbent assay for caffeine. J. Immunol. Meth., 129: 159-64. 1990.

Fleeker, J. Two enzyme immunoassays to screen for 2,4-dichlorophenoxyacetic acid in water. J. Assoc. Off. Anal. Chem., 70: 874-8. 1987.

Gee, S. J.; Miyamoto, T.; Goodrow, M. H.; Buster, D.; Hammock, B. D. Development of an enzyme-linked immunosorbent assay for the analysis of the thiocarbamate herbicide molinate J. Agric. Food Chem., 36: 863-70. 1988.

Haas, R. A.; Hanson, C. V.; Monteclaro, F. Development of immunoassays for detection of air pollutants in environmental and clinical samples. Proc. - APCA Annu. Meet., 79th(Vol. 2), 86/32.7, 13 pp. 1986.

Hack, R.; Martlbauer, E.; Terplan, G. Production and characterization of a monoclonal antibody to chloramphenicol. Food Agric. Immunol. 1: 197-202. 1989.

Harrison, R. O.; Braun, A.; Gee, S. J.; O'Brian, D. J.; Hammock, B. D. Evaluation of an enzyme-linked immunosorbent assay (ELISA) for the direct analysis of molinate (ordram) in rice field water. Food Agric. Immunol. 1: 37-52. 1989.

Harrison, R. O.; Brimfield, A. A.; Nelson, J. O. Development of a monoclonal antibody based enzyme immunoassay method for analysis of maleic hydrazide. J. Agric. Food Chem. 37: 958-964. 1989.

Heldman, E.; Balan, A.; Horowitz, O.; Ben-Zion, S.; Torten, M. A novel immunoassay with direct relevance to protection against organophosphate poisoning. FEBS Lett., 180: 243-8. 1985.

Hokama, Y.; Shirai, L. K.; Iwamoto, L. M.; Kobayashi, M. N.; Goto, C. S.; Nakagawa, L. K. Assessment of a rapid enzyme immunoassay stick test for the detection of ciguatoxin and related polyether toxins in fish tissues. Biol. Bull. (Woods Hole, Mass.), 172: 144-53. 1987.

Huber, S. J.; Hock, B. A solid-phase enzyme immunoassay for quantitative determination of the herbicide terbutryn. Z. Pflanzenkrankh. Pflanzenschutz, 92: 147-56. 1985.

Jung, F.; Meyer, H. H. D.; Hamm, R. T. Development of a sensitive enzyme-linked immunosorbent assay for the fungicide fenpropimorph. J. Agric. Food Chem., 37: 1183-7. 1989.

Kelley, M. M.; Zahnow, E. W.; Petersen, W. C.; Toy, S. T. Chlorsulfuron determination in soil extracts by enzyme immunoassay. J. Agric. Food Chem., 33: 962-5. 1985.

Kitagawa, T.; Gotoh, Y.; Uchihara, K.; Kohri, Y.; Kinoue, T.; Fujiwara, K.; Ohtani, W. Drug residues in animal tissues. J. Assoc. Off. Anal. Chem., 71: 915-20. 1988.

Kohli, K. K.; Albro, P. W.; McKinney, J. D. Radioisotope dilution assay (RIDA) for the estimation of polychlorinated biphenyls (PCBs). J. Anal. Toxicol., 3: 125-8. 1979.

Lapeyre, C.; Janin, F.; Kaveri, S. V. Indirect double sandwich ELISA using monoclonal antibodies for detection of staphylococcal enterotoxins A, B, C1, and D in food samples. Food Microbiol., 5: 25-31. 1988.

Langone, J. J.; Vunakis, H. V. Radioimmunoassay for dieldrin and aldrin. Res. Communic. Chem. Path. Pharma., 10: 163-75. 1975.

Li, Q. X.; Gee, S. J.; McChesney, M. M.; Hammock, B. D.; Seiber, J. N. Comparison of an enzyme-linked immunosorbent assay and gas chromatographic prodedures for the determination of molinate residues. Anal. Chem., 61: 819-23. 1989.

Luster, M. I.; Albro, P. W.; Clark, G.; Chae, K.; Chaudhary, S. K.; Lawson, L. D.; Corbett, J. T.; McKinney, J. D. Production and characterization of antisera specific for chlorinated biphenyl species: initiation of radioimmunoassay for aroclors. Toxicol. App. Pharm., 50: 147-55. 1979.

McDonald, J.; Gall, R.; Wiedenbach, P.; Bass, V. D.; DeLeon, B.; Brockus, C.; Stobert, D.; Wie, S.; Prange, C. A.; Yang, J.-M.; Tai, C. L.; Weckman, T. J.; Woods, W. E.; Tai, H.-H.; Blake, J. W.; Tobin, T. Immunoassay detection of drugs in horses. I. Particle concentration fluoroimmunoassay detection of fentanyl and its congeners. Res. Commun. Chem. Pathol. Pharma., 57: 389-400. 1987.

Mohammed, A. H.; McCallus, D. E.; Norcross, N. L. Development and evaluation of an enzyme-linked immunosorbent assay for endotoxin in milk. Vet. Microbiol., 18: 27-39. 1988.

Nagao, M.; Takehiko, T.; Wu, B.; Terazawa, K.; Gotouda, H.; Akabane, H. Development and characterization of monoclonal antibodies reactive with paraquat. J. Immuno., 10: 1-17. 1989.

Newsome, W. H. An enzyme-linked immunosorbent assay for metalaxyl in foods. J. Agric. Food Chem., 33: 528-30. 1985.

Newsome, W. H. Determination of iprodione in foods by ELISA. In: Pestic. Sci. Biotechnol., R. Greenhalgh; T. R. Roberts, Eds. Blackwell: Oxford, UK. 1987.

Newsome, W. H. Development of an enzyme-linked immunosorbent assay for triadimefon in foods. Bull. Environ. Contam. Toxicol., 36: 9-14. 1986.

Newsome, W. H; Collins, P. G. Enzyme-linked immunosorbent assay of benomyl and thiabendazole in some foods. J. - Assoc. Off. Anal. Chem., 70: 1025-7. 1987.

Newsome, W.H.; Collins, P.G. Determiniation of 2,4-D in foods by enzyme linked immunosorbent assay. Food Agric. Immunol. 1: 203-10. 1989.

Newsome, W. H.; Shields, J. B. Radioimmunoassay of PCBs in milk and blood. Intern. J. Environ. Anal. Chem., 10: 295-304. 1981.

Niewola, Z.; Benner, J. P.; Swaine, H. Determination of paraquat residues in soil by an enzyme linked immunosorbent assay. Analyst (London), 111: 399-403. 1986.

Niewola, Z.; Hayward, C.; Symington, B. A.; Robson, R. T. Quantitative estimation of paraquat by an enzyme linked immunosorbent assay using a monoclonal antibody. Clinica Chimica Acta, 148: 149-56. 1985.

Nouws, J. F. M.; Laurensen, J.; Aerts, M. M. L. Monitoring milk for chloramphenicol residues by an immunoassay (Quik-card). Vet. Q., 10: 270-2. 1988.

Rinder, D. F.; Fleeker, J. R. A radioimmunoassay to screen for 2,4-dichlorophenoxyacetic acid and 2,4,5-trichlorophenoxyacetic acid in surface water. Bull. Environm. Contam. Toxicol., 26: 375-80. 1981.

Schlaeppi, J. M.; Foery, W.; Ramsteiner, K. Hydroxyatrazine and atrazine determination in soil and water by enzyme-linked immunosorbent assay using specific monoclonal antibodies. J. Agric. Food Chem., 37: 1532-8. 1989.

Schwalbe, M.; Dorn, E.; Beyermann, K. Enzyme immunoassay and fluoroimmunoassay for the herbicide diclofop-methyl. J. Agric. Food Chem., 32: 734-41. 1984.

Singh, P.; Ram, B. P.; Sharkov, N. Enzyme immunoassay for screening sulfamethazine in swine. J. Agric. Food. Chem. 37: 109-14. 1989.

Stanker, L. H.; Watkins, B.; Rogers, N.; Vanderlaan, M. Monoclonal antibodies to dioxin: antibody characterization and assay development. Toxicol. 45: 229-243. 1987.

Stanker, L. H.; Watkins, B.; Vanderlaan, M.; Budde, W. L. Development of an immunoassay for chlorinated dioxins based on a monoclonal antibody and an enzyme linked immunosorbent assay (ELISA). Chemosphere, 16: 1635-39. 1987.

Stanker, L. H.; Bigbee, C.; van Emon, J.; Watkins, B. E.; Jensen, R. H.; Morris, C.; Vanderlaan, M. An immunoassay for pyrethroids: detection of permethrin in meat. J. Agric. Food Chem. 37: 834-9. 1989.

Turesky, R. J.; Forster, C. M.; Aeschbacher, H. U.; Wuzner, H. P.; Skipper, P. L.; Turdel, L. F.; Tannenbaum, R. R. Purification of the food-borne carcinogens 2-amino-3-methylimidazo[4,5-*f*]quinoline and 2-amino-3,8-dimethylimidazo[4,5-*f*]quinoxaline in heated meat products by immunoaffinity chromatography. Carcinogenesis, 10: 151-6. 1989.

Tyski, S. Radioimmunoassay of delta-toxin from B thuringiensis: correlation with bioassay. Toxicon, 27: 947-9. 1989.

Vallejo, R. P.; Bogus, E. R.; Mumma, R. O. Effects of hapten structure and bridging groups on antisera specificty in parathion immunoassay development. J. Agric. Food Chem., 30: 572-80. 1982.

Van de Water, C.; Tebbal, D.; Haagsma, N. Monoclonal antibody-mediated clean-up procedure for the high-performance liquid chromatographic analysis of chloramphenicol in milk and eggs. J. Chromatogr., 478: 205-15. 1989.

Van Emon, J. M.; Seiber, J. N.; Hammock, B. D. Applications of immunoassay to paraquat and other pesticides. ACS Symp. Ser., 276(Bioregul. Pest Control), 307-16. 1985.

Van Emon, J.; Seiber, J.; Hammock, B. Application of an enzyme-linked immunosorbent assay (ELISA) to determine paraquat residues in milk, beef, and potatoes. Bull. Environ. Contam. Toxicol., 39: 490-7. 1987.

Vanderlaan, M.; Watkins, B.E.; Hwang, M.; Knize, M.; Felton, J. S. Development of an immunoassay for 2-amino-1-methyl-6-phenyl-imidazo[4,5-f]-pyridine (PhIP) in cooked meats. Carcinogenesis, 10: 2215-21. 1989.

Vanderlaan, M.; Watkins, B. E.; Hwang, M.; Knize, M. G.; Felton, J.S. Monoclonal antibodies for the immunoassay of mutagenic compounds produced by cooking beef. Carcinogenesis. 9: 153-60. 1988.

Vanderlaan, M.; Stanker, L. H.; Watkins, B. E.; Petrovic, P.; Gorbach, S. Improvement and application of an immunoassay for screening environmental samples for dioxin contamination. Environ. Toxicol. Chem., 7: 859-70. 1988.

Ward, C. M.; Morgan, M. R. A. An immunoassay for determination of quinine in soft drinks. Food Add. Contam., 5: 555-61. 1988.

Watkins, B. E.; Stanker, L. H.; Vanderlaan, M. An immunoassay for chlorinated dioxins in soils. Chemosphere, 19: 267-70. 1989.

Wie, S. I.; Hammock, B. D.; Gill, S. S.; Grate, E.; Andrew, Jr., R. E.; Faust, R. M.; Bulla, Jr., L. A.; Schaefer, C. H. An improved enzyme-linked immunoassay for the detection and quantification of the entomocidal parasporal crystal proteins of Bacillus thuringiensis subspp. kurstaki and israelensis. J. App. Bacter., 57: 447-54. 1984.

Wie, S. I.; Hammock, B. D. The use of enzyme-linked immunosorbent assays (ELISA) for the determination of triton X nonionic detergents. Anal. Biochem., 125: 168-76. 1982.

Wie, S. I.; Sylwester, A. P.; Wing, K. D.; Hammock, B. D. Synthesis of haptens and potential radioligands and development of antibodies to insect growth regulators diflubenzuron and BAY SIR 8514. J. Agric. Food Chem., 30: 943-8. 1982.

Wie, S. I.; Hammock, B. D. Comparison of coating and immunizing antigen structure on the sensitivity and specificity of immunoassays for benzoylphenylurea insecticides. J. Agric. Food Chem., 32: 1294-301. 1984.

Wing, K. D.; Hammock, B. D. Stereoselectivity of a radioimmunoassay for the insecticide S-bioallethrin. Experientia, 35; 1619-20. 1979.

Wittmann, C.; Hock, B. Improved enzyme immunoassay for the analysis of s-triazines in water samples. Food Agric. Immunol. 1: 211-24. 1989.

Reviews in Non-English Languages

Arnold, D. Advantages and limits of the use of immunoassays in residue analysis of pharamacologically active substances. Lebensmittelchem., Lebensmittelqual., 13(Anal. Rueckstaenden Pharmakol. Wirksamer Stoffe), 61-75. 1988. German.

Fukal, L. Application of immunochemical methods to detect residues of veterinary drugs and anabolic hormones in foods of animal origin. Sb. UVTIZ, Potravin. Vedy, 6: 297-304. 1988. Czech.

Fukal, L.; Rauch, P.; Kas, J. Use of immunochemical methods in food production analytical chemistry I. Determination of nonimmunogenic low-molecular-weight compounds. Chem. Listy, 82: 959-77. 1988. Czech.

Galante, Y M.; Comitti, R. Applications of enzyme immunoassays in food analysis. Chim. Oggi, (1-2), 57-64. 1988. Italian.

Haberer, K; Kraemer, P. Availability of immunochemical detection methods for pesticides in water. Vom Wasser, 71: 231-44. 1988. German.

Hock, B. Determination of pesticides by enzyme immunoassay. Gewaesserschutz, Wasser, Abwasser, 106(Instrum. Anal. Pestiz. Wasser Boden), 330-54. 1989. German.

Hock, B. Enzyme immunoassays for the determination of pesticides in water. Z. Wasser Abwasser Forsch., 22: 78-84. 1989. German.

Kondo, F. Identification and analysis of remnant antibiotics by analytical instruments. High performance liquid chromatography, gas chromatography, and enzyme immunoassay. Chikusan no Kenkyu, 42(5), 579-84; 42(6) 705-9. 1988. Japanese.

Nishijima, O. Analytical methods for pesticide residues Bunseki, (7), 557-62. 1989. Japanese.

Tsuji, A; Maeda, M. Recent advances in immunoassay and its application to food analysis. Shokuhin Eiseigaku Zasshi, 30: 199-208. 1989. Japanese.

Non-English Language Articles on Specific Immunoassays

Chigrin, A. V.; Umnov, A. M.; Sokolova, G. D.; Khokhlov, P. S.; Chkanikov, D. I. Determination of chlorsulfuron by immunoenzyme analysis. Agrokhimiya, (8), 119-23. 1989. Russian.

Honda, T. Application of microimmunoassay to toxins and specific metabolites. Rinsho Kensa, 30: 1436-40. 1986. Japanese.

Huber, S. J.; Hock, B. Solid-phase enzyme immunoassay for the determination of herbicides from fresh water - polystyrene spheres in comparison to microtiter plates as antibody carriers. GIT Fachz. Lab., 29: 969-70, 973-5, 977. 1985. German.

Kindervater, R.; Schmid, R. D. Biosensors for water and wastewater analysis. Z. Wasser Abwasser Forsch., 22: 84-90. 1989. German.

Maertlbauer, E.; Terplan, G. Enzyme immunoassay for the detection of chloramphenicol in milk. Arch. Lebensmittelhyg., 38: 3-7. 1987. German.

APPENDIX 2
MYCOTOXIN   ANALYSIS

English Language Articles on Aflatoxin Immunoassays

Candlish, A. A. G.; Haynes, C. A.; Stimson, W. H. Detection and determination of aflatoxins using affinity chromatography. Int. J. Food Sci. Technol., 23: 479-85. 1988.

Chu, F. S. Immunoassays for mycotoxins. In: Mod. Methods Anal. Struct. Elucidation Mycotoxins, 207-37. Richard J. Cole (Ed.) Academic: Orlando, FL. 1986.

Chu, F. S. Recent studies on immunochemical analysis of mycotoxins. Bioact. Mol., 1(Mycotoxins Phycotoxins), 277-92. 1986.

Cole, R. J.; Dorner, J. W.; Kirksey, J. W.; Dowell, F. E. Comparison of visual, enzyme-linked immunosorbent assay screening, and HPLC methods in detecting aflatoxin in farmers stock peanut grade samples. Peanut Sci., 15: 61-3. 1988.

Dorner, J. W.; Cole, R. J. Comparison of two ELISA screening tests with liquid chromatography for determination of aflatoxins in raw peanuts. J. Assoc. Off. Anal. Chem., 72: 962-4. 1989.

Dragsted, L. O.; Bull, I.; Autrup, H. Substances with affinity to a monoclonal aflatoxin B1 antibody in Danish urine samples. Food Chem. Toxicol., 26: 233-42. 1988.

Fukal, L.; Reisnerova, H.; Rauch, P. Applications of radioimmunoassay with iodine-125 for the determination of aflatoxin B1 in foods. Sci. Aliments, 8: 397-403. 1988.

Groopman, J. D.; Trudel, L. J.; Donahue, P. R.; Marshak-Rothstein, A.; Wogan, G. N. High-affinity monoclonal antibodies for aflatoxins and their application to solid-phase Immunoassays. Proc Natl. Acad. Sci. U S A, 81: 7728-31. 1984.

Garner, R. C.; Dvorackova, I.; Tursi, F. Immunoassay procedures to detect exposure to aflatoxin B1 and benzo(a)pyrene in animals and man at the DNA level. Int. Arch. Occup. Environ. Health, 60: 145-50. 1988.

Garner, R. C. Monitoring aflatoxin exposure at a macromolecular level in man with immunological methods. Bioact. Mol., 10(Mycotoxins Phycotoxins '88), 29-35. 1989.

Groopman, J. D.; Donahue, K. F. Aflatoxin, a human carcinogen: determination in foods and biological samples by monoclonal antibody affinity chromatography. J. Assoc. Off. Anal. Chem., 71: 861-7. 1988.

Kaveri, S. V.; Fremy, J. M.; Lapeyre, C.; Strosberg, A. D. Immunodetection and immunopurification of aflatoxins using a high affinity monoclonal antibody to aflatoxin. Lett. Appl. Microbiol., 4: 71-5. 1987.

Morgan, M. R. A.; Wilkinson, A. P.; Denning, D. W. Human exposure to mycotoxins monitored by immunoassay. Bioact. Mol., 10(Mycotoxins Phycotoxins '88), 45-50. 1989.

Morgan, M. R. A. Mycotoxin immunoassays (with special reference to ELISAs). Tetrahedron, 45: 2237-49. 1989.

Mortimer, D. N.; Gilbert, J.; Shepherd, M. J. Rapid and highly sensitive analysis of aflatoxin M_1 in liquid and powdered milks using an affinity column cleanup. J. of Chromat., 407: 393-8. 1987.

Pestka, J. J. Enhanced surveillance of foodborne mycotoxins by immunochemical assay. J. - Assoc. Off. Anal. Chem., 71: 1075-81. 1988.

Rauch, P.; Fukal, L.; Brezina, P.; Kas, J. Interferences in radioimmunoassay of aflatoxins in food and fodder samples of plant origin J. - Assoc. Off. Anal. Chem., 71: 491-3. 1988.

Wild, C. P.; Pionneau, F. A.; Montesano, R.; Mutiro, C. F.; Chetsanga, C. J. Aflatoxin detected in human breast milk by immunoassay. Int. J. Cancer, 40: 328-33. 1987.

Wilkinson, A. P.; Denning, D. W.; Morgan, M. R. A. Analysis of UK sera for aflatoxin by enzyme-linked immunosorbent assay. Hum. Toxicol., 7: 353-6. 1988.

Wilkinson, A. P.; Denning, D. W.; Morgan, M. R. A. An ELISA method for the rapid and simple determination of aflatoxin in human serum. Food Addit. Contam., 5: 609-19. 1988.

Immunoassays for Other Mycotoxins

Casale, W. L. Inhibition of protein synthesis in maize and wheat by trichothecene mycotoxins and hybridoma-based enzyme immunoassay for deoxynivalenol. 168 pp. Avail. Univ. Microfilms Int., Order No. DA8722822. From: Diss. Abstr. Int. B 1988, 48(8), 2167. 1987.

Chiba, J.; Kawamura, O.; Kajii, H.; Otani, K.; Nagayama, S.; Ueno, Y. A sensitive enzyme-linked immunosorbent assay for detection of T-2 toxin with monoclonal antibodies. Food Addit. Contam., 5: 629-39. 1988.

Chu, F.S.; Lee, R.C. Immunochromatography of group A trichothecene mycotoxins. Food Agric. Immunol. 1: 127-36. 1989.

Fan, T. S. L.; Schubring, S. L.; Wei, R. D.; Chu, F. S. Production and characterization of a monoclonal antibody cross-reactive with most group A trichothecenes. App. Environ. Microbiol., 54: 2959-63. 1988.

Gendloff, E. H.; Pestka, J. J.; Dixon, D. E.; Hart, L. P. Production of a monoclonal antibody to T-2 toxin with strong cross-reactivity to T-2 metabolites. Phytopathology, 77: 57-9. 1987.

Hack, R.; Maertlbauer, E.; Terplan, G. A monoclonal antibody to the trichothecene T-2 toxin: screening for the antibody by a direct enzyme immunoassay. J. Vet. Med., Ser. B, 34: 538-44. 1987.

Kodama, A.M.; Garnier, L.; Katsura, K.; Miyahara, J.T.; Hokama, Y. Comparative immunological, pharmacological and biological characteristics of Ciguateric toxin found in Ctenochaetus sp. from Tahiti and Hawaii. Food Agric. Immunol., 1: 83-8. 1989.

Lee, R. C.; Wei, R. D.; Chu, F. S. Enzyme-linked immunosorbent assay for T-2 toxin metabolites in urine. J. - Assoc. Off. Anal. Chem., 72: 345-8. 1989.

Liu, M. T.; Ram, B. P.; Hart, L. P.; Pestka, J. J. Indirect enzyme-linked immunosorbent assay for the mycotoxin zearalenone. Appl. Environ. Microbiol., 50: 332-6. 1985.

Wei, R. D.; Chu, F. S. Production and characterization of a generic antibody against group A trichothecenes. Anal. Biochem., 160: 399-408. 1987.

Xu, Y.-C.; Zhang, G. S.; Chu, F. S. Enzyme-linked immunosorbent assay for deoxynivalenol in corn and wheat. J. Assoc. Off. Anal. Chem., 71: 945-9. 1988.

Human monitoring for mycotoxin exposure

Autrup, H.; Seremet, T. Detection of antibodies against aflatoxin-conjugate in sera from African and Danish populations. Eur. J. Cancer Clin. Ocol., 23: 1730-1. 1987.

Fukal, L.; Reisnerova, H. Monitoring of aflatoxins and ochratoxin A in Czechoslovak human sera by immunoassy. Bull. Environ. Contam. Toxicol., 44: 345-9. 1990.

Garner, R. C.; Dvorackova, I.; Tursi, F. Immunoassay procedures to detect exposure to aflatoxin B-1 and benzo-a-pyrene in animals and man at the DNA level. Int. Arch. Occup. Environ. Health, 60: 145-50. 1988.

Groopman, J. D.; Cain, L. G.; Kensler, T. W. Aflatoxin exposure in human populations: Measurements and relationship to cancer. CRC Crit. Rev. Toxicol., 19: 113-45. 1988.

Groopman, J. D.; Kensler, T. W. The use of monoclonal antibody affinity columns for assessing DNA damage and repair following exposure to aflatoxin B_1. Pharmac. Ther., 34: 321-34. 1987.

Groopman, J. D.; Haugen, A.; Goodrich, G. R.; Wogan, G. N.; Harris, C. C. Quantitation of aflatoxin B-1 modified DNA using monoclonal antibodies. Cancer Res., 42: 3120-4. 1982.

Haugen, A.; Groopman, J. D.; Hsu, I.; Goodrich, G. R.; Wogan, G. N.; Harris, C. C. Monoclonal antibody to aflatoxin B-1 modified DNA detected by enzyme immunoassay. Proc Natl. Acad. Sci. U S A, 78: 4124-7. 1981.

Hertzog, P. J.; Smith, J. R. L.; Garner, R. C. Production of monoclonal antibodies to guanine imidazole ring opened aflatoxin B-1 DNA the persistent DNA adduct in-vivo. Carcinogenesis (Lond), 3: 825-8. 1982.

Hsieh, L.; Hsu, S.; Chen, D.; Santella, R. M. Immunological detection of aflatoxin B-1-DNA adducts formed in-vivo. Cancer Res., 48: 6329-31. 1988.

Pestka, J. J.; Li, Y. K.; Chu, F. S. Reactivity of aflatoxin B-2a antibody with aflatoxin B-1-modified DNA and related metabolites. Appl. Environ. Microbiol., 44: 1159-65. 1982.

Wild, C. P.; Jiang, Y.-Z.; Sabbioni, G.; Chapot, B.; Montesano, R. Evaluation of methods for quantitation of aflatoxin-albumin adducts and their application to human exposure assessment. Cancer Res., 50: 245-51. 1990.

Non-English Language Articles on Mycotoxin Immunoassays

Chiba, J.; Kajii, H.; Kawamura, M.; Ohi, K.; Morooka, N.; Ueno, Y. Production of monoclonal antibodies reactive with ochratoxin A: enzyme-linked immunosorbent assay for detection of ochratoxin A. Maikotokishin (Tokyo), 21: 28-31. 1985. Japanese.

Goto, T.; Manabe, M. Application of ELISA to aflatoxin analysis: Results of preliminary tests of analysis kits on the market. Rep. Natl. Food Res. Inst., 52: 53-9. 1988. Japanese.

Isohata, E.; Toyoda, M.; Saito, Y. Studies on chemical analysis of mycotoxin. XIX. Enzyme immunoassay of T-2 toxin in foods. Eisei Shikensho Hokoku, (106), 117-20. 1988. Japanese.

Isohata, E.; Toyoda, M.; Saito, Y. Chemical analysis of mycotoxin. XVIII. Analysis of aflatoxin B1 by enzyme immunoassay. Eisei Shikensho Hokoku, (105), 93-9. 1987. Japanese.

Popova, T. Immunologic methods for mycotoxins determination. Suvrem. Med., 38: 3-5. 1987. Bulgarian.

Steimer, V. J.; Hahn, G.; Heeschen, W.; Bluethgen, A. Enzyme-immunological detection of aflatoxin M1 in milk: development of a rapid aflatoxin M1 screening test using polystyrene beads as solid phase. Milchwissenschaft, 43: 772-6. 1988. German.

Ueno, Y.; Kawamura, O. The detection and the quantitative analysis of mycotoxin - enzyme-linked immunosorbent assay. Kagaku to Seibutsu, 27: 318-27. 1989. Japanese.

Ueno, Y.; Ohtani, K.; Kawamura, O.; Nagayama, S. ELISA of mycotoxins with monoclonal antibodies. Tanpakushitsu Kakusan Koso, Bessatsu, (31), 80-9. 1987. Japanese.

APPENDIX 3
DNA-ADDUCTION   AND
PROTEIN-ADDUCT   IMMUNOASSAYS

One volume stands out above others because it contains so many papers on this subject. Rather than list each paper individually, readers are refered to the entire volume for a comprehensive overview. It is:

Bartsch, H.; Hemminki, K.; O'Neill, I. K., Editors. Methods for detecting DNA damaging agents in humans: Applications in cancer epidemiology and prevention. IARC Scientific Publications vol 89, International Agency for Research on Cancer, Lyon, France. 1988.

Primary Publications.

Baan, R. A.; Lansbergen, M. J.; De Bruin, P. A. F.; Willems, M. I.; Lohman, P. H. M. The organ-specific induction of DNA adducts in 2-acetylaminofluorene-treated rats studied by means of a sensitive immunochemical method. Mutat. Res., 150: 23-32. 1985.

Ball, S. S.; Quaranta, V.; Shadravan, F.; Walford, R. L. An ELISA for Detection of DNA-bound carcinogen using a monoclonal antibody to N-acetoxy-2-acetylaminofluorene-modified DNA. J. Immunol. Meth., 98: 195-200. 1987.

Bartolone, J. B.; Birge, R. B.; Sparks, K.; Cohen, S. D.; Khairallah, E. A. Immunochemical analysis of acetaminophen Covalent binding to proteins partial characterization of the major acetaminophen-binding liver proteins. Biochem. Pharmacol., 37: 4763-74. 1988.

Brylawski, B. P.; Cordeiro-Stone, M.; Kaufman, D. G. Ferritin-labeled rabbit Fab fragments for the single-step detection of benzo-a-pyrene-diol-epoxide adducts in DNA by electron microscopy. Carcinogenesis (Lond), 10: 199-202. 1989.

Chung, F.; Young, R.; Hecht, S. S. Detection of cyclic 1 N-2-propanodeoxyguanosine adducts in DNA of rats treated with n-nitrosopyrrolidine and mice treated with crotonaldehyde. Carcinogenesis (Lond), 10: 1291-8. 1989.

Degan, P.; Montesano, R.; Wild, C. P. Antibodies against 7-methyldeoxyguanosine: Its detection in rat peripheral blood lymphocyte DNA and potential applications to molecular epidemiology. Cancer Res., 48: 5065-70. 1988.

Fichtinger-Schepman, A. M. J.; Baan, R. A.; Luiten-Schuite, A.; Van Dijk, M.; Lohman, P. H. M. Immunochemical quantitation of adducts induced in DNA by cis diamminedichloroplatinum-II and analysis of adduct-related DNA-unwinding. Chem-Biol. Interact., 55: 275-88. 1985.

Fichtinger-Schepman, A. M. J.; Baan, R. A.; Berends, F. Influence of the degree of DNA modification on the immunochemical determination of cisplatin DNA adduct levels. Carcinogenesis (Lond), 10: 2367-70. 1989.

Fleisher, J. H.; Lung, C. C.; Meinke, G. C.; Pinnas, J. L. Acetaldehyde-albumin adduct formation possible relevance to an immunologic mechanism in alcoholism. Alcohol, 23: 133-42. 1988.

Foiles, P. G.; Akerkar, S. A.; Chung, F. Application of an immunoassay for cyclic acrolein deoxyguanosine adducts to assess their formation in DNA of Salmonella-Typhimurium under conditions of mutation induction by acrolein. Carcinogenesis (Lond), 10: 87-90. 1989.

Foiles, P. G.; Chung, F.; Hecht, S. S. Development of a monoclonal antibody-based immunoassay for cyclic DNA adducts resulting from exposure to crotonaldehyde. Cancer Res., 47: 360-3. 1987.

Foiles, P. G.; Miglietta, L. M.; Akerkar, S. A.; Everson, R. B.; Hecht, S. S. Detection of O-6 methyldeoxyguanosine in human placental DNA. Cancer Res., 48: 4184-8. 1988.

Foiles, P. G.; Trushin, N.; Castonguay, A. Measurement of O-6-methyldeoxyguanosine in DNA methylated by the tobacco-specific carcinogen 4-methylnitrosamino-1-3-pyridyl-1-butanone using a biotin-avidin enzyme-linked immunosorbent assay ELISA. Carcinogenesis (Lond), 6: 989-94. 1985.

Foiles, P. G.; Akerkar, S. A.; Chung, F. L. Application of an immunoassay for cyclic acrolein deoxyguanosine adducts to assess their formation in DNA of Salmonella typhimurium under conditions of mutation induction by acrolein. Carcinogenesis (London), 10: 87-90. 1989.

Foiles, P. G.; Chung, F. L.; Hecht, S. S. Development of a monoclonal antibody-based immunoassay for cyclic DNA adducts resulting from exposure to crotonaldehyde. Cancer Res., 47: 360-3. 1987.

Fram, R. J.; Woda, B. A.; Wilson, J. M.; Robichaud, N. Characterization of acquired resistance to cis diamminedichloroplatinum-II in human colon carcinoma cells. Cancer Res., 50: 72-7. 1990.

Groopman, J.D. DNA adducts of nitropyrene detected by specific antibodies. Res. Report/ Helath Effects Inst., 7: 1-17. 1987.

Hsieh, L. L.; Jeffrey, A. M.; Santella, R. M. Monoclonal antibodies to 1 aminopyrene-DNA. Carcinogenesis (Lond), 6: 1289-94. 1985.

Hsu, I.; Poirier, M. C.; Yuspa, S. H.; Grunberger, D.; Weinstein, I. B.; Yolken, R. H.; Harris, C. C. Measurement of benzo(a)pyrene-DNA adducts by enzyme immunoassays and radioimmunoassay. Cancer Res., 41: 1091-5. 1981.

Kriek, E.; Welling, M.; Van der Laken, C. J. Quantitation of carcinogen-DNA adducts by a standardized high-sensitive enzyme immunoassay. IARC Sci. Publ., 59: 297-305. 1984.

Leadon, S. A. Production of thymine glycols in DNA by radiation and chemical carcinogens as detected by a monoclonal antibody. Brit. J. Cancer Res. Suppl., 8:113-7. 1987.

Lee, B. M.; Santella, R. M. Quantitation of protein adducts as a marker of genotoxic exposure immunologic detection of benzo-a-pyrene-globin adducts in mice. Carcinogenesis (Lond), 9: 1773-8. 1988.

Lung, C. C.; Fleisher, J. H.; Meinke, G.; Pinnas, J. L. Immunochemical properties of malondialdehyde-protein adducts. J. Immunol. Meth., 128: 127-32. 1990.

Marini, S.; Citro, G.; Di Cesare, S.; Zito, R.; Giardina, B. Production and characterization of monoclonal antibodies against DNA single ring diamines adducts. Hybridoma, 7: 193-204. 1988.

Morita, T.; Ikeda, S.; Minoura, Y.; Kojima, M.; Tada, M. Polyclonal antibodies to DNA modified with 4 nitroquinoline 1-oxide application for the detection of 4 nitroquinoline 1-oxide-DNA adducts in-vivo. Jpn. J. Cancer Res. (GANN), 79: 195-203. 1988.

Nakayama, J.; Yuspa, S. H.; Poirier, M. C. Benzo-a-pyrene-DNA adduct formation and removal in mouse epidermis in-vivo and in-vitro relationships of DNA binding to initiation of skin carcinogenesis. Cancer Res., 44: 4087-95. 1984.

Nehls, P.; Rajewsky, M. F. Monoclonal antibody-based immunoassay for the determination of cellular enzymatic activity for repair of specific carcinogen-DNA adducts O-6 alkylguanine. Carcinogenesis (Lond), 11: 81-8. 1990.

Nehls, P.; Rajewsky, M. F. Monoclonal antibody-based immunoassay for the determination of cellular enzymic activity for repair of specific carcinogen-DNA adducts (O-6-alkylguanine). Carcinogenesis (Lond), 11: 81-7. 1990.

Nemoto, N.; Nakatsuru, Y.; Nakagawa, K.; Tazawa, A.; Ishikawa, T. Immunohistochemical detection of anti-racemic-trans-7 8 dihydroxy-9 10-epoxy-7 8 9 10-tetrahydrobenzo-a-pyrene-bound adduct in nuclei of cultured hela cells and mouse lung tissue. J. Cancer Res. Clin. Oncol., 114: 225-30. 1988.

Newman, M. J.; Light, B. A.; Weston, A.; Tollurud, D.; Clark, J. L.; Mann, D. L.; Blackmon, J. P.; Harris, C. C. Detection and characterization of human serum antibodies to polycyclic aromatic hydrocarbon diol-epoxide DNA adducts . J. Clin. Invest., 82: 145-53. 1988.

Paules, R. S.; Poirier, M. C.; Mass, M. J.; Yuspa, S. H.; Kaufman, D. G. Quantitation by electron microscopy of the binding of highly specific antibodies to benzo-a-pyrene DNA adducts. Carcinogenesis (Lond), 6: 193-8. 1985.

Perera, F. P.; Hemminki, K.; Young, T. L.; Brenner, D.; Kelly, G.; Santella, R. M. Detection of polycyclic aromatic hydrocarbon-DNA adducts in white blood cells of foundry workers. Cancer Res., 48: 2288-91. 1988.

Perera, F.; Mayer, J.; Jaretzki, A.; Hearne, S.; Brenner, D.; Young, T. L.; Fischman, H. K.; Grimes, M.; Grantham, S.; Et, A. A.comparison of DNA adducts and sister chromatid exchange in lung cancer cases and controls. Cancer Res., 49: 4446-51. 1989.

Perera, F. P.; Poirier, M. C.; Yuspa, S. H.; Nakaymama, J.; Jaretzki, A.; Curnen, M. M.; Knowles, D. M.; Weinstein, I. B. A pilot project in molecular cancer epidemiology determination of benzo-a-pyrene DNA adducts in animal and human tissues by immunoassays. Carcinogenesis (Lond), 3: 1405-10. 1982.

Perera, F.; Santella, R.; Poirier, M. Biomonitoring of workers exposed to carcinogens: Immunoassays to benzo-a-pyrene-DNA adducts as a prototype. J. Occup. Med,. 28: 1117-23. 1986.

Perera, F. P.; Santella, R. M.; Brenner, D.; Poirier, M. C.; Munshi, A. A.; Fischman, H. K.; Van Ryzin, J. DNA adducts protein adducts and sister chromatid exchange in cigarette smokers and nonsmokers. J. Natl. Cancer Inst., 79: 449-56. 1987.

Perera, F. P.; Poirier, M. C.; Yuspa, S. H.; Nakayama, J.; Jaretzki, A.; Curnen, M. M.; Knowles, D. M.; Weinstein, I. B. A pilot project in molecular cancer epidemiology: determination of benzo-a-pyrene-DNA adducts in animal and human tissues by immunoassays. Carcinogenesis (Lond), 3: 1405-10. 1982.

Poirier, M. C. Antibodies to carcinogen DNA adducts . J. Natl. Cancer Inst., 67: 515-20. 1981.

Poirier, M. C.; Lippard, S. J.; Zwelling, L. A.; Ushay, H. M.; Kerrigan, D.; Thill , C. C.; Santella, R. M.; Grunberger, D.; Yuspa, S. H. Antibodies elicited against cis-diammine-dichloro-platinum-II-modified-DNA are specific for cis-diammine-dichloro-platinum-II-DNA adducts formed in-vivo and in-vitro. Proc. Natl. Acad. Sci. U S A, 79: 6443-7. 1982.

Poirier, M. C.; Stanley, J. R.; Beckwith, J. B.; Weinstein, I. B.; Yuspa, S. H. Indirect immuno fluorescent localization of benzo-a-pyrene adducted to nucleic-acids in cultured mouse keratinocyte nuclei. Carcinogenesis (Lond), 3: 345-8. 1982.

Poirier, M. C. Antibodies to carcinogen-DNA adducts. J. Natl. Cancer Inst., 67: 515-9. 1981.

Poirier, M. C.; Liou, S. H.; Reed, E.; Strickland, P. T.; Tockman, M. S. Determination of carcinogen-DNA adducts by immunoassay. J. UOEH, 11: 353-67. 1989.

Poirier, M. C.; Nakayama, J.; Perera, F. P.; Weinstein, I. B.; Yuspa, S. H. Identification of carcinogen-DNA adducts by immunoassays. Environ. Sci. Res., 29: 427-40. 1983.

Poirier, M. C.; Stanley, J. R.; Beckwith, J. B.; Weinstein, I. B.; Yuspa, S. H. Indirect immunofluorescent localization of benzo-a-pyrene adducted to nucleic acids in cultured mouse keratinocyte nuclei. Carcinogenesis (Lond), 3: 345-8. 1982.

Poirier, M. C.; True, B.; Laishes, B. A. Determination of 2-acetylaminofluorene adducts by immunoassay. Environ. Health Perspect., 49: 93-9. 1983.

Potter, D. W.; Pumford, N. R.; Hinson, J. A.; Benson, R. W.; Roberts, D. W. Epitope characterization of acetaminophen bound to protein and nonprotein sulfhydryl groups by an ELISA. J. Pharmacol. Exp. Ther., 248: 182-9. 1989.

Pumford, N. R.; Hinson, J. A.; Potter, D. W.; Rowland, K. L.; Benson, R. W.; Roberts , D. W. Immunochemical quantitation of 3-cystein-S-yl-acetaminophen adducts in serum and liver proteins of acetaminophen-treated mice. J. Pharmacol. Exp. Ther., 248: 190-6. 1989.

Rajagopalan, R.; Melamede, R. J.; Laspia, M. F.; Erlanger, B. F.; Wallace, S. S. Properties of antibodies to thymine glycol, a product of the radiolysis of DNA. Rad. Res., 97: 499-510. 1984.

Reed, E.; Litterst, C. L.; Thill, C. C.; Yuspa, S. H.; Poirier, M. C. Cis-diamminedichloroplatinum-II-DNA adduct formation in renal gonadal and tumor tissues of male and female rats. Cancer Res., 47: 718-22. 1987.

Reed, E.; Ozols, R. F.; Tarone, R.; Yuspa, S. H.; Poirier, M. C. Platinum-DNA adducts in leukocyte DNA correlate with disease response in ovarian cancer patients receiving platinum-based chemotherapy. Proc. Natl. Acad. Sci. U S A, 84: 5024-8. 1987.

Reed, E.; Yuspa, S. H.; Zwelling, L. A.; Ozols, R. F.; Poirier, M. C. Quantitation of cis diamminedichloroplatinum-II cisplatin DNA-intrastrand adducts in testicular and ovarian cancer patients receiving cisplatin chemotherapy. J. Clin. Invest., 77: 545-50. 1986.

Rio, P.; Bazgar, S.; Leng, M. Detection of N-hydroxy-2-acetylamino fluorene DNA adducts in rat liver measured by radioimmunoassay. Carcinogenesis (Lond), 3: 225-8. 1982.

Roberts, D. W.; Benson, R. W.; Groopman, J. D.; Flammang, T. J.; Nagle, W. A.; Moss, A. J.; Kadlubar, F. F. Immunochemical quantification of DNA adducts derived from the human bladder carcinogen 4-aminobiphenyl. Cancer Res., 48: 6636-42. 1988.

Roberts, D. W.; Benson, R. W.; Flammang, T. J.; Kudlubar, F. F. Development of an avidin-biotin amplified enzyme-linked immunoassay for detection of DNA adducts of the human bladder carcinogen 4-aminobiphenyl. Basic Life Sci., 38: 479-88. 1986.

Roberts, D. W.; Pumford, N. R.; Potter, D. W.; Benson, R. W.; Hinson, J. A. A sensitive immunochemical assay for acetaminophen-protein adducts. J. Pharmacol. Exp. Ther., 241: 527-33. 1987.

Saffhill, R.; Boyle, J. M. Detection of carcinogen DNA adducts by radioimmunoassay. Br. J. Cancer, 44: 275. 1981.

Santella, R. M. Monitoring human exposure to carcinogens by DNA adduct measurement. Cell Biol. Toxicol., 4: 511-6. 1988.

Santella, R. M.; Lin, C. D.; Cleveland, W. L.; Weinstein, I. B. Monoclonal antibodies to DNA modified by a benzo-a-pyrene diol epoxide. Carcinogenesis (Lond), 5: 373-8. 1984.

Santella, R. M.; Lin, C. D.; Dharmaraja, N. Monoclonal antibodies to a benzo-a-pyrene diol epoxide modified protein. Carcinogenesis (Lond), 7: 441-4. 1986.

Santella, R. M.; Weston, A.; Perera, F. P.; Tivers, G. T.; Harris, C. C.; Young, T. L.; Nguyen, D.; Lee, B. M.; Poirier, M. C. Interlaboratory comparison of antisera and immunoassays for benzo-a-pyrene-diol-epoxide-I-modified DNA. Carcinogenesis (Lond), 9: 1265-70. 1988.

Shamsuddin, A. K. M.; Gan, R. Immunocytochemical localization of benzo-a-pyrene-DNA adducts in human tissues. Hum. Pathol., 19: 309-15. 1988.

Shamsuddin, A. K. M.; Sinopoli, N. T.; Hemminki, K.; Boesch, R. R.; Harris, C. C. Detection of benzo-a-pyrene DNA adducts in human white blood cells. Cancer Res., 45: 66-8. 1985.

Shuker, D. E. G. Detection of adducts arising from human exposure to N-nitroso compounds. Cancer Surv., 8: 475-88. 1989.

Stein, A. M.; Gratzner, H. G.; Stein, J. H.; Mccabe, M. M. High avidity monoclonal antibody to imidazole ring-opened 7 methylguanine. Carcinogenesis (Lond), 10: 927-32. 1989.

Sundquist, W. I.; Lippard, S. J.; Stollar, B. D. Monoclonal antibodies to DNA modified with cis or trans diamminedichloroplatinum-II. Proc. Natl. Acad. Sci. U S A, 84: 8225-9. 1987.

Tada, M.; Aoki, H.; Kojima, M.; Morita, T.; Shirai, T.; Yamada, H.; Ito, N. Preparation and characterization of antibodies against 3,2-dimethyl-4-aminobiphenyl-modified DNA. Carcinogenesis(Lond), 10: 1397-402. 1989.

Talbot, B.; Desnoyers, S.; Castonguay, A. Immunoassays for proteins alkylated by nicotine-derived N-nitrosamines. Arch. Toxicol., 64: 360-4. 1990.

Tierney, B.; Benson, A.; Garner, R. C. Immunoaffinity chromatography of carcinogen DNA adducts with polyclonal antibodies directed against benzo-a-pyrene-diol-epoxide DNA. J. Natl. Cancer Inst., 77: 261-8. 1986.

Tilby, M. J.; Lawley, P. D.; Farmer, P. B. Alkylation of DNA by melphalan in relation to immunoassay of melphalan DNA adducts characterization of mono-alkylated and cross-linked products from reaction of melphalan with deoxy-GMP and GMP. Chem-Biol. Interact, 73: 183-94. 1990.

Tromberg, B. J.; Sepaniak, M. J.; Alarie, J. P.; Vo-Dinh, T.; Santella, R. M. Development of antibody-based fiber-optic sensors for detection of a benzo-a-pyrene metabolite. Anal. Chem., 60: 1901-8. 1988.

Van Der Laken, C. J.; Hagenaars, A. M.; Hermsen, G.; Kriek, E.; Kuipers, A. J.; Nagel, J.; Scherer, E.; Welling, M. Measurement of O-6-ethyldeoxyguanosine and N-deoxyguanosin-8-yl-N-acetyl-2-aminofluorene in DNA by high sensitive enzyme immunoassays. Carcinogenesis (Lond), 3: 569-72. 1982.

Van Schooten, F. J.; Kreik, E.; Hillebrand, M. J. X.; Van Leeuwen, F. E. Reactivity of antibodies to DNA modified by benzo-a-pyrene is dependent on the level of modification Iimplications for quantitation of benzo-a-pyrene-DNA adducts in-vivo. Eur. J. Cancer Clin. Oncol., 23: 1809. 1987.

Van Schooten, F. J.; Kriek, E.; Steenwinkel M-J, S. T.; Noteborn, H. P. J.; Hillebrand, M. J. X.; Van Leeuwen, F. E. The binding efficiency of polyclonal and monoclonal antibodies to DNA modified with benzo-a-pyrenediol epoxide is dependent on the level of modification implications for quantitation of benzo-a-pyrene-DNA adducts in-vivo. Carcinogenesis (Lond), 8: 1263-70. 1987.

Wallin, H.; Jeffre, A. M.; Santella, R. M. Investigation of benzo-a-pyrene-globin adducts. Cancer Lett., 35: 139-46. 1987.

Wani, A. A.; D'ambrosio, S. M. Immunological quantitation of O-4 ethylthymidine in alkylated DNA repair of minor miscoding base in human cells. Carcinogenesis (Lond), 8: 1137-44. 1987.

Wani, A. A.; Sullivan, J. K.; Lebowitz, J. Immunoassays for carbodiimide modified DNA-detection of unpairing transitions in supercoiled cole 1 DNA. Nucleic Acids Res., 17: 9957-78. 1989.

Weston, A.; Newman, M.; Vahakangas, K.; Rowe, M.; Trivers, G. E.; Mann, D. L.; Harris, C. C. Measurement of carcinogen-macromolecular adducts and serum antibodies recognizing DNA adducts in biological specimens from people exposed to chemical carcinogens. Environ. Sci. Res., 36: 91-101. 1987.

Weston, A.; Newman, M. J.; Mann, D. L.; Brooks, B. R. Molecular mechanics and antibody binding in the structural analysis of polycyclic aromatic hydrocarbon diol-epoxide DNA adducts. Carcinogenesis (Lond), 11: 859-64. 1990.

Weston, A.; Rowe, M.; Poirier, M.; Trivers, G.; Vahakangas, K.; Newman, M.; Haugen, A.; Manchester, D.; Mann, D.; Harris, C. The application of immunoassays and fluorometry to the detection of polycyclic hydrocarbon-macromolecular adducts and anti-adduct antibodies in humans. Int. Arch. Occup. Environ. Health, 60: 157-62. 1988.

Wild, C. P.; Jiang, Y.; Sabbioni, G.; Chapot, B.; Montesano, R. Evaluation of methods for quantitation of aflatoxin albumin adducts and their application to human exposure assessment. Cancer Res., 50: 245-51. 1990.

Wilson, V. L.; Weston, A.; Manchester, D. K.; Trivers, G. E.; Roberts, D. W.; Kadlubar, F. F.; Wild, C. P.; Montesano, R.; Willey, J. C.; Et, A. Alkyl and aryl carcinogen adducts detected in human peripheral lung. Carcinogenesis (Lond), 10: 2149-54. 1989.

Wogan, G. N. Markers of exposure to carcinogens. Environ. Health Perspect, 81: 9-18. 1989.

Yang, X. Y.; Delohery, T.; Santella, R. M. Flow cytometric analysis of 8-methoxypsoralen-DNA photoadducts in human keratinocytes. Cancer Res., 48: 7013-7. 1988.

Yang, X. Y.; Deleo, V.; Santella, R. M. Immunological detection and visualization of 8-methoxypsoralen-DNA photoadducts. Cancer Res., 47: 2451-5. 1987.

Yang, X. Y.; Gasparro, F. P.; Deleo, V. A.; Santella, R. M. 8-methoxypsoralen-DNA adducts in patients treated with 8-methoxypsoralen and UV-A light. J. Invest. Dermatol., 92: 59-63. 1989.

Young, T. L.; Santella, R. M. Development of techniques to monitor for exposure to vinyl chloride monoclonal antibodies to ethenoadenosine and ethenocytidine. Carcinogenesis (Lond), 9: 589-92. 1988.

INDEXES

Author Index

Affiliation Index

Subject Index